Time Travel

Time Travel

Pop, Media and Sexuality 1976–96

Jon Savage

Chatto & Windus
LONDON

First published in 1996

1 3 5 7 9 10 8 6 4 2

Copyright © Jon Savage 1996

Jon Savage has asserted his right under the Copyright, Designs and
Patents Act, 1988 to be identified as the author of this work.

First published in Great Britain in 1996 by
Chatto & Windus Limited
Random House, 20 Vauxhall Bridge Road,
London SW1V 2SA

Random House Australia (Pty) Limited
20 Alfred Street, Milsons Point, Sydney
New South Wales 2061, Australia

Random House New Zealand Limited
18 Poland Road, Glenfield
Auckland 10, New Zealand

Random House South Africa (Pty) Limited
PO Box 337, Bergvlei, South Africa
Random House UK Limited Reg. No. 964009

Papers used by Random House UK Limited are natural, recyclable
products made from wood grown in sustainable forests. The
manufacturing processes conform to the environmental regulations of
the country of origin.

A CIP catalogue record for this book
is available from the British Library

ISBN 0 7011 6360 7

Designed by Margaret Sadler

Picture captions for the part title pages
Part 1: Linder, 'The Dream of the Sibyl Transmuted as a Household Appliance
and Cast Naked with Biting Breasts', montage, 1977
Part 2: Edwardians in a fashion show held at the Royal Ballroom, Tottenham,
1953
Part 3: Jon Wozencroft, 'Moebius', 1991

Contents

PART THREE: SPEED 1988–95

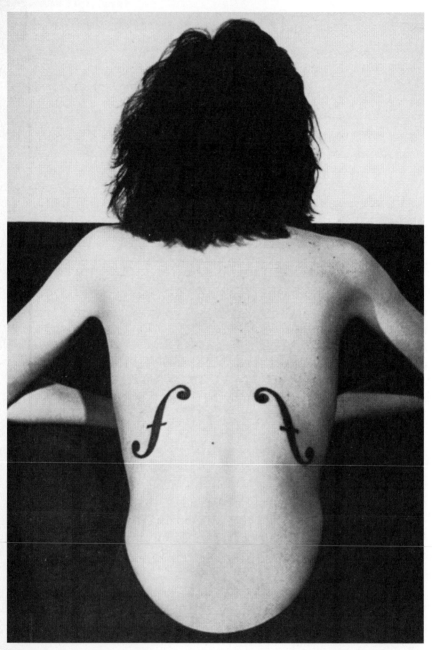

Ruth Marten: 'Homage to Man Ray': tattoo performed on Judy Nylon, London 1977
(photo: Jon Savage)

Introduction

I am afraid I cannot convey the peculiar sensations of time travel.
They are excessively unpleasant. There is a feeling exactly like that one has on
a switchback – of a helpless head-long motion! I felt the same horrible
anticipation too, of an imminent smash.

H. G. Wells, *The Time Machine*, 1895

2 September 1995: Labor Day weekend in Cleveland, and the population is out on the streets, celebrating a late summer holiday given extra zip by the Rock 'n' Roll Hall of Fame inauguration and concert. For a weekend, this normally closed city is open – amplified music blaring on every corner, a rock theme park. The music and media industries have gathered to worship at the shrine made in their image, I. M. Pei's $92 million museum, and CLE is basking, both in the unaccustomed attention and in the late summer heat – the sun zooms off Lake Erie in an speedy white glare.

The celebrations represent a considerable coup for a city which, according to local author Thomas Kelly, has suffered a 'forty year winter' of 'defeat and disaster'. The *Cleveland Plain Dealer* spells it out on its op-ed page: 'The hall promises to keep bringing tourists and dollars into this region for decades.' So far, so familiar. Like Manchester, a city of similar size and industrial archaeology, Cleveland is regenerating through culture and tourism, by capitalizing on its place in pop history: this is where local DJ Alan Freed first popularized the phrase rock 'n' roll (during 1952), where Elvis Presley first broke above the Mason-Dixon line (in September 1954) – a shift marked by packed out shows and rioting audiences.

Inside, the museum is airy and comfortable, with open space for five or six stories, leaving nooks and crannies for vitrines and TV monitors. Standout displays include Stephen Shore's 1966 photos of Andy Warhol's Factory, a reconstruction of Sam Phillips's famous Sun Studios and the Alan Freed exhibit. Here you can see the most extraordinary image from pop's dawn, Peter Hastings's photograph of the March 1952 Moondog Coronation Ball at the Cleveland Arena, which shows an all-black crowd in ceaseless movement, awestruck at finding its own power.

The meat of the collection is contained in an enclosed, theatre-sized area, the Ahmet Ertegun Exhibit Hall. The interior is broken down into a coherent sequence of vitrines which tell a chronological history of pop from the late 1940s to the present day, interspersed with video/interactive CD-ROM areas and mannequinned tableaux. Despite the problem inherent in displaying historical costumes, the charity shop syndrome, the curators have done a good job of assembling enough totems to keep pop-cult pervs like myself happy: Jim Morrison's school report card, a Dougie Millings Beatle jacket, mangled strips from the Otis Redding death plane, a whole display of early Hip Hop ephemera.

I'm here to check on my own pottery shard: a few pieces of memorabilia – handwritten Sex Pistols' lyrics, a Sid Vicious Action Man from *The Great Rock 'n 'Roll Swindle*, the blanked-out globe from Derek Jarman's *Jubilee* – which form one corner of the Punk display. Seeing this material behind glass is an unsettling experience. It feels strange to have the present that you once lived through so intensely become the past – even if, in my own case, I've contributed to the process. After a couple of hours, all this detail begins to accumulate into dead time: there's a whole history being constructed here, but I'm not sure that it's mine.

For all the research and the pizazz, the museum cannot, by its very nature, reflect contemporary pop. There is no House, no Techno, and nor is there any contemporary Rap, while 1990s white boy rebel music is represented by Seattle and a cheesy Stephen Sprouse tableau called 'All The Young Dudes'. Bearing in mind that senior members of the Time/Warner hierarchy are key players in the museum, this could be construed as a public lack of confidence in their subsidiary, Dr Dre's Interscope label – currently the subject of intense press and judicial scrutiny about Gangsta Rap – and thus artists like Snoop Doggy Dog, 2Pac and Nine Inch Nails. Dre certainly thinks so: together with Snoop, he pulls out of his appearance in the inaugural concert at the last minute, because his culture is not recognized here.

At one stroke, the event is skewed from the present into the past: the Gangsta Rappers occupy a position in 1995 America not unlike that of the Sex Pistols in 1977 – public enemies number one, national scapegoats. We know that history will be taken care of at this concert, and so it is, with incendiary cameos from Al Green, Jerry Lee Lewis, Aretha Franklin, Chrissie Hynde and Lou Reed – "Sweet Jane", dedicated with feeling to Sterling Morrison, whose funeral mass occurs today – but what the event is really about is shown by last-minute add-ons the Gin Blossoms. In place of Gangsta Rap's complex brutalities, we hear syrupy covers of the Beatles and the Byrds. This substitution cloaks music industry realpolitik: within a month of the show, Time/Warner will announce that they are selling their stake in Interscope, after lobbying by right-wing politicians and women's groups.

As an official baby-boomer – today is my forty-second birthday – I tap my toe to the Gin Blossoms' tepid "Feel A Whole Lot Better", then become uneasy. This flowers into resentment as archive clips of the Beatles and Stones intersperse the Radio Shack IDs on the video monitors – that's right, just another ad. As Greil Marcus writes in *The Dustbin of History*, 'The worry is that our sense of history, as it takes shape in everyday culture, is cramped, impoverished, and debilitating; that the commonplace assumption that history exists only in the past is a mystification powerfully resistant to any critical investigations that might reveal this assumption to be a fraud, or a jail.' Watching Sheryl Crow commit GBH on the Rolling Stones' "Let It Bleed", I realize which time I'm in: an eternity of 1969.

There are many pop 1969s, of course. There is the 1969 of Jamaican music, crashing into the UK charts as reggae for the first time; the 1969 of psychedelic funk – Sly and the Family Stone, the Temptations' "Cloud Nine"; the 1969 of David Bowie's "Space Oddity" and Lothar and the Hand People's "Space Hymn"; the 1969 of Terry Riley's *Rainbow in Curved Air*; the 1969 of "Sugar Sugar" and "Venus"; the 1969 of the Velvet Underground's third album and the Stooges' first – a blank negation seven years ahead of its time. Indeed, when Iggy Pop segues out of "Back Door Man" into "I Wanna Be Your Dog" at Cleveland Stadium, the hairs on the back of my neck stand on end and I lose control, for the first and only time this long night.

But the 1969 on offer here is an easy tour of the Woodstock myth: that moment when pop became rock, when the goal of feminine frenzy was replaced by musicianly machismo, when authenticity usurped artifice as a cultural ideal. Within the stadium, the festival ideal of hedonistic community has become a nostalgic ritual. The appearance of Bob Dylan is the final straw. He has *that* aureole of hair, *those* twisted mod clothes. I should be thrilled at viewing this icon for the first time. But the manipulations of the event are such that, as he grinds through "Just Like a Woman", it's as though he's in some scratchy archive film. For all his skills, he cannot step out of the Woodstock that never ends: I feel as though my own past has been frozen – rewritten in someone else's hand.

> I hang around 1995 for a week or so, and I realize that the texture of life is different from what it was in 1975. It's a subjective thing, and I try to put my finger on it. Nobody hitchhikes anymore. Nobody has hobbies. People use their phones, and phone-related devices like faxes and E-mail, all the time. Nobody seems to 'have time' anymore. People don't have 'lives' anymore.
> Douglas Coupland, 'Walk on the Wired Side', *Artforum*, December 1995

From birth to death, we are bounded by physical time. The revolutions of the earth, sun and moon impact on the daily and yearly rotations of light and dark, winter and summer, and on the body itself: in waking, sleep, eating and the prime meter of human time, the monthly female cycle. Ever since the invention of the mechanical clock during the fourteenth century, however, the measurement of time has become ever more exact, infinitesimal and merciless: from years to months to days to hours to minutes to seconds to nanoseconds. There is no excuse for not knowing what time it is at any instant: each moment is accounted for. And we agree to live by these rules in order to function in society, indeed to have a society: the world of timetables, appointments, and working hours.

This is what Joachim-Ernst Berendt, in his book on music and the landscape of consciousness, *Nada Brahma*, calls 'measured' or 'objective' time: 'considered to be steadfast and precise, the same time and obligatory for all people'. This public time is contrasted with what Berendt summarizes as 'lived' or 'subjective' time, which 'is the personal time of each single individual, passing by much too fast during moments of bliss and much too slowly in the hour of suffering'.

Turning to Greek mythology, Berendt posits this opposition in terms of their two gods of time: 'Chronos, the archfather, was the god of absolute, "eternal" time. For Kairos, however, the youngest son of Zeus, time meant the favourable moment, the "right time".' For him, 'rationalistic Western civilization has suppressed its Kairos. Its time is taken with clocks and watches, with *chrono*meters. Its time is from archfather Chronos. It is male, patriarchal, rational, and functional. In the more archaic cultures, as in the African continent, even today the time of Kairos is more important than clock time, which is "primal time" only for male-dominated thinking.' Indeed, 'clocks have to do with a lack of freedom'.

Music itself is ideally adapted to swim between these two times. With its varieties of rhythms and metres, it marks humanity's attempt to tell its own time, to measure the personal against the public. Post-war pop has been dominated by the 4/4 beat – clockwork writ large – in contrast to the varying metres favoured by Kairos cultures: the 6/8 of North Africa, for instance. It's no accident that the hourly grid of Bill Haley's "Rock Around the Clock" marked the emergence in 1955–6 of the modern music industry, which always seeks to lock pop into chronological time: the countdown of the Top Forty; the use of old hits in adverts; the accelerated BPM (beats per minute) played in boutiques to stimulate consumption, the concept of "Music While You Work" industrialized by Muzak.

The measurement of pop is part of this. Music is regularly charted in years, Top Forty positions and cultural movements. You can say that "Rock Around the Clock" is rock 'n' roll, that it went to number one for nine weeks and caused riots when filmed, that Chic's epic "Good Times" is Disco, that, as part of Grandmaster Flash's cut-up "The Wheels of Steel", it has had an incredible influence on Rap and dance music thereafter: thus validated, they are placed into the canon – a matrix of agreed, public recollection. Hits like these are a valuable tool for social historians, an instant access to the collective memory bank. Prime-time programmes like *The Rock 'n Roll Years* exploit this short cut, harnessing the private time of pop to the public time of news footage.

There is an apparent paradox here. Although it exists to heighten the present, pop is now saturated in the past. In this, the release of time capsules like the Beatles' *Anthology* has thrown up a wide variety of reactions, from exaltation through cynicism to psychotic hostility: how could they rewrite the past by recording over an old John Lennon vocal in "Free As a Bird" ? Much of the problem here – and in current arguments about nostalgia, retro, the question of authorship thrown up by sampling technology – has to do with the way that pop is still seen as a modernist medium – forever new, forever moving forwards. Its reliance on the past – whether it's Oasis cheekily sourcing Gary Glitter on "Hello" or the Bucketheads including a snatch of a fifteen-year-old record, Chicago's "Street Player", in the structure of their big summer 1995 hit, "The Bomb!" – is seen to indicate a lack of emotional authenticity, a loss of cultural power.

One of the reasons why the Beatles invite such extreme reactions is that they are seen as the embodiments of modernism, taking part in a collective, linear cultural motion. The 1960s were a time of great economic, social, sexual and spiritual gains: a healing of wartime wounds, the final overthrow – in private time, at least – of Victorian values. It is also the case that many of the liberations initiated during that period – consumerism, sex outside marriage, the social use of drugs – have since shown their darker side, but the measure of the pop 1960s lies in the unsolved social problems which they identified: ecological damage, obsolete puritan guilt, the lack of any spirituality in materialism. This is the journey the Beatles made in three years: from "Money" to "The Void".

On its release in August 1966, "Tomorrow Never Knows" (as "The Void" was later known) was an explicit assault on contemporary perception. The recording is like nothing else that the Beatles ever did. Built on a sequence of loops, whether Ringo Starr's drum thud – sounding like, in Ian MacDonald's words, 'a cosmic tabla played by a Vedic deity riding in a stormcloud' – or the five home-made tape loops which, overdubbed throughout, pull the rhythm out of everyday 4/4 into the eternity in a nanosecond offered by the LSD experience. If that wasn't enough, "Tomorrow Never Knows" also marks the moment when pop began to move out of linear and into serial time, when directional was replaced by circular motion, when the explicitly materialist was replaced by the spiritual.

In 1966, time went out of joint. The fundamentalist backlash began, triggered that summer by John Lennon's quote, 'We're more popular than Jesus now.' The loop became an accepted part of record production, whether in "Tomorrow Never Knows" or in John Cale's feedback instrumental which encodes form in title: "Loop". This time shift was also reflected in hallucinogenic lyrics looping on the words 'mind', 'circle', 'dream'; in a design change away from Pop Art to a series of Revivals – Art Nouveau, Art Deco, Arts and Crafts; in the move away from mod to antique styles – Victorian uniforms,

Edwardian moustaches, Depression-era Granny glasses; in the speed chaos of records like the Rolling Stones' "Have You Seen Your Mother Baby, Standing in the Shadow", or the Yardbirds' "Happenings Ten Years Time Ago" – the unitary, forward momentum of modernism accelerating to the point of catastrophe.

> There's a mickie in the tasting of disaster
> In time – in time you get faster.
>
> Sly Stone, "In Time", 1973

The impulse to speed is at the heart of post-war pop. In the words of famed producer Guy Stevens, 'All rock 'n' roll speeds up'. You can hear this within the tempo of punk staples like Lonnie Donegan's "Rock Island Line", the Beatles' "Twist and Shout", Patti Smith's "We're Gonna Have a Real Good Time Together", the Clash's "Brand New Cadillac", the Saints' "This Perfect Day". You can also hear it in the way that pop genres have evolved ever faster: Mod into the Ramones and Punk; Chicago House into Acid and Hardcore; Rare Groove and Breakbeat into the serious time damage that is Jungle. The cycles come and go, from motion to entropy, but the impulse to up the ante, to go faster than anyone else, is inherent in the twinning of technology and the adolescent psyche that occurs in Western consumerism.

Speed is vital because it is one of the few areas where teens are more powerful than adults, and time has been frequently used to express a rebellious attitude. You need think only of Sly Stone exquisitely cocking a snook at the world in the lyrics of "In Time" – Kairos as Staggerlee. In Punk classics like "The Last Time" and "No Time", the Rolling Stones and the Saints may well be directing their comments, as immature men are prone to do, at individual women, but the message is in the insolent drones of the music: a direct challenge to the established order, whether it's the nursery demons of the music industry, or authority figures like parents, teachers and politicians. I'm faster than you: I'm five years ahead of my time.

This is a generational war, expressed in time and perception. Pop is directly linked to the Second World War, both as a psychic purge and as an economic continuation. The modern music/media industry emerged out of the technologically driven, mass-production, fast-turnover nature of the post-war economy. Its speed-driven nature is epitomized by its inaugural icon, James Dean: in this struggle between life and death, the enemy is within. That's not to say that pop cannot also be an external war pursued by any other means. It can be argued that multi-national Western pop culture – MTV, Levis and Coca-Cola – has proved a much more effective method of strategic colonization than actual military conflict.

Pere Ubu's "30 Seconds Over Tokyo", recorded in 1975, takes this techno/teen speed to one logical conclusion. The title comes from a book by Captain Ted Lawson which tells the story of the 18 April 1942 'Doolittle Raids', a propaganda exercise where US bombers set out with just enough fuel to drop their bombs on Tokyo; the crew were then forced to bail out or crash-land. The song guides you through every stage of this 'suicide ride'. As the group goes into its final dive, David Thomas's badly amplified voice, as crackly as any RT, accelerates with it. Like Slim Pickens at the end of 'Dr Strangelove', he is bearing down so hard on the final nanoseconds that each one stretches into eternity. As the tape cuts dead, this human machine is frozen for ever on the point of impact, between death and the full intensity of life.

This intensity is both attractive and highly disturbing to adults: a devastating combination which accounts for the contradictory messages pumped at teens – commercial exploitation and real-life restriction. Mainstream attitudes to pop are always ambiguous: desirous of its attractions, hostile to its perceptions. One of the most persistent themes in this book is the conflict that occurs when pop speed comes up against the inertia of institutions, by definition locked in the past. These conflicts are played out in public scandals – whether involving the Sex Pistols, the Beastie Boys or Boy George; show trials, like the linked Modern Primitives and Operation Spanner cases, which briefly made illegal the body-piercing and S&M that had been a pop staple for over a decade; the suicides of Ian Curtis and Kurt Cobain, those most private of performers forced to enter public time.

In a famous allegory of the war between youth and age, Chronos swallows each of his male children – because it has been prophecied that one of his sons will dethrone him – until his wife Rhea manages to successfully conceal her third son Zeus, the eventual founder of Olympus. As Norman Cohn writes, 'the means by which the adults try to retain power is, precisely, cannibalistic infanticide'. This war against youth has reached a critical stage in the US and the UK, played out in headlines of teen suicide, teen sex, extreme teen violence. In this reaction, the very youthfulness of pop is deliberately underplayed. Having passed the fortieth year since "Rock Around the Clock", it is perceived as a middle-aged medium, acquiring all the traits popularly ascribed to middle-age: comfort, responsibility, a mild conservatism – the revenge of the baby-boomers who, having benefited so conspicuously from the ideal of youth, ought to know better.

It seems to me like I've been here before
The sounds I heard and the sights I saw
Was it real or in my dreams ?
I need to know what it all means.
The Yardbirds, "Happenings Ten Years Time Ago", 1966

Since the decline of pure modernism in 1966, pop has been passing through its postmodern phase, marked by overlapping revivals, looped time, the increased interconnection between music and the other media industries: advertising, newspaper and book publishing, film and TV. The music industry has expanded exponentially from its original teenage market to target all ages from childhood to death. You can hear music accompanying every human activity; you can walk into any megastore and find an incredible array of material stockpiled from the last hundred years – the result of a ten-year stockpiling of CDs. The effects of this can be bewildering (information overload can result in nausea) and maddening (if you can't afford it, or if your history has been tampered with).

Such overproduction encourages measurement, and this is what much pop writing has to do: assess the proliferating product and service a voracious industry. But this is to regard pop as nothing more than an industrial process – as chronological – and to ignore the fact that it is also organized sound, a complete sensory experience. According to Marshall McLuhan, 'we are enveloped by sound': through it, we can change worlds, from the public to the private, from measured into lived time. In *Ocean of Sound*, David Toop develops this through metaphor: 'On our watery planet, we return to the sea for a diagnosis of our current condition. Submersion into deep and romantic pools represents an intensely romantic desire for dispersion into nature, the unconscious, the womb, the chaotic stuff of which life is made.'

Indeed, the basis of pop music is that it provides a refuge from chronology: if only in its concentration on the moment, its insistence on pure pleasure. Play "Good Times" again: right from the phased fanfare, it remains as much in the present as it was in the moment it was recorded. Nile Rodgers's bassline has the vibrancy of lived time: an uptempo pulse that, locking into your heartbeat, compels you to dance and, in that physical movement, to lose time. Another irresistible stomper, James Brown's "There Was a Time", locks into a hypnotic groove that acts as the bridge between black music past and future: between the church and the sequencer.

It's also clear that, thirty years after the end of its modernist phase, pop has moved through postmodernism to something that does not yet have a name. Time-tripping is not a stylistic gambit but an industrial climate; a natural condition for the producers of music, another way of constructing their narrative. Contemporary pop exists within at least two or three separate speeds, held in balance by a conjuror's graceful sleight-of-hand. If, as Virilio states in *Pure War*, 'our consciousness is an effect of montage', then Rap, Jungle and Techno have turned perception into form – a different dimension, where past, present and future are not separated but part of the same continuum.

This applies to historical artefacts as well. Heard for the first time today, a record from 1973 (the Cosmic Jokers' "Galactic Supermarket"), 1948 (Nat

King Cole's "Nature Boy") or 1960 (Maxine Brown's "All In The Mind") can make as much sense of the present as any produced this year. Each of these time-scales has equal value, indeed exists on the same plane as the others. It's true to say that an archive release is locked in its own time: it will come with a set of expectations, a fund of shared, public knowledge. Yet it can also communicate directly with private time: although I know that, say, "Nature Boy" was recorded five years before I was born, that it was written by Eden Ahbez (an itinerant mystic who sported hippie length hair in 1948), what I zero in on is the melody and the lyric punchline – 'the only thing in the world/Is to love and be loved in return'.

By 1966, the Beatles excited comment when they were discovered doing at least four things at once: reading, listening to music, chatting, watching TV with the sound off. From this came many of their songs – like "Good Morning Good Morning", which was from a half-heard advert on TV. This is now a routine, mediated experience: today's musicians and teens have entered a more complex perception of time and space, one which may to outsiders appear like the maze of a computer game, with its constantly shifting perspectives yet definite rules. This could be the fifth dimension that the Byrds sang about in 1966: 'I will keep falling as long as I live, without ending/And I will remember the place that is now,/That has ended before the beginning.'

Today's varispeed pop encodes a harmony between the public and the private, the past and the present, between individual and environment: whether it's Techno in all its myriad forms (Ambient, Goa Trance), which pits expansion noises, suggesting space and transcendence, over machine rhythms and rapid fire film soundbites, or Jungle, which overlays hyperspeed breakbeat percussion (e.g. drum loops from 1970s records like James Brown's "Funky Drummer" accelerated to the status of computer-game soundtrack) on top of slow, rolling reggae basslines. Felt in every part of your body, the music offers a strong solution to today's perceptual conundrum: the fact that you can find an equilibrium between different speeds, the fact that you can live in several time zones at once without being torn apart.

This blurring of the classic distinctions between past, present and future returns us to the nature of the loop itself (somewhere in the world "Tomorrow Never Knows" is still playing). A loop can be closed – referring to nothing but its own circularity, like MTV – but it can also be the loop that resembles the mathematical symbol of infinity. (As disembodied as someone from a different time, John Lennon chants his mantra: 'Play the game existence to the end/Of the beginning.') Time has blurred into the everlasting present that has always been the hallmark of the teenage experience. This is at once the sound of instant gratification and a glimpse of eternity. With everything rushing at us, we can nevertheless tap into that eternity for an infinitesimal moment and, connecting with Kairos, find the 'right time'.

'We will travel not only in space but in time as well.' A Russian scientist said that. I have just returned from a thousand year time trip and I am here to tell you what I saw, to tell *you* how such time trips are made. It is a precision operation. It is difficult. It is as dangerous as the early days of aviation. It is the new frontier and only the adventurous need apply. It belongs to anyone who has the courage and the know how to travel. It belongs to you.

William Burroughs, "Time", 1965

This collection contains the first article I ever wrote for a nationally distributed magazine, which I stayed up all night to finish, and cuts off on the deadline for publication, February 1996. For much of this period, I've been fortunate enough to feel connected to something outside myself, a collective energy from which I drew strength: whether it be working for *Sounds* in 1977, *The Face* in 1982, the *New Statesman* or the *Observer* in 1986. Between 1977 (**Punk**) and 1988 (the end of **Style**) I felt completely in sync with the weekly and monthly deadlines of national journalism; it was a period of great excitement, intense rivalries, and, I now realize, unreckoned privilege. Since then, I've sought my own time: without the security of a regular deadline, I've had to learn to work at my own speed.

The book is in three sections. The first, **Punk**, covers the years that I was working for the weekly music press – *Sounds* and *Melody Maker* – where the prime object was to fill sixty or seventy broadsheet pages a week. Always concerned with the new, always living in the instant, the weeklies were, and remain, pure pop with rock window-dressing: great for learning in public, but you burn out quickly. Moving to *The Face* in 1980 was a liberation: writing at a shorter length, working closely with designer Neville Brody and photographer Chalkie Davies for a full integration of images and text. Following intuition rather than market research, Nick Logan produced a magazine that raised the game of everyone who worked there: each issue a wonderful object.

Going public with the June 1984 Fiftieth Issue Exhibition at the Photographers' Gallery, *The Face* came to define the mood of the mid 1980s. What had been fluid and playful became fixed – for me at least. At the same time, a sequence of disturbing public events – the Falklands War, the miners' strike, the onset of AIDS – pricked my pop bubble. Politics entered the equation, just as the full influence of *The Face* was felt in the mainstream media: by 1987, almost everybody who had worked on the magazine had been taken up by what was then still Fleet Street. The second section in this book, **Style**, is riven by the resulting tensions: tensions which increased to the point where, in 1989, the only logical conclusion was for me to stop full-time journalism and to concentrate on writing books. This decision was assisted by the increasing insecurity that has followed after the restructuring – many would say destabilisation – of Fleet Street .

Seven years later, I still miss the adrenalin addiction of weekly and monthly deadlines, that sense of being in a groove that you get with a mass market outlet – the sense of being *plugged in* – but I don't have to write stuff that makes me sick when I read it. There are other compensations: I can step off the PR conveyor belt and choose my own subjects. Writing for different publications in Japan, Germany, France, the US and the UK, I'm kept alert by the necessity of writing in different styles for different venues. I can also work in my own time: a change of speed which has allowed me to recapture the enthusiasm that I had when I started in 1977 – and of which I am still proud.

All the articles are arranged chronologically, in three sections: Punk, Style and Speed. They represent about a quarter of my total output during the last eighteen years: they are presented as published, except for the correction of minor typos, some tiny edits for pagination purposes and, in the first section, some small cuts to compensate for the almost total lack of sub-editing at *Sounds*. (It wasn't until I wrote for *New Society* that I ever worked from a proof.) On a few other occasions, where the cuts on publication were hurried and brutal, I have reinserted a few lines from the article as originally handed in. Nothing more has been added or revised.

The reappearance of these articles is a tribute to the editors and colleagues I have worked with, and I'd like to mention the following, in chronological and alphabetical order: Dave Fudger, Vivienne Goldman and Alan Lewis (*Sounds*); Vale (*Search and Destroy*); Ian Birch, Michael Watts and Richard Williams (*Melody Maker*); Neville Brody, Nick Logan, Paul Rambali, Neil Spencer, Steve Taylor and Lesley White (*The Face* and *Arena*); Tony Gould (*New Society*); Jon Wozencroft (*Touch*); Harriet Gilbert and Malcolm Imrie (*New Statesman*); Ann Barr, Andrew Billen, Dylan Jones, Michael Pilgrim and Nick Wapshott (*Observer*); Neil Spencer (*20/20*); Joe Dolce and James Truman (*Details*); Robert Christgau, R. J. Davis, Joe Levy and Doug Simmons (*Village Voice*); Philip Dodd (*Sight and Sound*); Jill Morgan (*The Times*); Stuart Jeffries and John Mulholland (*Guardian*); Paul Du Noyer, Mark Ellen, Barney Hoskyns, Susie Hooper, Jim Irvin and Mat Snow (*Mojo*); Amanda Sharp and Matthew Slotover (*Frieze*); Jack Bankowsky and David Frankel (*Artforum*); and Graham Fuller and Ingrid Sischy (*Interview*). The introduction contains material from articles published in *Artforum* and *Frieze*; Jim Jones, Jon Wozencroft and Ian MacDonald helped with ideas and research. Thanks to you all.

No this dream won't ever ever end
And time seems like it'll never begin

Pere Ube, "30 Seconds over Tokyo", 1975

Punk

1977–79

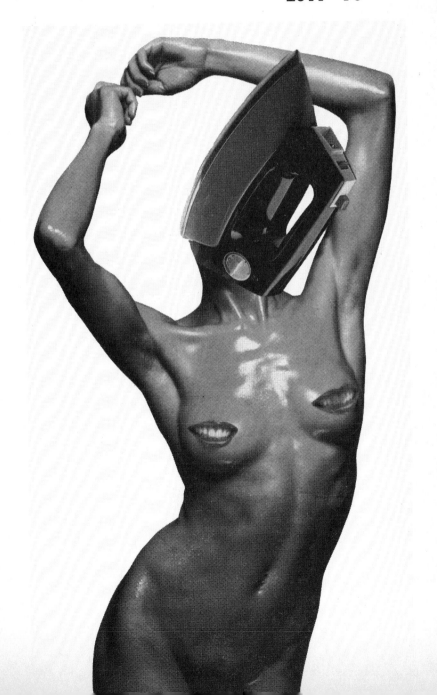

The Sex Pistols:
Live at the Screen on the Green

In which it must be conceded that Malcolm McLaren has a first-class media brain with a perfect instinct for theatre.

He can now have his cake and eat it – the media hype around the Pistols is so entangled that people will now believe anything. Always there are two or three different explanations for any given event or stroke pulled.

So, it isn't quite as simple as a band playing. Tonight, a fine set notwithstanding, theatre ruled. The Notre Dame incident was a good taste for the 'entrance game' – get in round the front, no, round the side, no, round the back, maybe *I'll* be shut out. Instant aristocracy – people fight to get in. And only the converted (or *very* curious) come to Islington on a Sunday night. But it is, at least, free.

After a longish wait, the audience of 350 get to watch a home movie collage of Pistol happenings – Swinging London camp (Young Nation), gigs, councillors, Derek Nimmo at Sex, *So It Goes* and the Grundy event. It's disturbing, cheap and nasty: the sight of Rotten filling the screen is chilling (someone's been watching *Privilege*). McLaren's shop gets plugged, the straight media is fooled, ritual bogey figures are there to be hissed at (press button). Us and them.

The Slits support – no competition. The fact that they are an all-girl band and that's great wears off into a headache after two numbers. But their musical ineptitude doesn't matter – having increasing presence and style as trash. Ari kicks a cymbal over/tempers flare – the circus bays for more.

More waiting, frustration and boredom. A good atmosphere for the Pistols to operate in. They've improved since last November. Even though the material is familiar – 11 songs plus two encores, the only newish material being 'EMI' and 'No Future'. In fact, they seem frozen in time: Rotten still berates the audience – not quite so wholeheartedly. It's only the rarity of gigs that allows them to stay still. Musically – Cook is the base, Jones runs through his guitar hero routines – Vicious stands legs astride, playing adequately, but he *looks* the part. Rotten is totally mesmeric: the lurker in derelict alleys, a spastic pantomime villain, with evil for real.

By now, to the audience, musical and stylistic considerations are all but irrelevant. The Pistols have become symbols – us against them – the songs anthems, inviolate from criticism. Just to see them is enough – it's a bonus that they played a good set.

So ultimately, the environment was totally controlled in favour of the

Pistols – no risks. I admire the media manipulation, but feel the sour taste of patronage and the exploitation of base brutality instincts. It's too easy. The eventual problem may be – who cares?

They're being overtaken. Fast.

Sounds, 9 April 1977

The Clash: Live at the Rainbow

I missed the Prefects, first on (sorry) – and caught the second half of the Sect. It doesn't make too good sense talking about 'halves' with the Sect as one song merges into another and you can't hear the lyrics. Style v. content. They're so like a movie of a band playing at the Rainbow, being ridiculously static, that I love them. Lead singer Vic looks as though someone's hung their old school-clothes on a peg and accidentally left their body inside them. Still, they don't seem to care too much and the audience don't either.

It's just strange seeing these bands at the Rainbow (with their titchy PAs) after being used to the view through an elbow at the Roxy, or the seedy funk of one-nighters like the Coliseum. The Rainbow is inhibiting: it's so plush and massive, and there are bouncers as obtrusive as the police at the Carnival last year. I mean, the stage itself is about the size of the Roxy basement.

At first, the Buzzcocks don't know what to do with it. Pete Shelley hunches over his Starway and spits out the lyrics; Steve Diggle just concentrates – bassist Garth tries a few lumbering runs that don't make it. But the music's great. The Great Lost Band – they aren't as assertive as other new wave bands, but another few months, plenty of gigs and (hopefully) record company investment and they'll be one of the best. Their sound is tight and controlled, carefully playing on their limitations. At the bottom is the drone of thousands of German bombers flying high – a grumbling growl – on top monotone vocals and rushed, desperate lyrics. Little inspired touches: the siren guitar in 'Boredom', and the car honks in 'Fast Cars'. Apart from the familiar 'Orgasm Addict' and '16', a new song, 'Whatever Happened To?', sounds excellent. Pertinent. They encore with 'Time's Up' and everyone begins to get loose.

There's plenty of action at the bar during all this, which only stops when the Clash come on. A social occasion, a gathering. A rarity in public as the supply of sympa places dries up – a new venue, please. And while we're on about it, the Seat Shock Horror was utterly unavoidable: people don't want to sit down, the music isn't about that at all. Movement and Energy. I mean, fixed seats are *totally ridiculous*. In the planning of venues for these gigs, it doesn't seem to be too much to ask that the few front rows be removed, as they were at Harlesden. They get removed anyway, very dangerously, so why not? And, furthermore, in contrast to the media idiocy, the atmosphere was very cool, relaxed even. They just aren't used to people leaping around and enjoying themselves actively.

As soon as the Jam arrive, we know that they're full of presence. They take that stage by the scruff of its neck and don't let go. The audience responds immediately. As usual, it's two-tone time – you could take these guys home to your granny. Very commercial, and hot with it – they're incredibly tight, flash and energetic. Non-stop bop: they revel in their and the audience's enjoyment. In fact the place starts going apeshit with a real excitement. Impressive. Only one real criticism: they steamroller 'Midnight Hour' and lose most of it – 'Batman' gets to be very tedious very quickly. (And please, not too much Conservative Party PR, hey, guys?)

And now we're in a different league. Simply, I thought the Clash performance here tonight was one of the best I've ever seen. Now – it's testament time. I last saw them in November 1976 at the RCA. A classic confrontation. And to me, they were so real, so raw, that I was totally turned around, provoked, galvanized into action. For that, if nothing else, my undying respect.

Six months on, they haven't lost that. They can communicate just as directly and devastatingly with 3,000 people, as opposed to 300. An amazing feat. Obviously, they've knocked off some of the rough edges, and what was once spontaneous has become a little more stylized. That's fine: to conquer the Rainbow, you just can't amble on – some elements of a *show* are needed. One of these is staging: at the back of the stage is a 25-foot backdrop, a blow-up of the back cover of the album or a similar shot. Next, lights flash – burning pink and orange, as well the more conventional colours. As soon as the band come on, there's an incredible electric tension – they're so much a part of London, England, 1977 that it's painfully intense. An awe-inspiring 'London's Burning': '1977' with Strummer framed for an instant in ice-blue for the last word – '1984'. Most of the material is from the LP: they didn't do 'Pressure Drop' (shame) and there was only one new song, 'Capital Radio'.

So, their performance. They've changed from their three-front-men days: Strummer is much more to the forefront. That leaves Mick and Paul much more room to play: and they do, beautifully. One neglected aspect, among the sociology and mythologizing of the album, was the playing. I mean, great rock 'n' roll, man! A sensibility second to none. 'Police and Thieves', where they stretch out, is a real moment. Strummer is emerging as one of the great front men. I could isolate it, briefly, to four moments. His involvement and encouragement of the drummer, the new kid, Nick, hidden almost behind the drums. His rush to the backdrop behind the band at the end of 'Police and Thieves' – mingling with photo-police, he stands apart yet *with* the rest of the band.

Suddenly, he holds the mike out to the audience, offering it to them. During the first encore, 'Garage Land', he reaches out into the audience, shakes hands and swaps his shirt for some guy's T-shirt. Look: the audience/performer barrier has been smashed in a rare moment of tenderness and solidarity.

A triumph … I'm thinking they could just have that once-in-a-generation thing.

Sounds, 21 May 1977

What Did You Do on the Jubilee?

Let's take all this sequentially: after an hour of waiting, the *Queen Elizabeth* left Charing Cross Pier at 6.30, and, after a moment's hesitation, decided to head downstream. If you aren't on the list, you aren't on. Nobody jumps … not even Palmolive. Bye-bye.

Begins very restrained – but come Rotherhithe, some booze and more food, and everyone gets mellow, if such a thing is possible. I mean, it's a nice evening (albeit a bit chilly) and there's space all around instead of tower blocks, so why be surprised?

The disparate crowd mixes surprisingly well – the only jarring note in fact is the refusal of the bar to serve doubles … never know what these notorious punk rockers might get up to. Downstream, we turn as a banner is unfurled along the length of the boat – red on yellow, it proclaims proudly '*Queen Elizabeth* – the new single by the Sex Pistols "God Save the Queen"', or something similar. Really low profile.

Inside, the conversation's covering some pretty recherché territory, but upstairs, in the covered area, the tapes start rolling. Dance. Great selection – moving from arcane dub to the Ramones, through Paul Revere and the Raiders. More boozing/dancing/yammering – general party patter – but expectation is heightened. They have to start playing outside the Houses of Parliament.

We repass under Tower Bridge, picking up a police boat on the way, but it falls behind; meanwhile, Jordan's telling me about this group she's managing called the Ants. Upstream it gets chillier – most take refuge in the downstairs bar (big boat this), ostensibly for a film that never happens. There's no pretence now: we're waiting.

More turns (Battersea funfair – for the detail-obsessed) and it's home-run time. The Pistols take the 'stage' – at the back of the raised covered area: the conditions are appalling, and it's amazing that any sort of sound comes out.

The main one is feedback – this delays their start, and is never fully resolved. Any blasé traces are swept away – pulses race/everyone rushed to be at the front. Pure mania.

Rotten gives up on losing the feedback, and the band slams into 'Anarchy', right on cue with the Houses of Parliament. A great moment. It's like they've been uncaged – the frustration in not being able to play bursts into total energy and attack.

Rotten's so close all you can see is a snarling mouth and wild eyes, framed by red spikes. Can't shake that feedback: he complains, won't sing for the first verse of 'No Feelings', but the others carry on. More frustration to explode.

By now the atmosphere is electric/heart thumps too hard/people pressing, swaying – it's like they have to play to blast them away. They're also playing for their/our lives – during 'Pretty Vacant' and the next song two police boats start moving around in earnest.

Now all adrenalin is flat out – do it do it do it now now now NOW. Suddenly in 'I Wanna Be Me' they get inspired and take off: 'No Fun' screamed out as the police boats move in for the kill is one of the greatest rock 'n' roll moments EVER. I mean EVER. (Think about that.)

Suddenly we're in the dock 'n' the power's off and Paul Cook's beating the hell out of the drums 'n' there are all those police and what's happening and what the fuck IS this …

We dock. The power is off. The bar is closed. Suddenly no more party. Suddenly a lot of police on the quay. Altercations begin. Nobody wants to leave. The police want us to leave. So does the owner. The owner can termi-nate the contract of hire at any time. Small print, baby.

Richard Branson loses his £500. Richard Branson doesn't want to leave. Tension. Indecision. People trickle off, slowly, after half an hour. Most stay on. More police. The police move on the boat. People move off. Nothing happens, bar a bit of pushing and shoving on either side.

Someone gets nicked. Now things start getting crazy. People are *aided* up the long gangway. Explosion of movement. Fear. Confusion. Flash/people running/ 'Get 'im'/crying faces/spin around/black marias/no objectivity/each for himself/ quick spurts of movement/hate/'You're shit.' And there's 11 people in the marias and we're on the pavement wondering what's been hap-pening. Very quick.

We leave. We go to Bow Street police station, via the Zanzibar (whose cutesy-poo decadence is sickening). No message. No bail. No press. No, not an IPC card. 'There are people we'd like to arrest but we don't know who they are.' A direct hint. Buzz off. And don't wait on the pavement. No help.

And then the seven of us *really* slip into *1984*: we move to this pub where everybody is enacting this weird ritual which involves the wearing of red/white/blue hats and 'singing' arcane folklore. They want us to join in/we can only make silly jokes out of pain. Zoom zoom/'I don't wanna grow up: there's too much contradiction.'

Chickens come home to roost, baby, you'd best believe.

Some jubilee. But look: I mean, McLaren's brilliant at the Theatre of Provocation, didn't he set all this up? To an extent. Provocation, yes, incitement, no. OK, I mean, all of us were expecting SOME interference, let's be frank – but not emotional over-reaction.

Whatever the rights and wrongs of the individual cases, objectively, yeah, responsibly, that's what it was. You know – nothing doing in the centre of town and these allegedly 'notorious-foul-mouthed-punk-rock-Sex-Pistols' … Image v. reality.

The charges run like this (approximately): Malcolm McLaren/'Using insulting words likely to provoke a breach of the peace'; Vivienne Westwood/ 'Obstructing a policeman'; Sophie Richmond and Alex McDowell/'Assault'; Dessert and O'Keefe/'Obstruction'; Ben Kelly and Chris Walsh/'Obstruction'; José Esquibel/'Threatening behaviour'; Jamie Reid/'Assault'. All have denied the alleged charges, and have been released on bail/surety until their case is heard.

No future, eh? Feelings are bound to run high. But wait. Neither 'side' is blameless, but there are a few things left to say: to a certain extent the barriers are down a bit more. That means if you look anything like a Sex Pistol, or a 'punk rocker', you're likely to get pulled in. Right: that means – no martyrs, no victims, no heroes, no stereotypes. No games on this score, no provocation. Things have gotten more serious. No escalation … ?

Uuuh. Um. Is it too late for me to say that it was a great party and that the Pistols were amazing? Oh, it is, but it's a shame, because that's a part of the evening too …

Sounds, 18 June 1977

Pulp! The History of Fan Magazines

In the beginning, there were stars. A lot of money had gone into the promotion of these stars (whom people 'worshipped'), and they didn't last long, on average. The people who'd put money into them wanted to extract the maximum money possible in the limited time-span available. It was also decided/discovered that the stars' audience wished to be reassured that their idol was accessible, normal, friendly, interesting, but entirely superhuman. The boy/girl next door (or even the neighbourhood tearaway, depending on image) but *unreachable*.

This formula lies at the heart of all the essentially exploitation (i.e. maximum profit/minimum time and effort) literature that is the subject of this piece. Sure, the emphasis changes from year to hear: in the mid 1950s, the thing was to be a family man/all-round entertainer, in the mid 1960s, a dedicated leader of fashion. Ephemera. Junk. And that's why it's fun, and (dare I say?) Important. There's no overlay of Art or self-consciousness (no time, when the quick buck's the thing): it's strictly contemporary, with all the follies/fads/attitudes of the times recorded in an instant for all posterity to see. For the flavour of a period, refer to Trash.

Hollywood Stars. In the 1930s, 'zines like *Photoplay* and *Picturegoer* (UK) set the format: lotsa pix, some 'candid', plus lotsa hot poop about the stars' likes/dislikes/lives, 'gossip', and frequent plugs. Vicarious glamour. Nothing is revealed although plenty of promises are made. Real Scandal? Nada: that was left to Hedda Hopper in the USA. Latest styles are displayed: in the absence of TV/youth culture, these mags were important transmission centres. Audience? Probably young marrieds/adolescents: there was no gap between school/job then: except for a *very* few, 14–21 year olds were expected to and did ape their elders. And no Cultural Respectability either: for the 'intellectuals', Film Mags were strictly 'shopgirl stuff'.

As Hollywood faded in the 1950s, its competitors, TV (specially), radio and pop/light music, hived off the leisure market between themselves. So – by the age-old laws of Exploitation – radio, TV and 'pop' annuals/periodicals emerge. We're mainly dealing with annuals here: an easy way of recycling material already used, they were also more attractive (colour pic covers and hard/cardboard binding) and could be guaranteed to sell lots over the 'silly' seasons – the Xmas/summer hols (and they tend to last, of course, which means you can still find 'em now).

So the TV and 'pop' annuals like *Discland* (1956) stick to the tried-and-

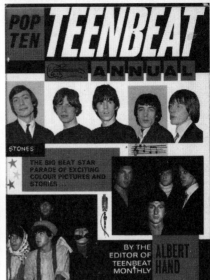

Helen Shapiro's Book for Girls, 1962   Teenbeat Annual, 1965

tested flick formula. In the latter, all the stars look middle-aged, and say middle-aged things: there's no hint of teen arrogance. I mean – pix of Ronnie Hilton at his work-bench in Leeds complement a piece in which he says how rilly grateful he is etc. … (no doubt); the only hint of a buzz comes in the cover shot of Dickie Valentine high-kicking, and an inside shot of Johnnie Ray getting his knickers ripped off. Don't worry, though: Dickie is snapped with wife and kid, and Johnnie? Well he jes' likes 'good home cookin''.

With rock 'n' roll, Teen Rebellion starts. Sure, but in Exploitationland the key word is still Whitewash: a sort of verbal equivalent of Elvis going into the army. Natch, the market becomes a little more specialized as the 'youth culture' becomes a potent spending force, but 'pop' is still subservient to movies. Just look at the way early rock 'n' rollers were shoved, mainly ill-advisedly, into low-budget quickies. Pop annuals and 'booklets' become more commonplace, but often they share a series or even the same book with film stars.

So, at the same time as *Elvis Monthly*, you have *Top Pop Stars* (early 1960s – sample headline: 'Vince Taylor the "Hot-Rod" boy who made good'), and the Fleetway (same house as those ubiquitous War/Battle/Air Ace Picture Libraries) Fan Star Library, in which the '*Amazing*' Elvis Presley rubs exciting glamour photos and behind-the-scenes peeps with Tab Hunter, James Dean, Pat Boone, Dirk Bogarde … Everyone is grateful, human (there are pix of Elvis as a kid – no immaculate conception, folks), fun and home-loving … both reassuring and distancing. Respectability was the ultimate goal: you had to play the game. (This changes of course: right now the way to make money is to be as offensive and snotty as possible).

This attitude continued up to the Beatles, and beyond. England started producing its own stars in the early 1960s: Adam Faith, Cliff Richard, Helen Shapiro. They're still robotic, loving their managers and jetting with their mums and dads. Like in Cliff's *Me and My Shadows* (*Daily Mirror*, 1961), in which he covers the last year of his life, in which – gosh – 'ever so many wonderful things happened to me' (obviously not ghosted). So it's Cliff as all-round entertainer (Palladium/movies) – loyal son – sex-pot (but clean): lotsa reindeer sweaters and greasy quiffs. This one's for Boys as well as Girls, which means there's an emphasis on Facts, Travel and Doing Things.

Now: *Helen Shapiro's Own Book for Girls* (*Daily Mirror*, 1962) assumes a propagandizing tone thankfully absent from Cliff's epic. The cover tells it – behind a lacquered Helen girls are drawn, knitting while watching TV, cooking, and, yes, making up dresses! Inside the theme is elaborated: after a pic feature about Princess Alexandra, we get strip cartoons which extol the virtues of Odette (heroine of the Resistance) and Amelia Earhart, short stories which revolve round Boys or being helpful to the dishy boss (in Publishing: must be Glamorous). At the end, 'Tony says: "I think you're something even better than a smart girl … I think you're a really *nice* one …"' True to the continuing but fast-waning (outside specialist markets) importance of movies at this stage (1962–3) are a series called *Top Numbers' Book of the Stars*, in which pop groups front a book of promo pics and barely rewritten publicity hand-outs which run the gamut of stars – right from pop to film, from TV to sport. Some are unintentionally hilarious, usually when you find an aspect of someone's past they'd rather forget: like Mickie Most as a rock singer, or Stevie Marriott as Buddy Holly's understudy.

With the Beatles, pop finally became respectable and took the place of movies as the mass leisure market, and the exploiters followed gamely. At first the change was in scale and emphasis, as movies started following 'pop'. More pure pop 'zines started to flood the market, no longer combined with movie stars. As an indication of the increased importance, pop started to appear in girls' comics, like *Rave* and *Petticoat*, alongside the trad comic-strip features. It was the era of Swinging London – Mod fashions – and comics played their part in the transmission of fashion changes and obligations.

Apart from the obvious Beatles monthlies, there was a flood of Beatles ephemera around 1963–4. Look at *The Beatles by Royal Command*: SEE Ringo bow as he shakes Princess Margaret's hand; WATCH them bow ironically; GASP as they breakfast with the stars. An even better one, by PYX productions (of course), is from really early days: 'At home with the Beatles' shows them still in their terraced houses making cuppas, ironing and collecting the milk bottles/interspersed with the inevitable pix of them in a barbershop. Just Ordinary Lads.

Another goodie from this era is *Give-a-Disc*; as well as centre-page shots of the Beatles and the Stones in their houndstooth jackets, and free disc ('Money/Swingin' on a Star') by the swingin' Leroys, you get the classic WOTS NOO?, or mainly for Mods by GREG, with bitchy comments like, 'Most of John Leyton's fan club are MIDS' – really, the worst. And there were so many others, like *Teenbeat* with Gerry's Mersey-side grin everywhere.

After this period of anarchy things settled down to being plain crazy. Through 1964–5 you have the girl's comics/*Top Pop Stars*/*Top Twenty*/*Teenbeat* annuals going strong, and that's just to name a few. There were also annuals tied to a specific show, or outlet, like the *Radio Luxembourg Record Stars Book*, or the *Thank Your Lucky Stars Annual*. And there were real fanzines like *Big Beat* (Pop 'n' romance magazine) wiv colour pop pin-ups, drama laughs and pop-packed articles. The tone isn't dissimilar to your 1977 fanzine: the authors are concerned to get to know the stars and to be on matey terms; breathless gossip is mixed with total trivia. Great! Best pic is one of the Pretty Things in their full rancid splendour, with the caption 'Those hair-raising Things!'

But really, the whole shooting match reaches its acme with the work of Albert Hand, the genius behind *Pop Weekly* and *Teenbeat*. Operating from Swingin' Heanor in fabby Derbyshire, Albert (veteran of *Elvis Weekly*) nevertheless managed to keep us readers up to the mark with the latest raves and all your own faves. Such was the (illusory) unity of the market that the Seekers could be struck next to the Who and nobody bothered. The writing is mandatory reading for all you would-be serious rocritix, cos Bert *knows* all the lads (behind the scenes news and views): with a few deft phrases he gets to the core – wottabout the Small Faces, then?

'Their name originated, quite simply, because someone pointed out that they did have SMALL faces. …' He puts the boot in the Who: 'They've got their reputation of being "mockers" of the scene and this is likely to hold them back when it comes to developing their careers on any sort of showbiz scene. Even though the business takes kindly enough to way-out characters …' And a flash of (haphazard) perception: 'Beatlemania can never be forgotten. It's like a world war: you can't live through it and not remember it.' Smaller wonder that come the arrival of 1967, drugs and the labyrinthine complexities of such as 'Crawdaddy' Albert got totally fazed …

Remembering fashion. The girls' 'zines were busy making as much mileage as possible from Swinging London. As well as sections in the weekly comics (down to *Rave*/*Petticoat*) and the more obvious fanzines (*Mod*/*Big Beat*), there were such goodies as '*Get Dressed*': (a useful guide to London's boutiques) edited by Millicent Bultitude with foreword by Mary Quant. If you spent your youth and money down the 'Gas Banana', or 'Foale and Tuffin', or even 'Hung on You' (where the Beatles bought theirs), this is for you, Lady Jane. *Fifteen* and *Trend Boyfriend 1968 Book* are similarly fashion-obsessed, but *Fifteen* carries the Helen Shapiro trend to 'nice girl' propagandizing to an extreme; quizzes about Good Behaviour, cartoon strips in the 'romantic' style, and subjects for discussion all tell you how to be a Good Girl, every inch of the dolly way. Should you forget the message, 'Write an essay of 150 words, and paste it in your journal'. Such folly represents a zenith of the concerned older generation's attempt to fight the war of decency on the enemy's turf.

1967, as ever, was the great divide. With the specialization of markets, the cross-over appeal of annuals became limited to the kiddipop bracket – *FAB 208* etc. – and the Monkees annuals (1967–9) come exactly at the time of change-over. Gushing and 'what's your fave colour' trivia became *unhip* as Pete Townshend was asked in the early *IT*s what he thought about his 'autodestructive' art, and the underground press became the interpreters of the new, messianic music.

So such exploitation quickies became the exclusive property of 10–17-year-olds, as each new rent-an-idol (Bolan/Slik/Bay City Rollers/David Essex et al.) was given his own annual. (Exploitation in the 'underground' zone worked in a different way.) Occasionally, the music press would shove out annuals – an easy way of getting an extra buck out of your writers over the 'sillies'. Still, in 1975, we find, by a neat reversal, ex-*Oz* editor Felix Dennis involved in the production of Kung-Fu quickies …

Mind you, in case you think we've lost that naïvety, look at the music press and, especially, many of the 1977 fanzines, which in their mixture of self-conscious outlaw approach and unselective gush suggest nothing so much as a marriage of *Oz* and *Big Beat*. Long may they flourish …

Sounds, 20 August 1977

Fanzines:
Every Home Should Print One

A few guidelines, for starters:

1. Something is mostly better than nothing. If someone has the energy/commitment to get up and produce a 'zine (no small effort), then it usually means they have something to say – and deserve our respect. Most people have something to say that's worth saying.

2. Now – if 'new wave' is an eezi-wrap package/culture, fanzines are the perfect expression – cheaper, more instant than records. Maybe *the* medium. A

democratization too – if the most committed/'pure' 'new wave' is about social change (a rhetorical question – it is), then the best fanzines express this too. Also: they're pluperfect ephemera, with no pretence to being anything else. Simply, anyone can do it.

3. But well? As 'Cells' say, 'It's what you say rather than how you say it.' Agreed, but such an approach has its pitfalls. The fact you can produce something, with a guaranteed audience, by yourself – no editorial *control*, no deadlines – holds the key to the best and worst in Xeroxland.

4. The best: a flexibility of format, which allows the freedom to experiment – i.e. one mag totally dedicated to one subject ('London's Burning', '48 Thrills 4', 'Kids Stuff 5', etc.), or guest issues ('Sniffin' Glue 11'). Editorial/commercial freedom allows extra musical/political comment given little space in the conventional rock press.

The effort required to produce a 'zine usually means that they're run by committed fans rather than professional observers. This (at its best, remember) gives an edge of enthusiasm/closeness both ways in interviews which often makes them more revealing than distanced music-press affairs.

Finally, and perhaps most importantly, outside blasé/saturated central London, they provide a vital function as a base/coordination point of the local scene. And that means Ilford just as much as Glasgow. That's where 'new wave' is spreading to – eventually new impetus, reinterpretation will come from there.

5. The worst: haste, the lack of any outside comment/advice, often means illegibility (the name of the game is *communication*), narrowness of vision and reduplication, both with the music press and with other fanzines. Most 'zines stick to a reviews/interview format – which tends to be a mistake, for several reasons. The music press itself is falling over to cover the 'new wave': it tends to be cheaper (15–18p, as opposed to 10–40p), quicker and better distributed. (Although it mostly doesn't have the attitude.)

By the time a 'zine comes out (usually after waiting/hustling for a printer), most of the news, etc. is out of date. With 40 or so fanzines flooding the market, and only so much punk product, reduplication is obviously inevitable – but the circle of Damned/Pistols/Clash/Heartbreakers/Clash/Pistols/Damned, etc. makes originality and choice real hard.

Another quibble is similarity of format, and limitation of vision. When Mark P urged in 'SG 5', 'Go out and start your own fanzines … flood the market with punk-writing', he didn't mean imitations. Just as when Rotten said early on, 'I want to change it so there are more bands like us', he didn't mean scores of paint-by-numbers Pistols.

The early 'SG's were startling – groundbreakers, secure in being at the forefront of a subculture ('We're the only mag who knows what's happening' – they were) – and remain interesting through the personality of Mark P, whose vehicle they (mostly) were.

Many fanzines seem content to have begun by copying the quickly hardened 'SG' format. The limitation of vision point is perhaps more obvious: a whole magazine devoted to the music within 'new wave' terms of reference ends up boring. To grow, survive even, the music and the fanzines must feed outside influences – they can't exist in isolation. And that needn't mean dilution … But 'begin' is the operative word.

6. Fanzines have already created graphic ideas (or re-presented them in a new context) that, in their best aspects, have been incorporated by the music press. More will follow …

7. They're still true to the 'Everyone will be famous for 15 minutes' spirit.

8. Maybe the word 'fanzine' is all wrong: 'fan' suggests a gushing, uncritical approach. Some writers get so excited (OK – a good thing) that the result is undiluted enthusiasm moving into sycophancy (crawling). This becomes tedious and does no one any good, ultimately. The best are frank and fearless, believing that if you're a fan you should criticize too.

9. The 'zines are usually better pictorially – suggesting that nobody has found a convincing way of writing 'punk' yet. Some of the tone of writing is forced – artificially tough . . .

10. 'It was easy. It was cheap. Go and do it.' Well, the first two may not be quite accurate, but …

Sounds, 10 September 1977

The Stranglers: Heroes or Zeros?

Ahhh – but these are testing times … now the very real euphoria has subsided, the scales have fallen from my eyes: not recantation, but re-evaluation. Timely sift and sort.

Oh, the Stranglers, such nice boys. But they *need* to be nasty, so squalid. And they do it so well. Look at Hugh Cornwell, standing on stage, posture saying, 'C'mon man, c'man get me, g'wan, I dare you …' They want to get up your nose. They want to shock. They want to confront you with the seamy white underbelly …

OK, OK, OK. So why did *Rattus Norvegicus* sell so well, then? Because they're bright and talented enough to translate their aggression and studied venom into direct musical terms: an instantly recognizable sound (which'll be hard to break away from) that scrapes under your skin and lodges there, even better as an irritant. You can't escape it. And of course they're heavy metal macho cross-over – perfect for the time when there wasn't much punk prod-uct and most were unconverted but … curious. And it was brilliantly pro-duced, and their constant playing paid dividends, and it was right in there with the then zeitgeist – all that stuff about rats and angry, suitably 'change'-oriented lyrix …

Well, here we are with a new product, all tarted up in a hideous – successful indeed as kitsch – chintzy chocolate-box-style sleeve. Inside, on cue, a rat appears – very reassuring. The themes of utter negativity, seediness, sleazo inputs continue, only, by the great law of Alice Cooper, a little more hysteri-cal, more strident, just nastier. Oh, look, more titles for 'liberals' to get fussed about: 'I Feel Like a Wog', 'Bring on the Nubiles'; and some creepy-crawlies: 'Peasant in the Big Shitty', 'School Mam'. Sort of like 'Plague of the Zombies'.

Oh, you guessed: I don't like the album. I've tried very hard (really, for all the 'right' reasons), but I still think it sucks. No, this isn't a critic's Set-'em-Up-and-Shoot-'em-Down exercise, nor a virulent manifestation of putative – new wave élitism. The Stranglers convinced me they had something when I heard 'Grip' thundering out over Portobello Road and couldn't rest until I'd found out who it was. I've got no axe to grind – but what I hear now turns me right off.

It'll sell. Half the album is full of very strong material: songs which are ridiculously catchy and well-constructed, and, oh yeah, they stay in the head … 'No More Heroes', 'Dagenham Dave', 'Bitching', and the best, 'Burning Up Time'. The rhythm section is simply very tight, relentless, while the organ that fleshes the sound out (and *does* bring to mind Seeds/Doors at 45 comparisons) holds some kind of magical power with its hypnotic swell, sinister undertone. Oh yes, they can do it …

But it sounds so *assembled*, somehow. And the material isn't as consistent as last time around; some of the songs, 'Dead Ringer', School Mam', 'Peasant in the Big Shitty', are plain awkward, embarrassing in parts … A problem is Cornwell's Lyrixstance, and the band's intrinsic and deep coldness. No amount of 'intellectual' rationalization can get round the fact that too many lyrics are dumb. Dumb – and Cornwell patently isn't. Like at the end of 'Burning Up Time', he goes into this 'Hello, little girl, want a sweetie …' rou-tine, and blows it. 'Bitching', with its 'Why don't you all go get screwed' refrain. Or the platitudes of 'Something Better Change'. Or the end of 'School Mam' … The rest of the band meshes so closely that his voice is given more prominence: under close security, it seems forced, trying to be tough, macho, too hard.

And the subject matter. 'Wog', 'Nubiles', 'Bitching'; point taken. Holding up a mirror, confrontation, etc. (although 'Nubiles' comes over most as being adolescent) – but who needs them as moralizers? Agreed that having your face rubbed in a cesspit can, on certain occasions, be salutary (shock/emetic).

Beyond a point, reached on this album, it seems more redundant, self-indulgent. I mean, we know already that England's 'going down the toilet', we've been told often enough. What to *do* about it? Because the Stranglers offer nothing positive, not even in their music. Look: the Pistols tell you we're being flushed too, but their music has a kick, a bounce, a tension that gives you energy, makes you want to do something. Some sort of life out of decay … The Stranglers rumble along relentlessly, zomboid, with sledgehammer blows driving their message home … they move, but they can be so wooden. Like a slamming coffin lid …

I suppose they got up my nose, didn't they? So they win in the end. Some pyrrhic victory, though. The music's powerful enough to get some reaction (always better than none) but what comes off this album, with its deliberate unrelenting wallowing, is the chill of death. No life force, nothing vital. Not so that it's frightening, just dull and irritating, ultimately. And it doesn't make it as a statement, even though it's all taken so seriously.

Oh, well – you can take it or leave it. They need this review like a hole in the head, so do you – no doubt you'll all buy it anyway. I know it isn't aimed at me, but it sounds as though everyone's intelligence is being insulted – yours, mine and that of this record's creators …

Sounds, 24 September 1977

Power Cut at the Electric Circus

Collyhurst Hill. A mile outside Manchester's gleaming inner centre. From the briefest of acquaintances this radius appears grim – low-density space, the gaps filled by disused Victorian industrial buildings, piles of rubble, anonymous housing projects … the Electric Circus stands near the crest of a hill. Opposite, there is a derelict 1930s estate, square acres of broken glass/graffiti/decaying stone, suggesting a post H-bomb no-life – on the other side stands a 1960s model, its forlorn attempt at modernity peeling. In all this, the punks standing outside the doors provide a brave splash of colour/togetherness: the Electric Circus is *needed* to a degree that's hard for Londoners, especially, to imagine. Tonight, the surroundings and music merge into a complete picture, indistinguishable to these eyes …

But – *let's rock let's roll let's deodorize/cauterize the night.* Walk through the door and the atmosphere cuts you like a knife. As the song especially made up for the occasion repeats, 'Eeeeeelectric'. Already there is the crackle/snap of a great night. Hip disco (ska/soul/new Virgin product – they're taping tonight for a possible album) segues into Warsaw. They look young and nervous. Desperate, thrashing, afraid of stopping/falling: 'What are you gonna do when the novelty's gone/You'll be back in the gutter where you came from' ('Novelty').

Quickly followed by the Negatives. At last! A group for whom the term 'minimalist aggregation' was coined. Led by local teendream/graffiti subject Paul Morley, fuelled by the sibilant saxophone of Richard Boon, driven by the powerhouse drums of Kevin Cummins, they honk and lurch their way through fifteen minutes (the concept of 'songs' is irrelevant) before eventually finding the lines from the Buzzcocks' '16': 'How I hate modern music, disco and pop ...' Of course, what they're making in fact is a Godawful joke racket, but that's the point: the feeling is such that they're called back for two encores and nearly get mobbed ...

The Prefects take some getting used to. As ever, the sound is terrible, but they're young (19), rough, but powerful. It's back to the old youth/growth potential argument. The music is utterly bleak urban noise, reminiscent occasionally of the Velvets' more monotonous passages – they are so *involved* that they demand some involvement back. Naturally, their approach holds problems, not the least the ideas/capacity ratio (far heavier on the former) – but capacity comes, and in the surroundings, their bleak, bitter approach makes sense.

Even more so the Worst. They look as though they've stepped right out of the industrial waste, totally uncompromising, blinking in the spotlight. No 'image'. They play not as though their life depends on it, but *because* it does. There's a hunger there. A three piece: the lead singer moves little, sings high – much is lost in the sound, and when his guitar breaks, they call it a day, with only one song, 'Fast Breeder', staying in the head. They're haunting, seeming to epitomize the evening's movie perfectly.

Maybe it's my energy lag, but The Fall don't move me. I register that the sound is better, that they're a five-piece with a female organist, but they don't prevent me from taking a quick breath of air. The hall is packed to the rafters, hot like an oven. Admitting that, you can take my recollection that they're competent but uninspiring with a pinch of salt. Or maybe I'm so hyped up on my *own* movie, and they don't fit it.

By this stage, it takes something special to retain enthusiasm/excitement, as one band begins to merge into another: the last three acts have it, in their different ways. Unannounced, Magazine play a short set of three numbers: 'Shot by Both Sides'/'The Light Pours Out of Me' (from the demo tape), and the old Buzzcocks number 'Big Dummy'. Their debut. Immediately, they're more musician than the other bands so far, capable of different textures – the sound isn't as clear or as confident as on the demo tapes, but this is understandable, and more than complemented by the visual presence of the band.

Still the centre of attention is Devoto: on stage he's a curious, compelling performer, awkward yet graceful, commanding yet ambiguous.

By now the bar's run out of everything except sherry. Upstairs, in the dressing room, they're singing through 'Nuggets' ('Dirty Water'/'You're Gonna Miss Me') with empassioned dedication, slicing through the sadness. John Cooper Clarke begins reading: small and skinny, reminiscent visually of 1966 Dylan, he declaims his poetry of wit and warmth fast, chewing gum, like wisecracking. The audience claps along to his rhythm. 'Daily Express' lists the horrors of that failing organ, with the punchline 'You won't get nipples in the *Daily Express*'. Or 'The Sad Tale of the Pest', where every other word begins with 'p': the explosive syllables fill the hall with popping phonetics. He goes down a storm: it's hard to imagine this in London, perhaps because up here there is no distinction between art/life. 'Art' or 'literature' is to be *lived with*, not revered or placed on a pedestal. A fundamental difference of attitude.

Well, the Buzzcocks have always been one of my favourite new bands and here they're on home ground and are the stars that hopefully they will be on a mass level. They've grown naturally and consistently, no hype, no fuss. Tonight, they are *on*. While they're tighter and faster, they haven't sacrificed their intrinsically understated and warm approach: in an age of hectoring, that merely adds to their strength rather than diminishes it. An offhand, dry resilience which will enable them to survive for as long as they want.

Their material is now very strong. Apart from the *Spiral Scratch* tracks, each song's got plenty to recommend it: 'No Reply' has a flash-fast call/response chorus like 'Time's Up'; '16' has heavily accented military drumming dinning in the frustration of the lyrics. During 'Pulse Beat', a long instrumental, they stretch out: John Maher keeps up the rhythm solo, leading to a couple of fine guitar solos wrenched out of Tony Hicks's axe (the Hollies – now there's Mancunian roots for you) by Steve Diggle, his face contorted in concentration. Another number, 'What Do I Get', is one of the finest pop songs I've heard this year.

From now on, things get heavily symbolic: the Buzzcocks end with 'Time's Up', and encore with 'Louie Louie' – remember the 'gotta go now' refrain? John the Postman and several others get up on stage to help, and then the audience goes berserk and rushes the stage … the plugs are pulled, lights turned off, 'This Could Be the Last Time' plays …

It had to end in lunacy. One of the finest gigs I've ever been to, with an atmosphere so charged it's stayed with me all week, given me strength. I felt honoured to have been there. It was hard, though, to escape the end-of-an-era feeling. The only way to look at it probably is to quote Shelley's introduction to 'Time's Up': 'If we get another place, let's make it better than this …'

I hope the 'if' isn't too hard.

Sounds, 15 October 1977

Howard Devoto: Heart Beats up Love

Whatever makes me tick
it takes away my concentration.
'Breakdown'

Being distorted selections from an interview (written and spoken) with Howard Devoto.

Who has, factually, among a lifetime of other things: formed the Buzzcocks, promoted the Sex Pistols at Manchester's Lesser Free Trade Hall July 1976, left the Buzzcocks unexpectedly early this year after contributing the words to the remarkable *Spiral Scratch*, and has been an important part of flourishing Mancunian musical activity. Still is.

The EP has become a classic: which could mean that it has entered the subconscious of enough individuals to enter that of the mass. Unusually, the lyrics are remarked upon as much as the music.

Devoto is now 'voice and occasional guitar' for Magazine. Who also are: Barry Adamson (bass), Martin Jackson (drums), Bob Dickinson (piano) and John McGeogh (guitar). Bob maybe will play some violin.

A demo tape of three songs, 'Shot by Both Sides', 'Suddenly We are Eating Sandwiches' and 'The Light Pours Out of Me', reveals music very different to the Buzzcocks period: open-ended, mature, haunting, commercial. Perhaps 'Shot' is nearest to the old sound (being in fact part composed with Pete Shelley): 'Eating Sandwiches' (the lyrix from Beckett's 'How it is') is jauntier, spry. 'The Light Pours Out of Me' begins with cool, metronomic drumming leading to a spare, deeper vocal, before reaching climaxes highlighted by high-register electric piano. A feeling reminiscent of Eno/motorik/disco, but more *definite*. Music full of light and shade, with gaps left for the listener to project into … the songs are still revolving within my head.

Record companies are *interested*.

Devoto is not easy to interview. Perhaps being aware that the media twist. As no doubt this piece has already. Or not wishing to give everything away. Why should he? The interview consequently takes place in two stages. Firstly, some questions are quickly thought up and related over the phone. At this stage, a shot in the dark. No meeting, little knowledge/surmise from scanty press and the EP. In between the questions and the return of the answer, an exploratory meeting. The demo tape is played. The answers arrive.

JON SAVAGE: Do you think we live in insane times?

HOWARD DEVOTO: Some of us do, some of us don't.

JS: On the EP, your lyrics are above beyond most punk bands … are you conscious of that?

HD: I'm not interested in being 'beyond' most people: i.e. over their heads. I get vertigo just like the rest. Sometimes it's better just to get under people's feet. What else can I say? I try to live up to my words, that is put my life where my mouth is and vice versa. Punk has made for a shift in perspective in what people look for in music and that gives me space to move. I want to move with the times but without going through the motions, if you see what I mean.

JS: Do you want to do something new with punk?

HD: Nobody invents new colours or new feelings. The first snowfall is fresh and in some sense new. But it isn't different necessarily from last year's snow. Nor is it old goods in a new wrapper.

JS: In retrospect your decision to leave the Buzzcocks was wise. Were you worried about being trapped in an image, and people's expectations?

HD: I don't understand how you could know it was wise. I don't want any credit for it. I didn't go through any pain making my decision: it happened naturally. At that time I don't think, and I can't think, I had an image to get maternal about. But I'm sure getting trapped in an image can be a big thrill. And expectations are useful things, like deadlines. You only get trapped in them if you let yourself.

JS: Do you improvise your lyrics?

HD: When I forget the words I make noises that sound like words.

JS: Do you change all the time?

HD: I forget.

JS: What do you want to do with Magazine?

HD: Improve people's memories.

JS: Do you want to be popular at a mass level?

HD: I can take it or leave it.

JS: Do you want to avoid being trivialized by commercialization?

HD: Being commercialized isn't necessarily to be trivialized. You can be on billboards and still go straight to people's hearts, and that is what you really want to do. It's not trivial. One of the reasons why rock is here and now so vital (at least as compared with other mediums) is because it's so commercialized.

JS: Do you remember dreams?

HD: I take pills to stop myself.

JS: Do you feel claustrophobic?

HD: I get in the same tight spots as anyone else. It makes no odds.

JS: Sometimes when you walk around the street do you feel alien?

HD: I have pills also to stop that happening. I give them to people who make me feel that way.

JS: Are you interested in Muzak?

HD: Not my type of question really (*commiserates with interviewer*).

JS: I'm fascinated by gaps in communication …

HD: I'm all for them. I don't believe in closing them up. I believe in trying but not succeeding. They've got to be big enough for an average-sized adult to pass through comfortably. There will be communication gaps until they've got the whole world bugged. There's something totalitarian about complete and perfect understanding. Do you see what I mean? They give you room to breathe, time to think.

JS: Do you listen to much music?

HD: Not my type of question really (*leaves the room*).

… These prompt a severe rethink and the desire for a further meeting.

That's fine – to my considerable interest is added the spice of a challenge. A further meeting transpires: we drive from Manchester through afternoon Oldham/Rochdale, the demo tape playing … ending at Birch Service Station, M62, to do the interview on cassette.

Devoto takes care over the answers, pausing frequently, face mobile as he searches for the right word. He is detached – glints of amusement appear sometimes around the eyes and mouth – but willing to talk. Some rapport is established – he relaxes. The conversation is undercut by taped Muzak/rattling cutlery.

HD: We were talking about gaps … I deliberately want to leave gaps, but it's also a matter of not trying to tell the whole story about something, when you can't. Not trying to make up the bits that are going to fit …

JS: Like writing about something and giving the impression you know everything about it …

HD: I guess so. But I feel really concerned about mistaken impressions, mishearing, and ambiguous experience. Just on straight sense things, when you think you've seen something, and it doesn't turn out to be what you've seen; I think there's a way of learning that as well, it doesn't always happen by accident.

JS: When we were talking about fanzines, you mentioned the phrase 'moral fibre' …

HD: Well … I don't understand what's *wrong* with the world, if there is anything wrong with it … but it might be something to do with what people call moral fibre. (*Quickly*) I just felt from reading an awful lot of things that I couldn't work out the stand that somebody has … like *Punk* magazine, or listening to the Talking Heads album … I just couldn't work out what this guy's angle was. It's a falling for style. But in fanzines there's a real opportunity for people to put in the little peculiarities of their lives …

JS: You seem to love language.

HD: It's once again where you think about phrases you've been using all your life, and you think, 'Where the hell did this phrase come from?' But I don't exactly love language, in many ways I've tried to get rid of it: like *Spiral Scratch* is *packed* from beginning to end with words, and I really want to try and get away from that.

JS: You said something about needing to burn out your rock 'n' roll obsession …

HD: I was just trying to get straight what Peter [Shelley] said to Caroline Coon [*Sounds*, 17 September]. This thing about how I wanted to feel what it was like to be a rock star, and once I found out, I wouldn't be interested any longer … which wasn't *quite* true. I mean, I suppose everybody wants to find that out …

JS: It's one of *the* fantasies of the age …

HD: But that was a very small part of the reason for doing anything.

JS: What were the others?

HD: Trying to find something to get excited about.

JS: Which often means you have to do it yourself.

HD: Mmm. Yes. I mean I've never been any good at sorting out what people meant by having a good time, having *fun*. That fascinated me for years … I don't think I ever found out what it is. It was like I had to substitute work for fun, going to parties, rubbing mean streaks with a lot of people. 'No fun' is a great phrase: I really don't understand this great plug that *fun* is having.

JS: You seem to be stating complex ideas in simple language in your range …

HD: Yes, but they're not worked out. I leave a lot of loose ends for me and everybody. Just taking certain phrases out of context and putting them together …

JS: 'I'm living in a movie which doesn't move me.' I love that …

HD: I think it's a very common feeling … it's just that I don't think a lot of people take notice of it. It's also: surely everybody has moments of wonder … This is a *free* country, surely everybody has them …

JS: The last question … why have you been wary of interviews?

HD: Because I know that this stuff, or some of it, is going down in black and white. I didn't write a song off the top of my head. Somebody might be trying to force me to shoot my mouth off about *anything* …

JS: I understand. Thank you.

Sounds, 5 November 1977

Never Mind the Bollocks Here's the Sex Pistols

Here am I – there are you – here's the Sex Pistols.

So I slip into it … It's very powerful. It's a very good rock 'n' roll album. It excites me. I want to dance. So will you. It's authentic.

It isn't aimed at *me*, at record 'reviewers'. The right perspective. Where it counts, that's the pockets, the pounds, the product shifted. As pop, what counts is that it's heard … parties, clubs, juke-boxes … I feel pretty … *redundant*. That doesn't annoy me. So why am I doing this? Well, remember 90 per cent of what is written about music is ego … Or, 'You wanna ruin me in your magazine … down with your pen and pad, ready to kill, to make me ill, wanna be someone wanna be someone Y'WANNA BE MEEEEE, RUIN MEEEEEE.' A great lyric.

From one of the songs that isn't on the album.

Oh, but you're so *trusting*: 125,000 advance orders? All those articles/reviews in fanzines saying, 'Haven't heard it yet but know it's amazing'? Got *you* sussed. Yeah, me too. To get factual: the track listing is – side one: 'Holidays In the Sun'/'Bodies'/'No Feelings'/'Liar'/'God Save the Queen'/ 'Problems'. Side two: 'Seventeen'/'Anarchy in the UK'/'Submission'/'Pretty Vacant'/'New York'/'EMI'. The singles are the same cuts.

Only three Vicious co-written tracks: 'Holidays'/'EMI'/'Bodies'/ The last the only one that hasn't been aired before. The older cuts – 'Submission', 'Seventeen' (a.k.a. 'I'm a Lazy Sod'), 'Liar', 'No Feelings', etc. have been changed around since earlier recordings. In the second, the now redundant *'You're only 29 …'* lyric (the generation battle's been won, what you gonna do with the space, kiddo?) has been dropped. I sorta liked all the messing around with release dates and all the bannings and record company swappings because it's pantomime and yet they're still out on the edge enough for the whole scam to cling nearish the 'Anarchy' byword. Ambiguous/double-edged so that nobody really knows what's going on till the last moment … nice. Anyway, 'conspicuous consumers always pay', that means whatever happens you and I will still fork out …

But come on: none of this would be taken seriously at *all* if they didn't play classic rock 'n' roll and weren't one of the best and most consistent *recording* bands this year. Look again at those advance orders – like the Beatles/Stones in the heyday. Or near enough.

It's been hard to avoid them in 1977 … I don't want to go on too much because it's really *all* been said ad nauseam.

But me? I'm here to offer *opinions*, aren't I? I mean, that's what this reviewing business is 'about', isn't it?

OK. I wanted to put in 'Bargain Bin' but they wouldn't let me. I wanted to write a one-word review. Y'know, write a book or write nothing. Why 'Bargain Bin'? Cos all double-think apart, it's a plain disappointment that there are four singles already released on the album. Can't get round that. K-Tel, etc.

'C'mon, I'm getting bored … How do they play, Savage?' Oh yeah … great, fantastic. Production's a wall of sound, Jones's guitar pure power chords with occasional subliminal overdubs, rhythm base hot, Rotten's vocal ever more stylish, ever more stylized. On 'EMI' and 'Bodies' his most sneering, obnoxious best. You know it all already; it's an all-out attack, the ultimate of rock as rock as outrage, as the big bad beautiful noise … very few albums ever top it for that – *White Light White Heat*, the first Ramones album, the only one coming to mind.

'OK. Tell us about some of the tracks … now …' – Well, 'Submission' has always been a great song and it's one of the best here maybe because it's taken at a slower pace than the rest and isn't a million miles away from being dopey. He wants to drown just like Jim Morrison in 'Moonlight Drive'. 'New York' has got some fun Dolls cops and piss-takes. 'Bodies' is 'about' an abortion and says fuck a lot and goes right over the top in being 'shocking' while remaining a strong song. 'Liar' isn't phased like on the bootleg but it makes no odds.

'C'mon, commit yourself … you haven't said what *you* really think …' Aaah, get off of my back. OK. Still think it doesn't matter as this is going to chart immediately but that's not true really otherwise I wouldn't be doing this jive at all, would I? A few general points: I'm not convinced that they've got enough impetus yet to produce stuff as good as the 1976 songs, of which there are nine on the album. But this is a good holding operation.

The publicity tactics? Unless cleverly handled (to date – brilliant) can become routine very fast (last year's rebels) but line has been walked with great skill. Interest still hyper-intense.

As rock 'n' roll, can't be faulted. As sound, dispels criticisms made by such as I about 'Pistols as media event'. Ashamed! I wuz wrong!

Noticed they haven't played live this year unless in an *adequately* controlled situation – abroad, late-night, private party, etc. Will they?

Still. Feel somehow that new songs smack of a creeping contrivance … and, yeah, what's fine on a single as blast wears me down with nihilism over an album. Believing in nothing, on plastic at least, makes you unassailable, but leaves a funny taste.

Sounds, 5 November 1977

New Musick

Program: present data suggest punk saturation/obsolescence in its present
form. Stagnation. Shock tactics used to gain space/attention now redundant.
Projex: post-punk projections, contrails. Print-out as follows.

Suggest that healthy future of mainstream punk assured. Already the term
redundant as new paths are charted. More overt reggae/dub influence, for
starters. Also fresh energy from regional centres, where they keep the faith
purer, and the USA. Any unity of the audience regretfully wishful thinking in
our private future, so the several strands delineated below (which'll all 'cross
over' of course) will be able to coexist happily or otherwise in the record
shops, 'zines and your hearts.

Your roots are showing, honey. Right now it's Iggy/Ramones/glam/ R&B/
1960s garage, to simplify. Major untapped (re-)sources for the future: the
obscurer side of the psychedelic explosion. Don't forget *they* all began in
1966–7 with their Bo Diddley mutated through the Yardbirds cops. Of course,
accelerated and more *knowing*: the ways we get fooled again are always *slightly*
different and maybe someone learns something and has a good time in the
process. Also (suggested by Greg Shaw) not as an experience but as a *form*.
Pere Ubu. C'mon, who needs LSD? We've all mutated that far in 10 years.
Look at what's in the air and comes out the TV set …

Clean teen-beat groups cleaning up and minting millions from where the
flamin' Groovies left off and starting where the Jam got boring. Synthemesc
nouveau pop and 'fun' along the way but so ultimately dire.

American reaction's already been mentioned but it'll be a lot weirder: we
don't *know* what we've started. The spectrum: from the clean teens above to
Satan's motorcycle slaves or whatever …

Aha! Somewhere in all there the 'New Musick'. Think of it rather as 'texture'. Sublimity. (As in sublime and subliminal). Nuclear nightdreams.
Waking thought sleep. Sensurround sound – the feelies. Roi Iggy is so much
older now than he was then: James Williamson to Eno/Bowie. Uh-huh.
Numb out. How's this? 'The absolute destruction of all the values it believed
in, the total loss of its position in the world, has put the UK in a psychotic
position – suffering from a mass nervous breakdown. One sign of that is a total
inability to face reality.' Oh. Now I'm blinded I can really see …

So catatonic bleakness. Withdrawal from our acceleration. Starting maybe
from integration of avant-garde techniques into the Velvets by John Cale –
from the opposite prism the 24-hour Technicolor dreams of Pink Floyd and
others that set the Germans on their course to the outer limits … Amon
Düül/Can/Kraftwerk/Faust (a few). Nico. Moving to mirrored isolation/mes-

merization – no feeling … Looking for a TV plug after the bomb's dropped. Or pure sensurround, but *active*: sound all-out to promote spontaneous physical reaction. Throbbing Gristle, Devo.

Or to harsh urban scrapings/controlled white noise/massively accented drumming – areas explored by the Subway Sect/the Prefects/Siouxsie and the Banshees/the Slits/Wire among others …

'It all sounds the same to me/manufactured in a factory.' Yes. That worries. Remoteness is fine but somewhere in the energy of punk lies some kind of life-force. *Love* the stone wall impenetrability erected by what this paper sees fit to term 'New Musick'. But there are enough cocoons around already to make the idea of more unnecessary. Who needs it? Oh yeah, remember that *I* don't know and if anyone else does they aren't telling: categories help lazies in their search for the nouveaunouveau. Lights on in your head: open the box. Don't take the money. Look where categorizations – punk/black/white/MOR/'England'/'Europe' get us …

Excuse me. It's too late to be late again. World awaits the *next* conceptual leap analogous to that one in the 1950s that shifted 'leisure' entertainment from movies into music …

Sounds, 26 November 1977

Penetration:
The Future is Female

Acceleration don't go to my head … London – a module, self-contained, trapped in an ever-accelerating time/style warp: a week seems like a month in our brave new world. Attitudes/reputations change in weeks, dissolve in days … Repetition, decreasing circles so close to the media encourage quick boredom, instant cliché … no doubt fun for the participants but to the out-side … well, confusing …

'It's hard to be trendy when you're 250 miles away … '

Caught somewhere between the spiralling capital and Newcastle – Ferryhill, 22 miles south, to be exact – are Penetration. Little local scene within which to put down roots and grow – 'People care but no one's opened a club – there's no focal point' – and forced to travel to London for manage-ment, record company, gigs, press. A young (19/20) punk band with remark-able freshness, excitement, invention within a form debased so quickly so deeply. 'It means we've been able to go our own way … '

The story is by now familiar: 'We'd gone to see groups for years – the New York Dolls in York, the Winkies, Silverhead – and I'd always fancied doing something … We saw the Pistols on their first northern gigs last May (1976). When I saw them, I thought "This is *it!*" I know it sounds clichéd now …'

Maybe in blaséd-out London, but outside – people keep the faith.

Pauline and Gary Chaplin, vocals and guitar respectively, had known each other for years; after several false starts which never got beyond the rehearsal stage, they formed Penetration in May, recorded a demo in June, have now played 40 gigs. They're at a crucial stage in their career: signed by Virgin on a one-off deal, further recording hinging on the sales of a recently released single: 'Don't Dictate'/'Money Talks'. The record company's hedging its bets, and the constant necessity for the seven-hour journey is proving a strain …

At Woolwich Poly they set up – hardly a break – after arriving: face an audience of 100 or so of the curious, eager, plain Saturday night apathetic – there's little to do. Vacuum tedium, in which they lose their edge, but persevere – arousing movement, an encore … Sound is basic punk, clear, confident, and adventurous: not believing it's dead, they keep it alive without thinking. On top: Pauline's voice – truly distinctive, able to warm and chill at the same time. Penetrating yet smooth, cuts to the marrow …

'It depends what the future brings as to what we do – it's a pretty crucial time at the moment. If we become pretty successful, then we don't really know what we'll do …'

Interviewed the next day, they're soft-spoken, determined, honest. Initial shyness turns to relaxation as the tape spins. Begins by mentioning that the single was a disappointment after the June demo: a collection of seven songs – 'Duty Free Technology'/'VIP'/'Money Talks'/'Silent Community'/'Firing Squad' and the single cuts – remarkable for Pauline's voice soaring above some sweet slicing James Williamson and some fine tunes …

'It's a bit watery.'

'If we'd had much more time to spend on it, maybe – I was happy with the recording of it, but when you're rushed straight from the recording to the mixing, and you're so tired after 12 hours' recording … it'd have been good to come back next morning and do it … we were just so tired – we didn't have any more time.'

Not to say that the single is *bad* – an oasis of promise and talent in the current product flood, but it could have been better … even down to the cover.

'We had no choice in the matter. Living right up there, everything happens down here and you just get something sent up. It's horrible – they'd already started getting them printed …'

'We'd like to do the next if we had the choice. We wanted to put mouths on it … don't dictate/money talks, you see.'

Torn between the two
Right around there is no answer
Don't tell me what to do
It's my choice I'm taking a chance
Don't dictate ...

A rare tolerance/integrity in a time of easily assumed hectoring and aggression. Matching their air of quiet determination – no modish toughness – and anxiousness to let the lyrics speak for themselves.

'It speaks for itself. It's not directed at anything, anybody – it's just general. You *could* read into that we're a very political band ... but it could be that it's about the mums and dads – person to person ...'

There's a lot of freedom
But not more choice
Habits never change
And there's only one voice
What will have to happen before the place will change?
Nothing to do and nothing to say
The Silent Community is here to stay ...

'It's funny though. We went into McLaren's shop – Sex then – and he had those Television posters and we started talking to him about the Pistols and he said, predicted, that there would be a scene in England, and it'd be massive. We couldn't imagine it. We wanted to go to New York – that's where we thought it was happening ... we knew the Ramones were *it* too as soon as we saw the pics ...'

New attitudes too. Like women in rock bands: it's for us commentators to find significance; for Penetration and Pauline it's natural.

'I just feel as though I'm a boy (*laughter*). Nooooo. You're with them all the time – I don't think of myself as being a *girl*.'

'We've never considered Pauline as anything different from just another member of the group – why *should* she be any different? It's person to person that's important ...'

So, after seven months they're still excited about being in a group/playing/something to say – still forming/planning for the future.

'We'd like to get our new material down now: we've advanced since we went into the studio. Got about as many new songs: "Life's a Gamble"/"Race Against Time"/"Lovers of Outrage"/"In the Future".'

The Future? well – it could be female ...

Sounds, 17 December 1977

The Jam: Live at the Hammersmith Odeon

Fragmentation strikes deep . . . as punk 'culture' is guided firmly into several easily categorizable (and therefore easier controlled)/marketable segments (divided we consume), it's clean teen night.

The disco plays Bad Company, Queen, etc. Safe. Making the point that the Jam are a neatly sealed cross-over product – energetic, 'modern', punkoid enough to sell to all those (thousands) who won't take the plunge but don't mind getting their toes wet. They've shown clearly that there *is* a market for tuneful fast 'fun' which'll grow in 1978. They fill it with panache and style. To these eyes/ears, it'll be totally dire – modern boredom – and I always thought one original aim of the Jam, among others, was to do away with the hideous 'product'/'market' syndrome – not to be fodder for another turn of the screw.

Well, the Jam have become stars and they let us know it. An hour-long wait for a 45-minute set. They run through much of their two albums: 'Change My Address', 'London Traffic', 'Carnaby Street', etc. – fast, proficient, professional. And so cold and sterile. Much of the material dissolves into a breakneck blur: the three-piece line-up (guitar/bass/drums) is used in a limited way – the pace is always frantic, the sound always treble, the vocals always shouted … ending up very two-dimensional. All shell and no substance. All on the surface …

The audience respond, of course, by Pavlovian reaction (i.e.'I've paid my money and I'll goddamn have a good time') *and* genuinely; but for all the atmosphere in the theatre the Jam might as well be wind-up dolls. For sure they do what they choose to do very well: leap around, put on a show, play dead sure. At no time do they break out of the predictability – no risks – or extend themselves. *They're* up on stage, unreachable, 'stars', hardly bothering to really communicate with the audience, *us*. At least the Clash, say, in a similar situation (big gig, 'image' to live up to) have the grace and humanity to try and break out of their image, to bridge the gulf between the audience/performer, even to *involve* the audience …

There are other, more disturbing elements. Looking hard at what they project/are on stage, the way it's presented, it becomes clear (no matter what they might say in 'interviews') that they are deeply conservative, if not reactionary. Dressed up, of course, in vague 'modern' clichés/form to snag the kids. OK, maybe it's too much to ask that they don't fall straight into the star trip, but there's more. Their image is of strict uniformity/blandness, directly. White shirts/black ties/suits. All the same. Cold, regular. The lighting is white. The only colours are the red of Weller's Rickenbackers – he has three, *exactly* the same – and the Union Jacks (one draped over a speaker, two on Buckler's drums). This last is ambiguous: fine as pop-art in 1965, understandable in the light of their unashamed Who revivalism (and why not, to a point?), but are they so unaware of its more sinister connotations in the year of the National Front media blitz? Of course. In these restricting times, though, any such chauvinism on a mass level, delivered so authoritatively/frontally, can only be dangerous in the long run, and I'm sure they are not unaware of that …

Sure they celebrate the modern world, but they leave things out: it's nothing so much as the bland world of drab concrete uniformity, security guards, carefully manicured intolerance, the C&A generation – everything unpleasant filtered out; and when it bursts through, others are the scapegoat …

Enough. Maybe they're just tired and naïve. I am fully aware that I've no doubt overstated, but I haven't disliked a concert so for a long time, for reasons coming from outside myself that are hard to define. I've tried to; maybe I'm wrong. One thing is sure: the Jam badly need the lay-off that they're about to get badly, because on this showing they've been spread thin …

Sounds, 31 December 1977

Tapper Zukie: *Man Ah Warrior*

An album full of dignity, grandeur and pride: smokey swirls intertwining to form a c(h)ord of steel …

Independent activity ensures that this is freely available in unadulterated form. Originally released in 1973, this is the real thing: essential.

There is variety in the sound: Zukie changes clothes fast: from the sweet through the playful to the diamond hard – talk-over voices vying – but no passion as a fashion. To be sure: fashionwear Zukie-style is the cloak of many colours, different guises through which to drive the message central to this album home: man as/is warrior, king. The strut/movement of the modern blues with a deep, serious intent. Unity, Dedication. Spiritual cleanliness. And yes, Rasta doctrines – but through the music, applicable to *all*. The musical crown and altar …

To isolate three tracks in sequence: the central 'I King Zukie' dominates the album, voices pitched against each other over a monumental, hypnotic bass riff. Rock solid. 'Simpleton Badness' mixes mysterious sound with eerie tape shrieks and moans – ritualistic, touching deeper/older nerves. 'Viego' makes a simple (and unfashionable) point tell through infectious catchiness:

Labour for learning before you grow old
Because learning is better than silver or gold
Silver and gold will fade away
But a good education will never decay.

Three out of 11: no significant quality lapse throughout.

Some general points. Opinion by way of explanation (thinking aloud). This is a product of totally different experience than you/I have experienced. It may be true (as has been suggested) that Rasta reggae shows superficial aspects of 'black hippiedom' – and much repetition. Mindless 'liberal' adulation hasn't aided understanding, let alone *feeling* – but there are fundamental differences. In the absence of any clear political leadership, music must provide much of that function – no doubt going some of the way to explaining why reggae is so didactic. And moreover, as such, it's an indication of a beginning: the first stirrings on a mass, concerted cultural level of a people, peoples rising up to take *their* place in history, rightfully. This history is inevitable. A thought.

What of us, offered this album, merely desiring a *review*, timorous products of a nation grown fat and spoiled (still), as yet still unrecognizing that we cannot live off the world any longer, as we have done for hundreds of years? *That* is one root cause of our deep malaise. We must take what we can generally there is much. An example: the current success of 'Uptown Top Ranking' (on an independent label, note) is merely the tip of the iceberg of reggae's 'commercial potential', so often unheard. This is to ignore deep strains of innocence, transcendence, poetry …

It should be mentioned here – to replay current pop-kid obsessions – that the MER label is run by Lenny Kaye, Patti Smith guitarist. This re-release is no doubt a payment in kind for strength drawn upon in the making of Smith's *Radio Ethiopia*. The central 'Ain't It Strange' – mysterious in its haunting power – reproduces the

Love is so sweet
Love is so pure
Love is something you must endure

lyrics from Zukie's 'I Ra Lion'. Just one example. For once, practical acknowledgement by white to black (our thanks).

Radio Ethiopia sleeve: 'Release (ethiopium) is the drug … an animal howl says it all … notes pour into the caste of freedom.' The greatness of this album makes me humble … and hopeful. The same greatness that makes the message, yes the *feeling* universal, that takes it beyond any 'market', any sect. Cliché turns to truth: one race, one human race. Savage will tread the thankless, shifting sands, dangers of 'white liberalism', 'woolly adulation', and in this case, perhaps, excessive praise, fired by the emotions/feelings that Zukie stirs. Courage, pride, conviction – so 'out of vogue' as spite and infantile debasement fester, and are rewarded in our 'civilized' society.

We are our own worst oppressors; we have much to learn …

Sounds, 14 January 1978

Power Pop!

Tick tock f—— the clock/the pendulum swings …

Oh it's so inevitable in post-punk letdown – the Pistols' 'split': sooooo symbolic – that attempts would be made to divert us from the more difficult task of facing the present and the future, into nice nostalgic alleyways. Play in *safety*, children! To be expected. Because punk as it began *was* and meant to be threatening, *did* and *does* offer *still* a context for change – not of fashions, but of attitudes and lives. This sort of thing, of course, is not to be encouraged, and, of course, isn't what rock 'n' roll's *supposed* to be about: I mean, it's all about fun 'n' barfing, eh, schoolgirls (corrrr!) and a piss-up, innit?

So anyway punk is finally put to bed (last year's t'ing) after being turned by countless shoddy imitations (inevitably) into another quik cheap fashion to gobbled up by the fast-fad industry and spewed out. Cut off dead, seemingly; any potential for growth or change to adapt to harsher, more realistic circumstances killed. Time for a new trend! But this time it'll be *nice*. So, suddenly, unbidden, largely unwanted except by threatened journalists not quick enough off the mark to understand punk and jump the 'trend', and record company execs bewildered as the punk tide swelled to an uncontrollable flood, appears 'power pop'! The Pistols had already made fools of the industry – so this time *complete* control!

Hello. Savage is this time around playing devil's advocate. Addressing himself more to the media/industry mechanics and implications of 'power pop' because it is in *these* areas that it lies, being *so far* largely manufactured out of thin air. As a 'trend': protests aside, it is being presented, accepted and swallowed wholeheartedly in certain quarters as such. True, it is created from elements that do exist – but as for any unity or real roots – forget it. IT WON'T WASH!

Constituent elements in this new 'sensation' appear to be a return to the 1960s golden age of pop – in today's context – and an insistence on being nice and 'popular' – wishing to be liked by everyone. The *Pleasers* … The previous editorial, in mentioning about 40 names, appears to be casting its net over a *ridiculously* wide area. Like going into a greengrocers', buying everything in the shop, and just calling what you've bought 'vegetables' … Categorizing the uncategorizable. And there's *still* very little full, clear or precise description what it is.

Well, anyway, 1963/66 for sure was the golden age you and I grew up with that but it can't happen again (regretfully). The very intensity of this golden age came from the compression of new ideas into a limited format: it was inevitable that this would fragment, as the limits shattered under the pressure. Pirate radio and the tranny culture was responsible also: come their shutdown and the control exercised by the studiedly vacuous and antiseptic Radio 1, what you heard as *pop* was severely limited and any division sharpened. It's no accident that EMI began their 'File under … ' system around the same time.

Neither had the record business or the 'youth culture' become the behemoth it is now. To that the incredible success of the Beatles contributed. Now it's division into markets, sections, 'cults': disco/MOR/Country/jazz-rock/funk/reggae blah blah – separate categories that are kept running together, the most popular or most radio-worthy of which feature in the charts. All the easier to control, offset one against the other, perpetuate. But no unity …

Point being that *now* most rock writers and record biz people were teen/sub-teen at the time of the Fab Four and *all that*. 'Carry me back … ' It's so damn easy and acceptable for all those old enough yet terrified/mesmerized by current problems, and for those who weren't there and want a surrogate, to float off into kitsch nostalgic fantasies – taking the Pleasers/Boyfriends/Stukas as indicative of 'power pop' as described here above all – of a time when the Beatles were kings of Swinging London and even Harold Wilson was fab, an era of no inflation as we know it, no crises, no unemployment, no acid writers … No bubbles burst, England still with empire. Aaaaah, but those were the days …

The past is always safe: it can't bite back.

Thus far, the groups who've allowed themselves happily enough to get involved with the whole schmeer have given little indication of any attitude other than to swallow/follow any lines possible, make ££££, and, yes, become another twist in the endless 'product'/'trend' turn of the screw. With a few concessions to 'modernity' thrown into the lyrics and packaging, and in their relations with the record companies.

It's not what they say but what they *do*, and so far there's little evidence of anything else. I'm quite happy to be proved wrong. And, of course, basically, good luck to them: it's better than being on the dole or working every day. What's offensive is the way they've been uncritically and excessively hyped as the *the* post-punk thing. IT WON'T WASH!

We have to deal with it/it will not go away. Yes. Retreats into nostalgia don't make it. And is it really so much 'fun' to be teenage? Mmmm. Maybe, but just *look* hard and you'll see the world isn't fun and is getting less so. Unless you choose to shut it out, blame others for the accelerating process of collapse. Don't hit me with all this updated and 'truly subversive' jive. It had to be admitted that punk was musically reactionary: fine as a shock to sweep the board, but further substance is needed to build it up again.

'Power pop' appears both musically and attitudinally reactionary – leading to a reflection of our political climate … and it *does* matter! Look: rock *is* the contemporary arena; for better or worse it's the only one we've got. You wouldn't be bothering to read all this otherwise, would you? To pretend otherwise is to do the ostrich and ultimately to evade responsibility. We know where that gets us, don't we?

Just a little more of your time. An alternative way of looking at things is to suggest that the promotion of 'power pop' is strengthening the divisions already present, defining ever more tightly, limits. OK – Devo/Pere Ubu/Magazine, let's say, have easily enough commercial appeal, given the plays, to be 'hits' of some sort and thus become, presumably, 'pop'. At the same time, they play fully contemporary music which happens to extend the medium – within the three-minute format.

To bracket these, and others, as esoterica is to do them and their potential audience considerable disservice (not the least by insulting the latter's intelligence), by limiting what they are supposed to view as 'pop'. Acting not unlike our fave Radio 1 'dee-jays'. Or you could consider John Cooper Clarke's twopence worth (which is as good as anybody's) quoted in *NME* 28 January:

> [The new wave] is the nearest thing that there's ever been to the
> working classes going into areas like surrealism and Dada … I think
> people in the new wave have done the smart thing and walked into
> those areas … I don't think I've ever seen a punk group that didn't
> have something imaginative about it …

So – room to move: maybe you all want to lose that. You're more than likely to by turning to 'power pop' as besuited clones try to turn the UK into the land of the bland. Confine yourselves at your peril! Just as *some* show signs of thinking for themselves …

But look, this is opinion, just like most of what you read: you don't have to believe me or anything. No *telling* you what to like. It's only possible to offer an alternative view to something that's been praised extensively, indiscriminately and uncritically. and if we're going to have 'power pop' (and there's room, c'mon), refer rather to 1970s models (mainly American) – Blue

Ash/Wackers/Big Star/etc. – rather than go back 15 years to hideous Merseyside mutations and 1962 Hardy Amies catalogues. Bah!

Such presumption! 'Power pop' has yet to prove itself convincingly as 'pop' – the main hit that comes to mind is the extremely attractive consumer's item 'Rich Kids' (red vinyl? It hooked me …) – and exactly where *they* stand in relation to 'power pop' isn't clear. And so far, so lightweight: as played in the UK, 'power pop' has very little 'power'.

Is this then the way the world ends: not with a bang but with a wimp … ?

Sounds, 18 February 1978

Throbbing Gristle: Tesco Disco

The interview takes place at Genesis P-Orridge's and Cosey Fanni Tutti's house in Hackney, east London. Away from the familiar, the capital is a concrete jungle – or concrete spiderweb. The approach east of the city is by degrees uninspiring, then threatening and terrifying, in a peculiarly English way, as National Front (NF) graffiti sprout on decaying low-density Victorian housing, the lot slowly subsiding into the industrial Lea Valley. The street where Gen and Cosey live is unremittingly grim: 1850s artisan housing – dirty brick façades, gaping wounds stretch the length of the street, broken only by a low railway, almost mathematically. Exactly the kind of street you can imagine Victorian murders of the cruellest, meanest kind committed, and no one ever knowing. Cobbles, grey sky.

Later Gen, Sheila Rock and I move to the Throbbing Gristle studio, the basement of a warehouse/factory at the edge of Hackney Fields. In medieval times, the particular area was used for plague pits! There's a chill in the studio so deep that even summer doesn't remove it. We talk about information, William Burroughs, the Image Bank. TG are nothing if not dedicated and systematic: the studio is filled with machines, tapes, filing cabinets. It might be possible, also, to underestimate their sense of humour …

SEARCH & DESTROY: What do you think the reaction will be to your music?
GENESIS P-ORRIDGE: *Confusion*. A few people will try and dismiss it as post-psychedelic trash. And some people will think it's a *bit* like the German groups, or a *bit* like ENO, or a *bit* like Donna Summer (*laughs*). It isn't really – that's like saying anyone who plays electric guitar is like Eric Clapton. It's a *cop out* – people don't want to think any more. It's the same old thing: people *do* want things in a little cell, categorized. …

S&D: You want your LP to blow that away?

GP: This first one is deliberately difficult. It's not easy to listen to, it's not easy to assimilate. And that's partly because of the situation at the moment. I usually dismiss it by saying to people that it's propaganda, but a lot of people (because of the name, and what they hear second-hand) think that we are actually a punk group – y'know, expect us to come on, be outrageous, play the drums, do lots of quick numbers … At least now, if they want to find out what we sound like, they'll buy the record, and they don't have to waste their time or ours getting something they don't expect. That's why there's no picture on front: because any picture at all would link you with a *type* – people go by record sleeves and quite often they make assumptions about the sound. The kind of images we like anyway could seem to be *heavy punk* – a bit heavier than the punks do but linked, which'd mislead people, or so bland it'd be a bit artsy-fartsy. So we thought this way: the only way you could guess what we might sound like is to buy it and listen to it.

The sleeve's done like an ICI Report … Also, with everyone pretending they're not interested in the *business* and all that, we've done it like a report for shareholders, only barer … Or else it's a bit like what you'd get at a museum – when you get an LP of New Guinea tribal chants or something. It's meant to be those two things, really …

S&D: How many did you press?

GP: About 750 – that's all we could afford. It's also actually a report: side one is taken at almost six-weekly intervals through the first year we were playing. So you can actually hear a survey of what we sounded like, how it changed and developed. Side two is a film soundtrack.

S&D: Sandy Robertson told me you played for one hour exactly …

GP: We have a digital clock and as soon as it hits the hour everything switches off and that's it. We want to get a little micro-processor to do that for us so we don't make mistakes.

S&D: What you're doing seems another forward conceptual leap …

GP: It's inevitable, really. What we're doing – it isn't that we're any more *talented* than anyone else – we just worked it out logically. It seemed inevitable … assimilate all that everyone knows this far, and then you think of what might happen next. We've been very lucky because Chris is totally into electronics and buys all the specialist magazines, and as soon as anyone even does research on something, we know about it. We know as quick as a 'scientist' what's going on, and what's *possible* … He gets circuits and builds them, tests them, and then finds whether they're of any use, and sometimes he alters them so that he's inventing new things for us to use immediately. So we're like a workshop as well …

S&D: You changed in 1976 – that ICA show – from Coum Transmissions to Throbbing Gristle, didn't you? Are you now concentrating totally on TG?

PROSTITUTION

INSTITUTE OF CONTEMPORARY ARTS

LEFT Genesis P. Orridge: collage, late 1977
ABOVE Throbbing Gristle: logo, autumn 1977

GP: Oh yes, completely. That's another thing: at the ICA, the opening night was meant to be the statement: we are now Throbbing Gristle and not Art; this is a retrospective of what we did as an art group but now we're going to take the perception of what we did as *active performers* (and all our ideas) but we want to use *sound*. It's what my mother told me to do! – I said to her, 'I'm pissed off really with all this art crap' (and I said we'd gone to Milan and done all this stuff and I don't feel very pleased about it). And she said, 'Well, sit back and think, *What do you really want to do?*' I told her I'd always liked music, and that's true – the only thing that's been continuous has been writing and and playing – cos my father was in a dance band and I was playing the drums at three. And also the only thing that always has a physical or an emotional effect on me is *sound* … Rock 'n' roll is metabolic music: it's a *global* event as well. Usually everybody invents a drug – everybody invents *sound* – but they don't all invent *writing* cos they can pass things verbally (like, through storytellers) so I thought, Why are we fucking about with all this sort of –

S&D: That *art world* can be –

GP: It's sick. The rock 'n' roll world is sick, but it's a blatant kind of sick … The art world is exactly the same as the rock scene, but much more pretentious and snotty – and it doesn't get through to people! The reason we were performers as opposed to painters is that we met real people and did things directly to an audience.

So anyway, we decided to concentrate on Gristle. Also Chris and Sleazy [Peter] are interested in that area and it wasn't fair to try and pull them into an area that wasn't natural to them – they just didn't like it – thought it was a waste of time.

S&D: What about punk?

GP: You read *File*, didn't you? Well, a year ago punk was interesting, now it's not important, it's become unnecessary. The most interesting people who happened to be involved in it will stay interesting, but that's because they're interesting people, not because they're *punk* ... There must be some: Howard Devoto sounds as though he *could* be – although I don't know enough. The best story I heard about him was that he'd left to do a thing on Burroughs and somebody said he was doing an LP of all cut-ups – but that apparently was a myth ... But it's just rock 'n' roll – I've got a big T-shirt which I've had for a long time, which says ROCK & ROLL IS FOR ARSE-LICKERS! because rock is based on popularity, and that the rock music scene – as big companies, contracts and tours – is an arse-licking scene. It *does* happen that some things that are interesting come through it ... the ones that they're all idolizing now, the Velvet Underground and Iggy, all sold really badly when they came out, and had awful trouble with the record companies ...

S&D: Do you think that things are accelerating?

GP: That's *general*, isn't it? Information being pillaged ...

S&D: What do you think'll be the end result?

GP: I think it works in favour of the people in control of the planet, because images are getting so blurred that everyone's helping them destroy the actual meaning or effect of *anything*. They don't even think, they just say, *Oh, great: somebody with a swastika, let's slam it in our magazine.* And that swastika syndrome, for instance, is working in favour of the people on the right wing – simply because it's making the swastika lose its symbolic power of offence. The more you see something the more you get used to it – after a few months you don't even notice – which is why/how TV is meant to operate of course: that's why the commercial channel is more important! The control system: they break it up and they switch it around so you get dead bodies equivalent to washing machines equivalent to comedy programmes and you don't know which is the real thing any more ...

S&D: It's all homogenized to the same level –

GP: You can walk in and think it's the news and you don't know whether it is or not. ...

S&D: Some of punk's general imagery – do you think it goes to kids' heads or just washes over their heads?

GP: I think it becomes another fashionable style and I don't think it goes deeper than that: that's what's terrifying. And that's why we're being much more careful about the imagery we use. The last song we did was just called 'We are Very Polite'. And we were saying:

> We are very polite
> It's very nice here tonight
> Everybody's being very quiet and waiting for us to say something
> important and there isn't really anything at the moment that we can
> think of that is important so we're not going to do any more songs.

That was the end of the song. Because there wasn't anything I could say with words that I could say with safety, without it being very dangerous to use anything *specific*. And we're very conscious every time we play to make sure that we're not wanking over something and reinforcing the *dilution of information* – which is what's happening.

S&D: Have you seen the Clash?

GP: No. I haven't met them either, but I have doubts. From the design of their clothes – the Rauschenberg look – and from the fact that they're a lot more sophisticated than they pretend – I think it's dangerous for the kids. You see all these people who are actually sophisticated intellectuals bullshitting the kids with their 'suffered in my 'igh-rise right flat innit'. That guy from the *Daily Mirror* said to me that the difference with punk was that all the kids were from high-rise flats. It's not true. In the Merseybeat days, in the Liverpool sound, an awful lot of those groups must have lived in council houses. Y'know, there must have been quite a lot of people in rock 'n' roll who at some point have lived in council flats. So why the big deal? It's just a fact of life, a certain percentage of the population happens to live there. By all means refer to it, but don't make it into a bloody religious symbol. It's probably better to say, *I lived in a council flat, it was pretty horrible, but look, I still managed to get something together and do something, so don't worry*. Instead of saying, 'Worry about it' say, 'No look, I'm having a good time, YOU can … ' It's become another hype wallowing in self-pity – and most of it is false self-pity – glam-rock glitter-rock, *slum rock …*

S&D: Are you optimistic or pessimistic about our future?

GP: I don't mind, you see: I find it *interesting …*

SHEILA ROCK: Yecch, these song titles: 'Slug Bait' –

GP: Out of 12 songs we had at the beginning … Sleazy liked it the best as the backing was so strange. I made the lyrics up – I have a title and a subject and I make them up on the spot. If you're doing improvised music, you can't have a fixed lyric, you can only have a subject. It's a true story …

SR: Is it?

GP: It's from the *Guardian*. In Rhodesia one of the first murders by the guerrillas was a young couple on a farm and the woman was pregnant. The guerrillas came and they cut the husband's balls off in front of the wife. The wife is actually Sharon Tate – they didn't kill the wife … It's a combination of the two … a condensation of that attitude, of that kind of human behaviour that you can get. Historically it's been going on all the time. It's a *standard response* – for some reason, as soon as they get into that fighting … a technique. The Russians did it in the last war …

S&D: Don't you feel it comes over as being very ambivalent? Whether you approve of it, or really enjoy it, or whether it's really terrible – isn't it open to misinterpretation?

GP: It is now! It wasn't when we did it – nobody else was doing songs like that. This is the *old problem*: if you refer to anything specific, you've this awful problem like the Fugs had – that it becomes outdated lyrically, and even if it's funny and quite good to listen to, you end up being bored and fed up with hearing about Richard Nixon … It becomes much more of a historical thing than something that's relevant … This song is about something that people will Keep On Doing: the guy next door might do it to me if the NF got in, y'know.

S&D: Is there much NF around here?

GP: Hackney got the highest NF vote at the last election, they reckon it'll have the first NF MP. This is where Mosley started in the 1930s …

S&D: Do you see the connections I'm trying to make? If a punk band says Violence Is Fun, or if the Pistols say 'Bodies' is fun – what happens if the NF comes along?

GP: They're being conditioned *not* to be surprised or horrified.

S&D: Don't you think you could be helping that along?

GP: I don't think so for two reasons. First, we don't do those songs any more because it's too dangerous, and because they no longer make the point we were making at the time. At the time they were contradictory to what everybody else was doing – in 1976, before the Pistols made records, the punks at the ICA thought the songs were really frightening. John Towe [ex-Generation X drummer] said he thought I was going to hang myself at the end – he'd heard a rumour … he was scared to death of us. There's all these weird stories going round … What do you think Death Factory means?

S&D: On one level I can take it as a metaphor for what's going on all around us – our consumer society –

GP: We're trying to create a sound that is equivalent to the experience – not a sound that is normal – at the end of one hour listening to us you feel like you've been through a condensed version of it – hopefully you might actually – people do actually come up to us and say they feel physically affected. Not sick – they just feel they've been through something special and strange that they don't understand – they want to talk about it.

S&D: Do you want them to go away and think about it?

GP: Well, I'm not worried because it's their problem. But I want us to project what we think as exactly as possible … We respond to the situation, which is why we play different music every time. Now the last time we played at Winchester, the people were 'nice' all the way through, so helpful, that we felt nice, they felt nice, and we played nice music, which they enjoyed … One bloke came up afterwards and said to us, 'I'm really pissed off with that!' and I said, 'Why?' 'Oh, I wanted you to be nasty to everybody and spit at them and insult them and beat them up.' I said, 'Why should I do that? No one's done it to us.' He said, 'Uuhhh – you should go out and annoy people …' and I finally said, 'Well, why annoy them if they're not annoying you? Why alienate them when they're happy to listen to what you've got to say?' At the ICA it was quite different: the people there were very hostile to start with, so we just said, 'You think you can be hostile, dear, you just wait until we start – we'll throw it straight back.' We did.

S&D: How many gigs have you done?

GP: We really try to do one a month: it must be about 15 … Cosey was interesting at Winchester – she said, 'People at first expect us to play music with melody lines and riffs and so on and they're a bit confused when we start and they realize that it's not built that way – usually if they realize that, after a while – you're dealing with sound as *sound*, you're not interested in dealing with whether it happens or it doesn't happen to be a *note*, but you're actually building sound like somebody builds a car – then they're all right again because they start listening to it. They rediscipline their brain to listen to it as sound, blocks of sound, units.

S&D: Do you want to move into video, movies?

GP: We've done a film … we've got a 16mm camera but we can't afford to use it … That's what was nice about the record; you've got total control over what comes out. At the end of the LP, all the sound on it is what's *chosen*. Live – there's an element that means it can never be exactly right.

S&D: Have you heard Eno's *Music for Films*?

GP: That's one I'd like to hear.

S&D: It's very fragmented – 14 tracks a side – a sampler – but it comes across almost as being a cut-up …

GP: Side one of our album was assembled from tapes over the last year … Y'know that Bowie didn't meet Burroughs until *Rolling Stone* brought them together – he'd never read any of his books until he did the interview. I actually asked W.B. about it. He said, 'Oh, I don't like all this modern jazz stuff …' (!) – that's what he calls rock (*laughs*). Burroughs? Oh, he'll hang around anywhere he can get free drinks. All the young people appreciate him more than his own generation …

S&D: I've been thinking a lot recently about those circles – Burroughs/1930s montages – Grosz/Heartfield …

GP: Well, the Dada thing was all about cut-ups and collages and all the type-setting that's now punk cliché …

S&D: And ransom note graphics …

GP: And ransom notes – it's a nice double-think: they've got this awful problem: now that 1984 will be here soon, what will they do then? I don't mean the effect, I mean the year …

It's like science – you get one person who goes so far, and he's reworking as far as he got, and younger people take it up … Burroughs is like that – his concepts of control through the word and media have got *so far*. You know what he does for money now? He lectures to the CIA on control systems, in the Midwest. I'd have thought it was obvious – nobody else understands what he's talking about, except the people who are doing it. The CIA – they know anyway, cos he worked it out from what they were doing.

S&D: I've been thinking about subliminal messages programmed into TV shows …

GP: It's actually processed into the way they say things and the context they say it in, but there's no subliminal cut-in … I've got a friend who works in BBC news planning – where they plan what the news is going to be – he took Cosey along to watch the 5.40 news – what they choose NOT to put in. They actually do *conspire* to do it – it's not just paranoia!

S&D: There's that Burroughs' paraphrase: 'A paranoid is someone who knows what's going on …'

Search & Destroy 6, February 1978

Devo: Are We Not Ready?

The first thing you notice on arrival at Köln airport is the modernity and organization. No baggage queues. No pre-fab ramshackle buildings. Instead you get sweeping architecture and (by comparison) lavish decoration: glass walls, marble, straight lines, expensive consumer items in glass display cages. If you're bored, you can find a chair with a small TV set on one arm: just insert pfennigs and watch awhiles. Immediate impressions indicate a clean dream …

Köln itself was largely flattened by Allied bombing in WW II: we thought-fully left the central cathedral intact. The reconstruction of the town around it results in grey, bland uniformity. Not depressing, just hardly there. Not dis-similar to Britain, except Germany seems further on down the Affluence road: 'A fake chandelier in every living room!' The shops parade the usual,

Devo in San Francisco, September 1977 (photo: Ruby Ray)

perhaps more blatantly, more confidently. Furniture, cameras, wiener fish bars, pastries, all the excess baggage of a country richer than us and which is in some respects five/ten yards on in time.

These eyes hunt hard for any cracks in the prosperous façade – Baader-Meinhof traces especially (Savage's naïve terrorist chic – Eno's 'RAF' runs through his head) – but it takes them two days to find any.

Conny's Studio turns out at the end of the taxi-ride to be a converted Victorian farm on the outskirts of Wolperath, a small village 30km ESE of Köln. High up and snow-bound: as a greeting, the weather turns and remains its coldest for many winters. The studio itself is a converted stable, its unpretentious façade hiding what could be one of the best computerized desks in the world.

Conny – Conrad Plank – has worked since the late 1960s producing such as Kraftwerk, in their earlier days, Neu, Cluster (with and without Eno), La Düsseldorf, Harmonia (Michael Rother and Cluster), Fritz Muller … his knowledge abut this end of German music is virtually encyclopaedic. Here Eno mixed down four tracks of 'Before and After Science' last summer: both are now working on Devo's first album.

Devo themselves are, to say the least, in an *interesting* position. No record contract, no production contract (as everyone is at pains to emphasize), no manager – Jerry Casale handles all that – and apparently little finance, yet they're in the middle of recording an album in an excellently equipped

German studio with Brian Eno producing and David Bowie expected to appear.

Meanwhile in England Stiff picks up the rights to two older singles, an astute and badly needed prestigious move, and promotes them heavily with ads that heighten mystique. The media guns are trained: word-of-mouth and oblique (admittedly) articles by such as yours truly have already started to spread the word. Information aplenty but still little insight: the mystique remains. This piece also, by devoting three pages to them, indicates an importance placed upon them that, in concrete terms, they could be considered to have done little to earn. Thus far, in the UK at any rate, Devo amount to two hard-to-find import 45s – 'Satisfaction'/'Mongoloid', the latter now released by Stiff – occasional adulatory press and a hefty cult. As for now, a media phenomenon, a gimmick almost, rather than a *band*.

Remember Magazine? Sometimes too much (uncritical) press can be counterproductive.

These guys are well ready for it, all the same. Jerry Casale and Mark Mothersbaugh have been working together for about five years, while the group as it is now – Jim Mothersbaugh/guitar, Bob Casale/guitar, and Alan Myers/drums – have been together for about 18 months. Hardly overnight sensations.

But there's a *lot* of pressure all at once. The pressures of moving out from Akron and Cleveland, Ohio, playing other American cities, emerging into the global spotlight as Bowie takes them under his wing. And apart from the simple acclimatization from the USA – this being their first time outside – they're here working what amounts to 12 hours a day in freezing conditions. While the UK media gets the hots, record companies continue to buzz round the band like wasps round a good fat honey-pot. Most of this devolves to one person, Jerry, but the rest must feel it. Hardly the most relaxed situation. Be stiff …

Into which *Sounds* blithely flies.

But who are these people, what do they eat for breakfast, what's their favourite colour, etc? Devo, learning fast about media manipulation, aren't exactly going to let it all out. This fits in with their chosen image: a corporate unit – Devo – with the individual members and their history unimportant: for instance, no pictures were to be taken of them without their Devo suits. It made sense: the main strength of Devo as a phenomenon so far has been their consistent, brilliant presentation of the group as a total package – music, visuals, image, ideology, language, films, each referring to itself and each other, and solidifying the circular links …

Although I tend to feel like the fly in the ointment rather than the fly on the wall (*my* paranoia), it's enjoyable enough. Most communication is with Jerry and Mark, the others being friendly but low profile. Jerry is the principal organizational force – he's taken on the chores of a manager – while Mark could be the spark at bottom, being responsible, if nothing else, for the pin-head routines and the synthesizer that's at the root of their sound. 'It's one of Mark's specialities, projecting insanity.' Both are creators of the band's visuals: familiar elements wrenched out of context, or once-used images represented in a different form. They've both been in contact, in various degrees of involvement, with the Image Bank, a Canadian art organization who could be superficially described as working in similar areas. The bands are *very* American, clever, and a paradoxical (but calculated) mixture of sophistication and naïvety.

In all this, the music is easily forgotten and shouldn't be. Conny's Studio is the first time where the band have been let loose in a 24-track studio. The two singles were recorded on a four-track, 'Mongoloid' in particular on a Revox in their garage in December 1976. No heating: the weather was so cold that Mark played with his gloves on. It could be why it sounds slow. The 45s are being re-recorded for the album – and even in their unmixed state, the versions are very different, 'Mongoloid' for instance featuring a drum snap/slap nowhere to be found on the original, where the drums are buried in the mix. The album will probably contain 12 tracks, including stage favourites 'Uncontrollable Urge', 'Too Much Paranoias' and maybe 'Gut Feeling'. Studio time is booked until early March: the group plan to come to Britain to play at least one date, probably the Roundhouse on 11 March. The album is scheduled for release in May or June.

The studio process in itself is simply unglamorous and very hard work. The group had gone through the first flush of getting most of the basic tracks down, and were in the middle period of getting the tiny elements right, adding overdubs, before the final remixing could begin. The picky bits. Remake/remodel, sift and sort, match and mismatch. This involves constant listening and relistening, constant decisions as to the prominence the various elements are to take in the mix, quite apart from the choice of the elements themselves.

It's a fickle business; some tracks seem to lag behind. Eno's role as producer is that of intermediary between man and tape, an interpreter almost; with 24 tracks also, organization is all-important. It's a difficult task, to balance the almost scientific quality of running through a tape for the hundredth time with the (gut) feeling that must remain. So far, the results were impressive ...

The interview took place in between breakfast and lunch in the studio. It was the hardest I've ever done. Barring Jerry, the group didn't (doesn't) want to talk in an interview situation, and the atmosphere of unwillingness and suspicion was strong.

JON SAVAGE: Can we start with why you came out to Germany to record?

JERRY CASALE: We were told to. We didn't know what we wanted. It was just as easy to be told where to come. It was through the Bowie connection, but we could go further than him, right now, if you know what I *mean*.

JS: Can we talk about the production on the singles?

JC: It would seem fairly obvious what the production was …

JS: Not really.

JC: It was – a combination of our degree of organization and the amount of money we had with what was available. So it represented really a random point in time. 'Mongoloid' was recorded in December 1976, and 'Satisfaction' in August/September 1977.

JS: I was puzzled by the cover art of 'Satisfaction', which was no doubt the idea …

JC: Yes. It was a parody of slickness. Those glasses were 3D glasses. Just Hollywood. A parody of sexuality – plastic tits …

JS: So what's the situation now with your record contract?

JC: Mmmm. We don't want to go too far into it, but don't be surprised if you see a big WB on the album jacket.

JS: There's that whole argument that runs that when you enter the business you get sucked in by it …

JC: I don't even think that's a question: you get sucked in. But then if the choice is between being sucked in or not being sucked in, I'd rather be sucked … I really think that's up to us. That's what becomes the creative process at that point – the creative process then is so inexorably connected with business that it's impossible to separate them.

JS: Can you explain the idea of Devolution?

JC: Devolution's a big idea about the way things are. Everyone has a big idea about the way things are whether they admit it or not: a lot of people's ideas masquerade themselves as non-ideas, which we find the most dishonest. Devo just has the biggest, best and most interesting idea about reality (!) that allows people to discover things, which is exactly what other ideas don't allow. Other ideas begin by ignoring what's there so their idea doesn't account for the whole picture. It's like when people thought that the earth was at the centre of the universe but the movement of certain planets didn't really match up to that idea – they couldn't make it match because their idea of what was happening was at basis wrong. And when the premise is wrong, everything else that follows is *sick*.

JS: Do you feel that our culture is accelerating; accelerating almost to the point of implosion?

JC: Implosion, right. Critical mass. Who knows how long it will take. In the meantime we're just providing the wake, a big party …

MARK MOTHERSBAUGH: Read-outs …

JC: In other words, rather than being uptight about it, we're just like … bringing the good news. The quicker it happens, the more we'll like it, personally.

JS: Normally you're conditioned to view the unknown with fear.

JC: Right. We got over that because we lived in Akron, Ohio. That area, that experience of living in America, made it easy to overcome that, because if we didn't, we wouldn't have merely imploded, we'd be invisible now, we wouldn't exist as far as you were concerned.

JS: Why Akron?

JC: It's in the centre of the most highly industrialized part of the United States. It's hilly, grey, like culturally stripped. There's one thing different about Akron though, and that's that it's safe. It made it really easy to just watch everything happening that was going on everywhere else but not really to be in it, but be aware of it. It wasn't so isolated that we didn't know what was happening.

JS: Are certain people Devo?

JC: What do you think of the guy holding a potato? Would you say he's Devo?

JS: Umm. About 100 per cent.

JC: Devo means everything good and bad at once. High and low Devo. When Booji Boy says it in the movie 'We're all Devo', he means it. But some people don't know it and are uptight about it – some people don't know it at all, like the guy holding the potato, and just act it.

JS: I couldn't believe that image, thought you'd made it up.

JC: Right. We don't need to make anything up. We just point things out. We select things and recognize them, and use them. Selection is probably the only artistic process – originality is a corny idea in the 1980s.

JS: With Devolution, what you're saying is that we've reached any limit of expansion?

JC: Right. The consumer attitude can only go so far. When you've eaten everything on the plate, what's next? Goo. Yeah. – Evo/Devo, consume/shrivel up. The idea that people have of themselves and their purpose on the planet has got to change.

JS: Devo to me is an example of a strong undercurrent, a wish to express 1978 disorientation, to break down the way we think …

JC: The breaking down musically has occurred – punk – and Devo are here to mutate. Devo's just the clean-up squad of the 1980s, the Smart Patrol.

JS: It occurs to me that people will be able to take the album on several levels: it's accessible enough for people to cut out the parts that threaten them. Many people are going to be threatened by what you're saying.

JC: I don't know about you, but I've always liked being pleasantly threatened.

JS: But you have to accept the fact that some don't.

JC: The people that need it most always hate it, but they come around.

JS: How are you finding working with Eno? Are you getting what you want?

MM: I think so. We weren't really looking for a *producer*.

JC: So we're getting what we want. He's sensitive to us, OK. I think it'd be hard for anyone here, unless they got some idea that I don't know about, to say that there'd be anybody who'd be any better. He gets good drum sounds, doesn't he, Alan?

ALAN MYERS: Yes.

JC: It's exactly what he was playing in his head, but that's very hard to transfer. I mean, to sound real, it can't be real. Eno understands that. It's just like a total movie set, recording and mixing. Making all your props look incredibly real so the finished scene transfers the illusion that it's happening.

JS: The comparison with the released singles works in your favour.

JC: Right. If people can like that, we can certainly surpass it. Because those were like black and white sketches. (*Pause.*) If we could add odour to our records.

JS: The Raspberries did.

JC: Oh no, oh please!

BOB CASALE: Put scratch 'n' sniff round, right on the vinyl, and let the needle do the work.

JC: Yeah. You have video disc and the odours come out different for every song.

JS: How would 'Mongoloid' smell?

JC: Pablum and bacon frying. Hospitals.

JS: 'Jocko Homo'?

JC: Is there any question? The zoo!

JS: 'Too Much Paranoias'?

JC: I think that'd have to be a chemical smell. Something to ring the alarm bells in our system.

JS: When did you start playing outside Akron and Cleveland?

JC: When they wouldn't let us play any more. April 1977. We went to CBGB's New York.

JS: What was that like?

JC: Perfect. We got on stage at two o'clock in the morning.

MM: Got into a fight with the Dead Boys.

JC: The crowd loved it. It had nothing to do with music – it was the aliens against the spuds. The Dead Boys attacked us on stage during 'Jocko Homo' …

JS: You must have really got to them?

JC: Sure. They took it personally. 'If the spud fits, wear it.' And the crowd loved it … we continued to play all through the fight and ended up looking good. Mark offered himself up first, being in the front line.

JS: What about the latex bags?

JC: They're in the movie, during 'Jocko Homo'. It's everything, like maggots, paramecium, foetal things.

JS: Disgusting but comfortable.

JC: Yes. That's the nature of research or birth or discovery. It's pretty disgusting.

JS: Another thing that strikes me about Devo is that there's a whole area of American life, the kitsch, that we just don't know about, and that you draw from.

JC: There is too much. More than you can think of. That's why Devo would happen in America, because it's like puking up the hairballs after eating too much. It's there you can reach the saturation level, the critical mass of consumerism first, because it's happening at such a rate … I mean, families will buy *eight* bags of groceries, and in there will be wiener wraps, cheese dogs, banana marshmallows, hostess hohos.

MM: Stove-top Stuffing! It's real surrealistic.

JS: Is there any music you get inspiration from?

MM: I like TV commercials. The musical content. Usually they're much more creative than anything you hear otherwise, because it's a free format to work in.

JC: No. I'm not the same. I hate TV commercials, that's maybe why I watch them at all. Personally I can't separate the *function* from the techniques or sounds. What'll happen to me is just like when you're in a daydreaming state I'll combine things. I'll take one thing I hate and another thing I hate with something I like. I don't ever like anything unless its purpose has been mutated. But there's no doubt that most of the best things Devo do start as a joke.

(Tape 2. With Jerry Casale the next day.)

JS: What was your involvement with Iggy?

JC: I was probably a superficial involvement. Iggy's always in a plane slightly obtuse to any kind of tangible relationship. He drifts in and out of focus.

JS: Could you tell me more about the origin of Devo?

JC: Devolution was a combination of a Wonder Woman comic book and the movie the *Island of Lost Souls*, the original, with Bela Lugosi, Lon Chaney, Charles Laughton. That was various things I'd been thinking about – devolution, of going ahead to go back, things falling apart, entropy. It grabbed every piece of information and gave it some kind of cohesive presence – it was a package. Just as our music and our identity exist as a technique rather than a style.

JS: There seems to be a new way of approaching rock 'n' roll: a few bands are emerging with their own ideology, package.

JC: Yeah. It's the next logical step. It will ensure the existence of vital rock 'n' roll. If rock 'n' roll's going to maintain its position, its purpose, then the emphasis has to switch, otherwise it'll become a vestigial organ, meaningless.

JS: You're in an interesting position now.

JC: Yes. We're like stored energy about to become kinetic.

After the interview(s), the movie – *The Truth About Devolution* – is shown a few times. Pleasantly threatening. Very friendly. Made in spring 1976 by Chuck Statler on location in Akron, it features a four-piece Devo, Jim Mothersbaugh having since been replaced by Alan Myers, and Bob Casale joining later. It runs for about nine minutes. The soundtrack consists of two songs, 'Secret Agent Man' and 'Jocko Homo', with synthesized siren whoops as fillers. Less of a plot, more of a mood, the visuals muck up your reality in the way that, say, the best Monty Python could: to the extent that, on switching on German TV afterwards, it was impossible to tell whether the movie had ended. Maybe it still hasn't. It turns media/rock clichés on their head in a dry, alien, cruelly witty manner. A good taster for the band: as such it is used to open their set ...

In the end, Devo appear sure *they* have the answer. By casting a loose enough net, they can make adjustments to suit any circumstance which will still fall within their scheme of things. There are still considerable areas of doubt: most importantly, we haven't yet seen them live. Reports vary. Although much of what they have to say is impressive, and makes good, clear-headed sense, they aren't entirely blameless of being wilfully obscure and clever for the sake of it, of using ten words and an oblique idea when five would do.

Some of what they say when broken down away from their (powerful) presence isn't so omniscient as it seems, and veers on occasions towards arrogance, and sweeping generalizations. It could be the arrogance of pressure, paranoia, or everyone bidding for you on a world scale, or it could be merely to provoke, to polarize, or to screen.

It doesn't matter now. The album so far signifies that Devo are well putting their actions where their mouth is, and more. Like the film, the album, as it was, was already shaping up as an attractive, yet disorienting mixture of the familiar and the cliché, mixed around and stripped to sound like nothing you've heard before, yet ... Exactly right in its remoteness. Still the most powerful music to come down the pike since ... Time will tell. They could be *the* transitional band as records give way to video discs – they're already waiting ...

Returning to Britain with its overt decay, rows of crumbling Victorian housing and cramped ribbon development ... its overwhelming poverty in comparison, it's only possible to wonder how Devo, with their neat yet convincing paradoxes of order out of chaos, and chaos out of order, will take us, and how we will take them.

Sounds, 4 March 1978

Cabaret Voltaire: Blue Religion, Grey Outside

Inside the house, an hour to kill before going into town. Hungover. Sit on the sofa and watch TV with the sound off. A tape plays. Flick the channels irritably – the synchronicity doesn't work – until the TV comes to a rest at a Disney dog show. Cut in, cut out. Fluctuating reception causes fading patterns in the electronic images. Time passes. No feelings. Blue religion, grey outside. The terrace opposite stops short in the grey air, thick with moisture, revealing vistas of factories, tower blocks, endless tightly patterned semis ... Hills in the distance. Sometimes the factories work at night – the noise can be heard in the house, filtering through dreams: dull, percussive, hypnotic.

Cabaret Voltaire – Richard, Mal and Chris – have been working together since 1974. They've played about a dozen gigs in two years, mostly in home town Sheffield. Their first gig ended up with Mal going to hospital; their first London date recently (at the Lyceum, supporting Buzzcocks) ended in a hail of feedback and flying glasses from the new conservatives after four numbers ...

'We got together as a mutual interest in sound: it was only later that the idea of gigging came up ... '

Their equipment runs to two synthesizers: one EMS, which is used for treating Chris's voice and organ, the other treats Mal's voice and bass. Richard plays guitar, and occasionally clarinet. Percussion is provided by a rhythm generator: sometimes live, sometimes prerecorded. Tapes are prerecorded, from any source, and run through various treatments ...

'I think we try and do things as immediately as we can. It's a fault in a lot of ways but it's the way we work – we try to be as spontaneous as possible and get down as quick as possible what we're feeling ...'

A Demo of eight songs recorded last October gives a fair idea of what the results are. Recorded on two-track in Chris's attic, it's satisfactory to them as a document of the time, but they could produce better now. Immediate impressions are of haziness and blandness: the songs are short, remote and synthetic. A cool yet harsh throb, slivers of sound slowly edging their way under your skin. Itchy. Few melodies, more a concern with sound as texture, with the possibilities of sound within the instruments themselves.

The synthesizers are used as instruments with tonal qualities of their own rather than to reproduce the sound of another. Nothing remains the same, nothing is as it seems: vocals, more recitations (think of 'The Gift') swim in and out of the mix: instruments appear, collide and fade … They're fascinated, among other things, by Kraftwerk and dub – yet the mix isn't as blatant as that: reference points. The constantly shifting sound, with instruments, random sounds dropping in and out of focus, is reminiscent of dub, the cool repetitious rhythm of German motorik …

'I'll be your mirror/reflect what you are … '

'Repetition's what we work on: that Warhol/Velvets thing – repetition becoming hypnotic … '

Lyric subjects are comments rather than statements. Preoccupations with the random violence and psychosis lying under the bland homogenized façade of our present mesmerization. The gaps between electronic reproduction – TV/stereos et al. – and the reality outside. The things that people do to each other, tortured by the psychic demands made on them.

You can read it in the papers every day. Beneath the superficially bland façade of the music, the cool texture, these are presented obliquely: sometimes you have to strain to hear the deadpan voices say, sometimes it's lost in the shifting mix. 'Capsules', 'Control Addict', 'Loves in Vein', and a version of the Velvets' 'Here She Comes Now' where the original is shattered into fragments, while the fragile beauty is retained …

Frozen, fragmentary, but always the throbbing, synthetic beat … This last makes it difficult to categorize them simply as experimental.

Stripped of such a daunting tag, the insistent pulse beat makes them accessible to anyone who wants to listen … Avant-garde smokescreens can put people off needlessly. CV don't see themselves as being anything, particularly (except maybe outsiders); they just keep on doing what has been coming naturally, as an extension and reaction to their environment and resources …

'In some ways staying up in Sheffield has helped us, because we've developed at our own pace and how we've wanted to, but in other ways we've not got recognition for what we've been doing as nobody's known about us, really.

'It's an urban environment. Full stop. I don't think any other city would have produced anything different.'

The interview takes place in Chris's attic, the four of us are barely able to fit in between equipment, tapes and stored data. The group are unused to interviews, and are understandably wary of the turning tape and dangling microphone. It's here that they've got the art of recording to a fine art: one corner is filled with taped records of what they've been playing over the last two years …

'We used to come up here, three times a week, say for two hours, and just keep churning ideas and numbers out – all those tapes there. Hundreds of them. About 70 per cent are still valid.

'It's our form of notation. We're that tape-oriented that everything goes on tape for our own use like people write diaries or whatever – tapes are the medium that we work in. I wouldn't say that everything we've got down is great: I think that at a certain stage I'd like to go back, not to repeat what we've been doing, but to use some of the ideas that we used perhaps once or twice …

'That's why we recorded them all …

'If you limit yourself, you can experiment within that scope – to set limitations is always an advantage.'

Live, they also use films as a back projection: the success of this depends on sympathetic venues but it indicates their desire to present a total environment …

'We hate the idea of one-dimensional stuff, purely aural …

'We've got two main performance movies that we use. They're just like barrages of images coming at you: there's no theme to them, they're just a source of images – some abstract, some figurative, some taken from TV …

'The latest one has got a 1960s French porno movie added into it, which worked out well in Sheffield last time. We'd just begun 'Here She Comes Now' when the porno flick started up … '

The consistency of their bleak vision and their uncompromising execution of it on stage only adds to hostile reactions from those conditioned to expect any performing group to behave in a certain way. Unlike the German groups, they don't celebrate the machines, but rather make clear their dreadful and ambiguous possibilities.

Maybe, in reflecting too closely and too accurately the grey featurelessness of our electronic privatization they do too good a job, in an industry working for escapism, to the extent that many people can't take it, like they can't take their own boredom …

'Control Addicts', from the tape, was composed after reading and digesting Burroughs's *Dead Fingers Talk*, futuristic nightmares of total control which may yet be avoidable but the unlikelihood increases every day … another song is 'Do the Mussolini (Headkick)' …

'That was from a piece of newsreel where Mussolini had just been killed off, right, and there's all the peasants standing around, and the corpses on the ground, and some geezer kicking the corpse about … '

Maybe such fascinations and the intention and meaning of the song *could* be ambiguous to a wide audience and be misinterpreted with unpleasant results? The group are taken aback, and spend some time in mulling it over …

'We don't try to make clear statements and set ourselves up … People can interpret it how they want. I suppose, though, it might be our playing with things beyond our control …

'I don't know what our reaction would be if that were misinterpreted. I suppose if that happened we'd make some statement … '

Right now CV are treading water. Sheffield and their isolation there, once productive, is by now frustrating. Even though they've been coopted into the small yet flourishing (and under-exposed) local scene and the demand is there, places to play are hard to find. They see little point in continuing to develop as fast as they have been as there's no one to hear them.

People's attitudes don't make it easier – the climate is retrogressive. Concurrent revivals. 'In ever decreasing circles'.

Later we all go to see Siouxsie and the Banshees play some 1978 music – their studied yet liberating movie of provocation and isolation. And good songs, fuck it! A brilliant show to a smattering of applause, while the disco changes gently yet firmly from 'acceptable' punk to computa-disco, as the disco kids got more money … get drunk and sod it …

Sounds, 15 April 1978

Siouxsie and the Banshees: Live at the Music Machine

Sweet Siouxsie and her boys in black play loud, angular, claustrophobic. Batter batter into submission: make you want to do bad things …

Play a full-set midnight Music Machine to a fullish hall: two encores. 'Helter Skelter', 'Mirage', 'Nicotine Stains', 'Make Up to Break Up', 'Metal', 'Hong Kong Garden', 'Carcass', 'Overground', 'Suburban Relapse', 'Love in a Void' and 'The Lord's Prayer'. A couple of others, a couple twice. If not a triumph, at least a headlined show of strength: is anyone listening?

'Metal is tough/metal is clean … ' – Banshees are a clean machine. With

Siouxsie and the Banshees, early 1978 (Bryn Jones)

disturbing yet liberating overtones. They don't actually *do* very much; it's what they *are*. Tonight, sartorial correctness in black and white. Guitarist stands right, stares and pouts sanpaku: bassist, bleach blond on black, provides spasms of bursting motion. Drummer concentrates on playing massive. Siouxsie swings and pivots in 90 degree arcs, legs and arms scything. Stylized, angular, full of latent violence and emotion, a perfect visual complement for …

The music: the Banshees aim for something grand, massive, structured: (mostly) they pull it off. Something always lies just outside their reach: that's good. Always first the pounding deep drums, then the interplay of controlled discordant guitar and booming melodic bass, Siouxsie's staccato vocals. At very first forbidding, most of the material quickly becomes addictive and commercial: Banshees have carved out their niche and stand alone …

'I was minding my own business/when my string snapped ...' ('Suburban Relapse') their most fully realized mix. Words, music, visuals. Very English, very suburban. Paraquat parties behind the privet hedges, the pebble-dash prisons that keep the occupants *in* just as much as they keep out the outside world. English remoteness, that stiff upper lip, emotional and physical isolation turning ever inwards into psychosis, unnameable perversions in deep closets ... Banshees tap that feeling, somehow, like the Pistols, strip off an outer layer, the façade, present it in a stylized and controlled (horror) show ...

Obverse, inverse, reverse and perverse. The Banshees don't do what you want them to do: you don't get what you want. Expecting nothing in particular, I was mesmerized at Sheffield two weeks back (which they'd prefer to forget): tonight, wanting them to be amazing, I was distanced by the Music Machine, its attendant ghouls, and the extraneous floor cabaret. Concentration wavers, the band seems on and off: sometimes they're almost boring, sometimes it all comes together with considerable power – 'Suburban Relapse', 'The Lord's Prayer' ...

Maybe this wilfulness, and the fact Siouxsie won't suck her thumb, makes people 'scared'. The unique is always uncategorizable and therefore 'difficult' – but then it *lasts*. When they're *on*, the Banshees are one of the best contemporary dance bands: all it needs it vision. Meanwhile, the band seem set to slip into a vicious circle of being 'underdogs' – it'll be interesting to see how they'll react to the acclaim and success that's to come ...

Sounds, 13 May 1978

The Psychology of Pere Ubu

I need a car that can get me around
Night in the city where the air can shine ...

Limits. Mostly, limits. Here, the black and white of print sets limits. Occasionally words can fly out and spark something off, but mostly ... The boundaries of this magazine (as to space, imposed by advertising revenue; as to content, shall we say, uh self-imposed?) set limits. A rock interview HA HA. Oh, you know, you know: a hurried selection of quotable quotes from a random (at best) conversation, mummified on chromium dioxide and then twisted and turned to fit time/space limitations. Any attempt to reproduce, recapture the effect of music (bar video sensurround) has to be at best approximate. But in print … It's just a joke, mon …

So we get back to the car and the air can shine. Pere Ubu see themselves as a pop band and there's no reason why you and I shouldn't do the same. They are also much more … An easy enough way to begin approaching them is through the images that line their songs: not that they're intrinsically 'difficult' to enjoy or understand or anything like that, it's just that the initial impact of their music as a whole is so strong as to make the 'Oh, here's a Chuck Berry lick' approach totally, irrelevant.

The quote at the head of this piece is a personal favourite and there for no other reason: it's not that important *per se*. But then they all are. And maybe it transcends the black and white limits: that's also much of what Ubu are about – in their own words 'expanding the borders of expression' *and* pop(ularizing).

DATAFIX: Right now Ubu have just finished their first tour of the UK. It's been vital for the band in some ways. An illustration is the fact that, of the thousand or so copies of their last 45, 'The Modern Dance'/'Heaven', pressed on their own label Hearthan, 700 were sent to England. The previous experience of other bands (take Blondie as an extreme example) has shown that the UK is a good place to bounce back off with enough momentum to even begin thinking about cracking the vast and notoriously difficult American market.

Ubu don't find it particularly easy to play in their home town Cleveland even after three years together, let alone shift enough 'tonnage' or even 'megabucks' (to use two hideous terms currently favoured by the American rock biz as the conveyor-belt process *really* takes off) …

In a previous communication, David Thomas mentioned that the only way that he knew how to present Ubu was in performance. That's why this piece has to be at best a poor substitute. That's why it was delayed until after seeing them three times, but … On stage Ubu deliver and surpass: in a studio they seem to be freer, more extravagant. And their control and strength are the more apparent; one example is the quality of their sound, which is excellent – no crackle no hum proper mixing – and allows them to transmit exactly what they want (an object lesson to others) …

… the name Pere Ubu (Ubu for convenience). Beyond the direct association with Jarry's plays, it's unspecific enough for you to project into it what you want, even on the level of pronouncing it in different ways, shortening it, whatever. Ubu as a group entity is also strong yet unspecific enough to allow the five grown, very different, individuals that form it room to move and yet contribute what they wish. Similarly the music created on record and live is loose yet strong enough for you, again, to project what you will into it: it's not *necessary* (I mean, Ubu rock out amply) but they do repay attention. There is depth there: they do think about what they're doing. Ubu: a rare mixture of

Pere Ube: generic tour poster, 1978

intellect, emotion and instinct …

The last is also 'important'. One reason why Ubu aren't an analytical band. They don't have their approach verbalized and mapped out in easy quotes – no Ubu by numbers. Neither are they particularly enamoured with the star-making machinery or the whole gossip/marketing superstructure erected over what is after all our finest growth industry. This is fairly obvious from the way they dress on stage – they don't change their everyday clothes – and their reluctance to pose for any photographs outside a concert. It's not that they're 'difficult' as people (none could be friendlier or more polite), it's just that to them it seems fairly irrelevant to what they're doing, which they tend to see as part of a life's work rather than a quick way to 'stardom', chicks and quick megabucks.

Neither are they at ease during interviews … This time, David Thomas is relaxed enough by his standards, but it's plain that he finds the process frustrating, as he talks at a rush, in tangents, hums and hahs, buzzes – fighting to get the words exactly right. Of course, it's impossible, but he makes an attempt. Bassist Tony Maimone is also present, balancing, more poised and precise. An overcast day on the Gloucester Road:

JON SAVAGE: On stage you come across as being very disciplined …
DAVID THOMAS: Mmhmm. Discipline in *any* art is one of the most vital
things, y'know – it's important to us because … we try to allow room for
accidents, for a certain amount of change … from performance to
performance and to be able to maintain. To be able to do that, to be able to
change and still maintain a viable, translatable thing, you must have
discipline, y'know.
JS: And if you're that way on stage maybe the audience will react in the same
way …?
DT: To have fun, there must be work first, y'know, and to work there must be
discipline – so 'fun' is discipline in one way. (*Pause.*) That's not true! But
there are some performers [here an American artist is mentioned] … the
thing is when you incite something and then when it happens you act real
hurt and, uh, like 'Why are you doing that?', that's bullshit. If you incite it,
then you have to be able to take the consequences … That's why discipline
is important. There *is* a difference between us and the audience and that
difference must always be made clear – I'm not saying it's like a rock star
thing, but … they've paid their money, we're going to perform, we're gonna
do our music … emotional incitement is easy …

JS: That Ubu are very aware of this is clear. The studio version of
'Sentimental Journey' begins with bottles smashing, which the live version
omits entirely. This extends even to their refusal to put 'Final Solution' out
on the 'Datapanik' EP, or even play it live often, because of any potential
associations with the Nazi programme which, it is to be admitted, hadn't
even occurred to this writer …
DT: We thought of ways that we could put that in 'Sentimental Journey', but
there were no ways it could be done … You can't break glass and expect no
consequences. You can't control that … We've found that in Cleveland our
audiences have become very self-disciplining – like if there's some arseholes
or something, the people in the audience take care of it.
JS: And anyway Iggy defined that particular number for all time ten years
ago …
DT: Oh yes, yes. That whole violence thing is just old, it's just old. I'm not
interested in it. It also doesn't lead anywhere. Like also, self-destruction is so
easy – it's the easiest thing you can do …
JS: It's still surprisingly popular …
DT: It's very bad. It's just too easy. To destroy yourself all you have to do is let
it happen, y'know. It's much harder to work for discipline, to work not to
destroy yourself …

Here is as good a place as any to mention former Ubu guitarist and writer Peter Laughner (responsible for 'Life Stinks' on the album, and, in collaborations with the others, both sides of the first 45, 'Tokyo'; he also wrote the brilliant sleevenotes – mood – for the same 45 …), victim of the drink 'n' drugs-as-glamour culture. David doesn't particularly want to talk about it, feeling that the subject has been put to bed in a recent interview in *Search & Destroy*, no.6:

I think it's a mistake to romanticize it … Peter was an extremely talented musician; he was also an extremely talented artist. He was also a fool, and the fool killed him. That's all that *should* be said about it.

TONY MAIMONE: There's the energy thing too: to really channel energy, coordinate energy, it takes time. You have to work at it.

DT: It takes a machine. Like a machine made of humans, not like mechanistic … which is what science fiction literature, a lot of the ideal literature is always concerned with – the ideal interaction between a group of humans, y'know. *That* takes discipline.

JS: The vision is …? [This causes difficulty.]

DT: … It's just what everybody – there is no … I mean, I can't say it's this and this and this. The vision is just to push further, y'know … is to, is to, uh, one of the favourite things that was ever said about us was when somebody said that we were expanding the borders of expression, y'know, and that's what I want to do, that's what everybody wants to do. We never talk about this stuff … so … but I know that's what I want to do.

JS: I was thinking about the songs and they all seem equally 'important' …

DT: They're all part of a pattern, but if there is one that is the most important it *is* 'Sentimental Journey', because that's the final … I don't know what the implications of 'Sentimental Journey' are, but I know that it's important. Nothing is really endorsed, we don't say anything … this is good or this is bad, y'know: it's just pictures – it starts in a house, goes out on to the street, goes back into the house, goes to sleep. Obviously there's various things … the implications are drawn from them.

To do 'Sentimental Journey' really well for me personally I have to be real tranced out and I have to really prepare for it, which is why 'Over My Head' is real good because it goes down like that …

JS: There's a real dream-like quality to 'Sentimental Journey', and to those three songs in the middle of the second side of the album ('Real World', 'Over My Head') …

DT: The second side is paced real nicely. It's (*pause*) a sort of dream-like quality – it's a sort of waking dream, that sort of thing, it's not like a sleeping dream. I don't dream in my sleep for some reason – I haven't dreamt at night for three or four years, except for occasional dreams. I don't know anything about dreams, I don't know if anyone does really: it's too hard.

What the band is doing is concerned with views of perception.

On stage they play every song they've recorded (and indeed most aren't recorded until they're worked out live) except '30 Seconds over Tokyo' and 'Chinese Radiation'. 'Cloud 149' and 'Final Solution' are changed around some. 'Chinese Radiation' is the only song that was worked out in the studio, and is one of their elliptical numbers …

DT: Weeeeeeeel. There's various themes in there that have to do with various things. I personally feel that I could have done my part of it better, but I don't … but that's past. One thing that might help is that the crowd noises were taken from Shea Stadium – Grand Funk Railroad unreleased tapes – that's almost trivial, but there's a certain implication in there that I found real relevant, personal to me, real funny … there's a whole series of things in there involving rock 'n' roll involving American culture involving Chinese culture – the original title from the song, the association came, because every … when the Chinese blow off their nuclear bombs in China, the radiation for some reason always comes over Cleveland.

Obviously, Cleveland is important to them as their home town, shaping their attitudes and music: what do they think about the current 'Cleveland' media hype? They must feel pretty weird about it …
DT: Oh yes, because the reality of it and what they're talking about is so far apart.
TM: Downtown Cleveland is a very dangerous place at night if you don't know where you shouldn't go. I mean all this mystical shit about Cleveland. It's just a regular city.
DT: There *is* a mystery about Cleveland but it's really something that's not easy to communicate. The whole thing … I think Cleveland has a real chance, a real future, if somebody with a vision gets in there in charge and hangs everything together – because it's hit bottom, absolute bottom, stripped to the bone.
TM: Standard Oil has like huge refineries there, Standard Oil, US Steel and Public Steel, those are really big.
DT: All the industry is real industry, that's all. All you have is industry, a place to live, and places to buy your food, places to buy your entertainment: it's just stripped down to that, that's all there is. We saw more people here on a street corner last night at midnight than you'll find in *all* downtown Cleveland at nine or ten practically. From the centre of downtown to the Steel Mills is maybe a mile … But Cleveland – as opposed to other cities, you never see a skyline here (in London), you never see the city versus the

world. But in Cleveland it is, the city versus the world …

JS: There's a lot of space in your songs …

DT: Yes, there's a lot of space in Cleveland, a *lot* of space …

They talk a good deal more about Cleveland, which they obviously care about, to the extent that they've seen beyond merely reflecting the undoubted industrial squalor and desolation there. Now to *do* something about it … This colours their music and especially their live performance: their ontrol and discipline, and David Thomas's unpredictable buzzings and explosions are, apart from providing great dance music, a triumphant assertion of the power *now* out of any time, as things are breaking down, of the individual to shape the course of events. With just as much application to use, as things are breaking down have …

1984 in America is going to be real neat – because everybody's getting so paranoid about it that it's going to slip right in. It's happening right now. Things are going so fast that people – when terrorism starts in America that's going to be real heavy. Because none of the American industries are prepared for it; US Steel – get serious! – any of those things …

There are these huge gas tanks and things down there, no security, no fences, even …

Another thing, the whole thing about failure is central to Cleveland psychology, because all there is … it's just like the baseball team. The Indians haven't had a season where they've even been respectable since 1958. Every season, every season, in spring training, they're going to be contenders this year, they're going to finish third, be real good, and every season, by the third week, it's Bam! down into the basement for the rest of the year. In Cleveland … everybody's been lied to for so long, y'know. It's going to get better, everything's going to be OK!, and then it's just … everybody's not pessimistic, just withdrawn.

TM: It's just like OK – we're not going to leave, you're going to do what you're going to do so obviously just get on with it. But the thing is, most of the people have abandoned the city and are out in the suburbs now … The thing is: there is a chance for Cleveland, there *is* a chance for Cleveland and it's not as bad as it may sound today with us talking like this …

Another aspect of their adjustment to this world is the sense of humour. 'Humour me', the last cut on the album, is a pivotal song, its refrain *'Well that was fate'* may be reflecting the band's attitude to Cleveland's decay: no point in getting *depressed*. That's where the discipline of their stage performance comes in, their sense of being a unit, even down to the 'Red Guard' in 'Chinese Radiation': somehow, it all fits … In Cleveland's 'blank spaces and empty places' change and growth can and will occur. We in the UK could do well to think on that ... Certainly, if nothing else, Ubu are one of the most invigorating stage bands in a long while, *giving*, at the same time as transmitting, energy ...

We come to the Alfred Jarry plays, from where the band's name. The refrain in 'The Modern Dance' is 'Merdre Merdre', the same refrain (Shit with an extra R in French) as in the Jarry play *Ubu Roi* ...

DT: I don't know if there's an explanation that I could give, that anybody could ...
TM: Because it would be so easy for it to be misunderstood.
DT: The whole thing about the character Ubu versus the band is that it's used ... sometimes it can be used as a counterimage to what the band's doing and sometimes it can be used – various elements of it – there's lots of elements to the character Pere Ubu – it's used basically as the Dadaists used to put opposing things together sometimes, anything together. It's a perfect name for the band: there's a number of ways to perceive it as being in connection with the band – that's up to each person to do.

That's where your input comes in. My knowledge of the play is limited to a cursory reading, but one aspect worth mentioning is that in relation to their time, the plays had an effect analogous, not dissimilar to that of punk in the rock matrix: a polarizing force, almost an enema, or explosion if you like, clearing away dead ground, dead attitudes, to give room for reconstruction. Ubu see many current attitudes as being irrelevant, and are certainly interested in reconstruction; a new world out of the old ...

An example is their attitude to the rock 'n' roll superstructure mentioned already ...
DT: As you know, we're not really into interviews and that kind of stuff, and the whole thing: there's just no reason for it. Our job is to make music as a steelmaker's is to ... whatever his particular function is. What's the difference? We just happen to deal with different media, with a different sort of raw material.
JS: There's still all that gossip and sycophancy ...
DT: Well, that's old wave. I mean if this is supposed to be 'The New Wave' you know then it should genuinely be 'The New Wave' – let out all that worship of personality and all that sort of stuff – it's partly – because you know that most rock stars are dumb!

Ubu look beyond what was ever 'the new wave', whatever *that* was or is. In *CLE* magazine, David and Johnny Dromette put together a 'Beyond the new wave' project ...
DT: That was just a series of experiments ... when John and I get together,

did you read that thing in *Search & Destroy?* Psycho-bulk? Things like that are just … that's all from watching TV. TV is real important to the band, not because we *like* it or anything, it's really horrible, but it's *the* best indication you're gonna get of how far things have gone, how much people are going to accept in the name of consumerism and entertainment …

This piece has been an attempt to pull a few threads together. There's a lot to Pere Ubu, or, just as much as you want. If it's too 'serious', too 'heavy' – well, don't forget the album title: *The Modern Dance* – that's what it is, and Ubu are a modern dance band.

DT: When I look at Ubu I don't just see … I mean, I see this as part of my life, that doesn't mean it'll be over in five years. Maybe Ubu will be over in five years, maybe Ubu will be over next week, next month, two years from now, who knows? But the whole thing is – I could see Ubu going for ever until we're all dumb old dead men, y'know – but you got to look at it as part of your life …

> And another lot of young people will appear, and consider us completely outdated, and they will write ballads to express their loathing of us, and there is no reason why this should ever end.
>
> Alfred Jarry

Sounds, 20 May 1978

Subway Sect: Life Underground

Everyone's a prostitute
Singing in the song in prison
Moral standards the wallpaper
The wall's a bad religion

Media teach me what to speak
Take my decisions
It's how to find your inner self time
On the television

No one knows what they're for
No one even cares
We'll should publicity handouts
Nobody's scared

So take me talking with clichés
Betray yourself for money
Having is more than being now …

'Nobody's Scared'

In a business whose norm is hyperbole about mediocrity – may the loudest, the most packageable, win! – Subway Sect come over as an oasis of undersell. Their talent and quality are beyond dispute; the fact that their clothes, actions and personalities – the usual raw material for an image – don't shout it out has meant that they've been cruelly underrated, even worse, ignored …

There have been other things too, I guess. Their first ever performance was in the full media glare of the 100 Club 'Punk' Fest, September 1976, which though excellent for publicity proved dangerous for typecasting. And, by now, nothing but a red herring. While other contemporaries have entered the glare and become shrivelled to a crisp, the Sect have stayed in the shade, developing at their own pace, refusing to be steamrollered into making a move before they were ready. It wasn't entirely of their own volition: they were inactive for a period of about six months last year, seemingly playing second fiddle to co-Bernard Rhodes managees the Clash, as they worked and were manoeuvred into becoming this generation's Rolling Stones (necessary rock 'n' roll rebellion) …

During the last three months or so they've come into their own. Early 1977 they were oblique, uncompromising, still forming: at the Roxy Harlesden and the Rainbow making no concessions, confusing, alienating. And planting

Subway Sect in Rehearsal Rehearsals, Autumn 1976 (photo: Sheila Rock)

some irritant seed that stuck in your brain. Rob Ward was added on drums in September – his solid drumming making them more accessible by degrees.

The development graph became steeper: a single was released on Rhodes's own Braik label – 'Nobody's Scared' and 'Don't Split It', which although fine to these ears (the group aren't particularly happy with it) seemingly confirmed the impression that neither Rhodes nor the producer knew what the hell to do with them. Recorded six months before its release, it represents for them at best a point in time: their brooding, scratching, almost hypnotic intensity comes through better on the flip, 'Don't Split It', which even so is only a hint of how it's played live now …

Recent dates supporting the Patti Smith group at the Rainbow, where they played a set superior to the headliners in almost every way to a smattering of applause, and at the Fulham Town Hall have shown their true capabilities. Sect songs mix intricacy with a trebly insistence resulting in a freshness and bite parallel to early 1950s rock 'n' roll. In the way they're heard, rather than the form, the Sect have developed their approach and music to the point where they're nearing a stylistic breakthrough.

On stage their very normalcy of dress and uncompromising rejection of the establishment 'rock' techniques (Love me! Worship my instrument! Worship ME!) provoke reactions of at first alienation and then stimulation. Bassist Paul Myers and guitarist Rob Simmons flank singer Vic Godard, their faces never altering as they chop out angular rhythms – Simmons stares to a point on the ceiling, holding his guitar to his chin, all the time playing rhythm guitar as lead … Godard hunches himself and contorts around the mike stand, his hand clenching his hair in a paroxysm of tortured introversion, seemingly oblivious to the audience feet away … (beware of your senses) … squeezing out pointed yet personal lyrics in a high-pitched quaver …

If you don't know tomorrow then
tomorrow's never going to know you …

Subway Sect provide another relevant contemporary adjustment which holds seeds for the future. Ways out of the (artistic) cul-de-sac that good old rock 'n' roll finds itself in as the end result of punk prolongs the active life of recycled licks and attitudes and provides a new generation of packageable fodder. They cast a quirky, sharp eye on our nation's slow downfall.

Like the music, interviewing them took time but it was worth it in the end. The first was a classic exercise in mutual non-communication, any dialogue falling into the abyss of mutual suspicion and lack of rapport. Dumb questions and dumb, sullen replies. Both sides wondering why they were bothering at all. Try again. This time easier with Vic Godard by himself at his domicile in Mortlake, one slow Sunday afternoon …

JON SAVAGE: Sometimes it comes across that you don't need an audience …
VIC GODARD: I wouldn't like it if no one liked us at all every time we played, but on the other hand I wouldn't like it very much if people liked us every time we played … I'd just prefer there to be a good reaction one night, and the next night a hostile reaction, as long as it changes.
JS: Your music seems intricate yet fresh, an interplay between the four of you …
VG: As we go on the songs do get more intricate and complex, but we don't practise songs very much. Once we know it roughly, we don't aim to get it perfect, we just get it so that it can be played, and then move on to the next song …
JS: You've recorded an album which is coming out soon. Are you happier with it than the 45 … ?
VG: Oh yes, much happier. When it's made I hope that it's going to be … one of my favourite albums of all time, but I know that Television's first album is better, I know that now. That's the trouble, you always rate someone else better than yourself.
JS: What weren't you happy with about the single?
VG: The thing is, it's a song that we did in a period when we were just getting our ideas together, before it finally sort of *happened*. It was recorded in October.
JS: When I saw you last year, there was some irritant factor that made you

stick in my mind, and there's something of that on the single …

VG: Yeah, that was a Fender Mustang. Very cutting. It used to remind me of a dustbin. Dustbins clanging. It wasn't really any good live because it used to get drowned out by everything. But that's what used to annoy a lot of people – we used to have the guitar set like that – they just couldn't really take it, their ears …

JS: You've developed a good deal …

VG: Everything's become more 'sophisticated' in the music – we've changed so much we've probably become more 'palatable'.

JS: After that initial flash of publicity at the 100 Club you didn't do much for a long time: were you worried about getting lumped in with the punks?

VG: Yes, it wouldn't have been a very satisfying thing: we're more interested in musical development. We didn't do any gigs because we were just practising, writing more songs. We didn't have enough at that stage.

JS: That was your first gig, wasn't it? Were you pleased with it … ?

VG: Quite good, because there was a lot of improvisation in it. We did some straight songs, and in the last song I was just making up the words as I was going along. I don't ever do that now: it's a bit self-indulgent, improvising.

JS: You sing 'Don't want to play rock 'n' roll': don't you feel that the album will put you into the mainstream of the marketing business, something you've avoided so far?

VG: I don't think it will. It's hardly as though we've got the resources to have adverts on the tube. That quote was about …rock 'n' roll as a sort of stance that doesn't need anything else to support it – you know, like Generation X …

JS: Rock 'n' roll as a way of life …

VG: Yes. Rock 'n' roll. It also has to do with a way of writing songs that's based on a riff – like most groups, they get a riff, and then they put the words around it, whereas we might start out with the lyric, and then set the guitar to it.

JS: You seem also to be bringing in quite a lot of words and ideas that aren't normally used …

VG: We use *English* words as much as we can, also English words that aren't normally used in conversation, let alone in songs. Like in 'Eastern Europe' '*I shall take no more*' instead of, oh something American like '*I ain't gonna take it*' or something like that. Sometimes America seems like everything that's bad in England ganging up on you.

JS: The Clash's 'I'm so bored with the USA' …

VG: We used to do a song called 'USA' …

JS: Do you think that punk has mainly perpetuated rock 'n' roll?

VG: Nearly all of it. The only things that seem to be getting away from it are things like Television and the Voidoids. I'd have minded less if punk had got absorbed more into pop than rock 'n' roll. It'd have been much better. The sound of it – instead of going into pop which I thought it might do, I thought great! Loads of singles in the charts, all poppy sort of stuff, and it never really happened.

Most of the punk things that get into the charts are mere rock things that have just taken the place of the occasional rock 'n' roll thing that used to get into the charts. It seems so funny: when it first started, I couldn't believe that was going to happen. I didn't think anybody was interested in that sort of thing any more. I think, when it first started, a lot of people were hiding what they were really after.

JS: It's difficult to be original these days: I suppose it's a question of pooling influences and stripping them down, as there's so much information …

VG: It's impossible. It's a question of pooling influences from any period, I reckon, before 1950. We've got quite a lot … some of the melodies from Frank Sinatra – I reckon he could do versions of our songs, the slower ones … and the things that we do get from rock 'n' roll are from the eccentrics that have always been on the outskirts since it started.

JS: Like Beefheart?

VG: Yes, and the Velvets. We get something from classical music: all it is is a sort of freedom of playing – playing guitar, instead of being influenced by rock noises, and being taught to go to certain places on the neck, playing the way that's closer to piano playing or Spanish guitar, something like that.

JS: In the interview we aborted you said you wanted to do something 'timeless', without reference to the age we're in …

VG: Yes, I'm very interested in that. I'm concerned about not writing songs that will be dead three months after you've written them because they won't be relevant any more, or they've become fashionable. The things that I write about aren't really topical anyway, they tend to deal with age-old things …

JS: Did you ever think about what's going to happen to England in the future?

VG: One of the songs is about that. It's about education: what it's going to do to England – 'Enclave':

> The idiot of all is no amusement hall
> He's just dressing up ready for the great downfall
> He maybe an exception to the rule
> But he's a very special kind of fool with too many heads to do the basic stand
> Eyes in the wall, he stands six feet tall
> He's the idiot of all.
> He says: 'My garden holds no interest to me
> The system's builders have all gone to tea

Your life is too much for me to bear
All there is left for me to do is stare – '

– it's about people who go to college and learn things that aren't geared for life. It's not really that college should be geared to life, but that life is geared wrong anyway. But then you get the thought that can life be geared correctly anyway? It's really celebrating what old people regard as the biggest malady in modern life.

JS: Will you keep on changing?

VG: We've come to a difficult point now as we're writing songs we can't play – what we're going to have to do is work very hard at instruments, because in the past I used to think it was great to avoid being taught because you wouldn't become like anyone; but now I realize to really do anything new you have to know all the old stuff inside out, and to know what to take from …

JS: Have you been happy with your management?

VG: Not all the way along, because there was a period – this is quite a long time ago – when the Clash were just taking off and we obviously didn't get much attention, but it was all right, because it gave us time to catch up. Now it's fine.

Track listing for the album seems to be (in no particular order): 'Chain Smoking', 'Birth and Death', 'De-railed Sense', 'The Ambition', 'Staying (Out of Touch)', 'Imbalance' (a brilliant instrumental), 'Eastern Europe', 'The Exit of No Return', 'Forgotten Weakness', 'Enclave' and 'Rock and Roll, Even'. Even? *'Rock and Roll … is the opium of youth, for people who've got boring jobs, go and see a group on Friday night.'*

The Sect have it all. What happens now depends on them and their management. Stay tuned.

' *…the true escape into timelessness …*'

Sounds, 1 July 1978

The Clash: *Give 'em Enough Rope*

Waiting for God (oh) we get the Clash. Faced with the weight of expectation and a well-aired critical backlash, it'd be very easy to over-react here – one way or the other. Why should expectation be so intense? Other groups can release albums two, three years apart, and no one bats an eyelid; in this case, beyond the rescheduling, beyond the acceleration of our media-dominated society accurately reflected by the latest youth cult, it comes down to the Clash's 'cultural significance' to a generation.

The mantle that they sought frankly from the word go fell fairly and squarely on their shoulders. The weight is enormous – more than anyone could bear – and, from recent reports and the evidence of this album, it fits them ill. Timed for them to relax and for the audience to relax on them.

Sure, the Clash have suffered from doing all their growing up in public, but then, they always wanted it that way. If the Sex Pistols were hoisted to fame by astute media exploitation, then the Clash weren't far behind; no slouches in that game. From 13 August 1976, when they played their first gig-cum-rehearsal in front of the then relevant journalists – thus ensuring glowing lead write-ups and their place high in the pecking order – they never looked back as the quickly emergent and expanding 'Punk Movement' took every word as gospel (ignoring any element of humour) and crystallized them into *the* attitudes, *the* catchwords, *the* myth. Perhaps even more than the Pistols: for every kid who scrawled 'Anarchy' on his ripped shirt, another read the press and took in all that stuff about 'high-rise living' and 'the Westway' and 'dole queue rock' and and and and.

For a few months, for the idealistic, young and needy, it seemed as though the Clash could combine the functions of fashion plates, politicians, artists and rock stars in a heady mixture (copied by the noticeably less talented – Generation X to name one) that would change the world.

When the world spun a few times and very little changed in real terms, the Clash started to buckle under the pressure. Their press coverage wasn't (and isn't yet) matched by record sales; much of that very coverage was sycophantic, written by camp-followers, or plain silly: the statement 'The Greatest Rock 'n' Roll Band on the Planet', indefensible on any level, was said too often by those old enough to know better.

As the disparity between the ideals expected from them by the fans (and reinforced by the press) and the reality of the situation grew, they themselves relapsed into backbiting, self-justification, double-think and dull petty crime. From an all-encompassing revolutionary package into just another rock band: suddenly, to many they become ordinary.

That strain is all over this album.

From the sub-Warholian grainy Day-glo of the militaristic, spaghetti western cover, the general mood of the album, as hinted by the title, is pretty depressed. The Clash's view of the human condition, while imprecisely expressed, isn't very sanguine this time out. The sharp, direct attack of the first album itself holding out hope by the accuracy of targets selected and hit, has been replaced by a confused lashing out and a muddy attempt to come to terms with the violence of the outside world, which the Clash plainly see as hostile through and through.

Flicking through the titles, you catch the words repeated – 'Drugs', 'Guns' – and the general themes of gangs and fights, all too rarely enlivened with the humour that marked the first album. They sound as though they're writing about what they think is expected of them, rather than what they want to write about, or need to. It's as though they see their function in terms of 'the modern outlaw' – obligatory 'rock 'n' roll' rebellion similar to the Stones – and conservationists of the punk ethos they so singularly helped to create.

None of all this would matter at all if this was a great album: that it isn't in the end, is down to the music.

The first side is by far the superior, both in the quality of the individual songs and its playing as a whole. 'Safe European Home' begins the album in confident and rousing fashion, where the Clash's penchant for singing about their own mythology – viz. 'Complete Control' – is continued. This time, however, they tell the story – the visit to Jamaica by Strummer and Jones – against themselves with wit and with no attempt to justify themselves. And perception: 'Where every white face is an invitation to robbery'; the song itself is lilting and memorable, with great gabbled vocals from Strummer, a well-integrated reggae break, and irresistible motion. A broadening that's entirely successful.

'English Civil War', while hypertense, works in the context of the surrounding cuts, and because it reworks closely a song which can't fail to rouse. Strummer is one of the few rock vocalists who could carry it off successfully: the range and depth of emotion expressed by his voice are two of the few consistent things about the album and are where much of the Clash's undoubted greatness stems from.

'Tommy Gun' is another emotional *tour de force* and a natural 45: in the break near the end the band works itself into a wall of sound as Strummer fights to be heard over the holocausts of noise and the band achieves the massive general statement they want.

The side sags with the amusing yet lightweight 'Julie's in the Drug Squad' – referring to the recent huge LSD busts in Wales – where the sound is thin and some of the asides gratuitous. Things pick up again with 'Last Gang in Town': the Clash's sometimes ambiguous fascination with gangs and gang warfare is laid out for all time. Built around a basic rock 'n' roll riff, the song misses making a statement, but is remarkable for Strummer's deep compassion (already in evidence on the 'Hammersmith Palais' 45) and the genuinely taut and chilling chorus.

Side two is more problematic. 'Guns on the Roof' unhappily mixes one of Strummer's best vocals with the Clash at their most tendentious. The intro rips off 'Clash City Rockers', which rips off 'I Can't Explain', which ripped off some old Kinks song I can't remember. While most pop is about plagiarism, in this instance it seems to signify lack of inspiration and an over-reliance on the tried and trusted. The lyric extrapolates from their conviction for shooting pigeons – very 'creatively violent' – to take in an attack against the legal system and then world-wide violence. How this is managed is unclear, even from the lyric sheet, but the song continues this habit the Clash have of justifying their actions by making tenuous general externalizations thereon. 'Drug-Stabbing Times', however, is another great moment: beginning with a flashy rock riff, the band break fast into the chorus, driven by cowbells and fat sax; the lyric tells you what happens when the car comes to take you away.

From here on in it's downhill. 'Stay Free' is embarrassing in the failure of Mick Jones's vocal to carry the weight of the song, overburdened by 'tasteful' arrangements and the attempt at personal dedication. Both 'Cheapskates' and 'All The Young Punks (New Boots and Contracts)' show the difficulty (and pain) of the Clash's adjustment to the pressure place don them and the failure of 'punk': both are depressing and sad – both unintentionally and intentionally – and not much fun to listen to. 'Cheapskates' is irritating in the assumption that the lyric ('We're cheapskates/Anything'll do/We're cheapskates/What we s'posed to do?') absolves them from any mistake or any failure; 'All the Young Punks' attempts a hymnal 'All the Young Dudes' and ends the album on a very downbeat, if touching, note.

In some ways, much of the musical and social development of the last 18 months might never have happened as far as this album is concerned. Musically, the Clash stray rarely from rock 'n' roll roots and basic Who/Kinks/Mott models. Jones's guitar work is best when concise and using drones: several times on the album he stretches out in a manner which suggests he thinks he's better than he is. Simonon's bass work at times seems to drag rather than pump, adding to the leaden air of cuts like 'Cheapskates': it's best on moments like the chorus of 'Last Gang'. The only consistent elements are Strummer's singing and Headon's drumming; this is occasionally over-emphasized by the curious production, tampering as it does with the general sound, without performing a general overhaul throughout – appended rather than integrated. An unsatisfactory compromise, illustrating the well-publicized vacillation that went into the album's making.

Socially … It's hard when you define a period so accurately. The Pistols broke up and neatly avoided the issue. Here, the Clash seem locked in time, stranded on their conception of what the problems are, where solutions are to be found, and what problems face their audience. They have an audience which is loyal to a point of fanaticism; enviable but dangerous – they often seem to relate to each other on the basis of mutual reinforcement: trapped in this circle, the Clash's solution is to rock 'n' roll. From being radicals, they become conservatives.

It is *not* being cynical to say that punk does not rool OK: what's needed now is a recognition that the problems are more complex and must be met not with facility, but with adaptability: often the refusal to adapt is justified by strict loyalty, to some ethos whose time and circumstances have passed as an end in itself. It now takes a medium less tainted, sharper than pop music, to define the problems that face us; the solutions can only be worked for in the real world.

Here even the Clash's function as basic consciousness-raisers can be called into question: it's often hard here to work out what it is exactly that they're articulating, even more so when they aim at statements.

Vague yet constant allusions to guns, violence, drugs and militarism in the lack of clear articulation can appear, if taken wrongly, to smack of terrorist and militarist chic – which they already have, remembering the Belfast pix – been misunderstood on – and which can be seen as simply irresponsible.

And still the double-think: the promo pic is a classic example – time-honoured cute 'rock star' poses in front of a Russian propaganda poster or painting, Mick Jones with a Red Star appliqué – which as far as real politics goes just seems kindergarten.

And so on and on: signing to CBS and then bitching non-stop, going to the USA to finish an album which, in its allusions to drugs, four-letter words and determinedly English patois, would seem to have very little hope of American airplay. So do they squander their greatness.

Melody Maker, 11 November 1978

Public Image Ltd: Live at the Rainbow

If Rotten (with the latter-day Pistols) excelled at the theatre of alienation – that gulf between performer and audience which, when stretched, could be intensely provocative and destructive of many of the false 'object relation-ships' endemic to entertainment – then he crowned it on Christmas night.

The man looked at the crowd – everybody's favourite nightmare: the audi-ence looked at the band – three nervous musicians and a nervous charisma. The vortex between them sucked everything out.

The video of the Hit 45 suggests a fair likeness of PIL on stage. Wobble sits throughout, 'skiving'; Walker holds the bottom down at a nod from the oth-ers. Levine stands stock still, occasionally turning his back or executing ran-dom passes with Rotten, at a stroke shifting the whole visual focus. This, however, as expected, is mainly Rotten's show.

Tonight he pulls some of the old SP stunts – insulting the audience, being deliberately lazy – while using the space that PIL give him to move around a lot. Except now there is a new element of preaching and self-justification: he whines about the Pistols, the press, the audience, Glitterbest, etc. more than is necessary. It sits ill with the traces of his previous persona, and subtracts from the still fragile freedom that PIL have provided. And it's all about him: pointing at himself, megalomania rampant, I Wanna Be MEEEEE!

With thousands of eyes bleeding him, he is still magnetic and mesmeric, but his gestures fall into the gap between the raw band and the audience, wanting only, if not to be loved or even led, to watch.

A year – did you miss him while he was away? Unsurprisingly, he finds it hard to carve out a new relationship between himself, the audience, and the stage.

PIL play the album except 'Fodderstompf,' with an added 'Belsen was a Gas', the lyrics thankfully changed. The set starts brilliantly with 'Theme': the album begins to make sense, played live, as the massive Germanic struc-tures boom and Rotten whirls, but bad timing, the harangues, equipment problems and a couple of crudely manipulative touches reinforce the alien-ation. From then on, it's all downhill.

The longer they play, the more the clapping decreases. Unable, unwilling to dance, the audiences seethes and the fight starts. Rotten stops the song, yells in disgust: 'You skins – wankers – you got the wrong enemy again; will you never learn?' He's still in control, but it's a bit late to start housetraining the audience now.

After a perfunctory reading of the hit, the group splits. No encore. The audience feels had. The atmosphere is ugly: idealism perverted into inept nihilism, liable to move into more dangerous areas.

PIL have great and undoubted potential; as their album is an infuriating, amusing mixture of the brilliant and diabolical, so was this performance. Behind obvious first-night nerves and inexperience, aggravated by PIL's insistence on getting their feet wet in the Rainbow, lie more fundamental problems: PIL and Rotten in particular seem to treat the business of making statements so seriously that they neglect to get their base right.

Rotten is quite right to pin the blame for much of what has happened over the last year on the biz, exemplified as much as anything by the vampirous mood of the crowd. It's an area where he has the power to effect change, but the music just isn't up to carrying the load.

PIL are also too prone to blame everyone except themselves: that tonight was far from a success can be laid fairly and squarely on them, just as much as on the audience. They will, no doubt, improve – time will tell.

Meanwhile, the final frame is of Rotten, alone (the group has fled) on the stage. In front of him a dozen burly bouncers face the crowd. They start to peel off, attack … Oh, go, Johnny, go.

Melody Maker, 31 December 1978

The Sex Pistols:
The Great Rock 'n' Roll Swindle

'What needs understanding is the state of paralysis everyone is in …'

The Sex Pistols and Glitterbest, after shattering old myths as effectively as anyone can, and often brutally clearing new ground, find the tools used becoming the new myths – 'Anarchy!' 'Pretty Vacant' – now inviolable, hallowed, stuck in time. This album – at once very funny and deadly serious – smashes those myths (setting up more in turn) and with stunning and sure cheek savages the hand that feeds it – the industry of human happiness.

In early 1977, the Pistols publicly made fools of two major record companies all over the front pages of the press – striking directly at the root of discreet respectability sought by a business which, like many others, extends its tentacles into some very shady areas – and walked away with £125,000.

This album rubs that in something rotten, and with a vicious (and obvious) hindsight makes out the whole scam was a swindle anyway. In a brilliant twist,

McLaren turns a position of apparent weakness into strength, and keeps everyone guessing, entertained and annoyed. Nobody is spared.

A most potent weapon is humour. (This album makes me laugh out loud.) The shock tactics, so often understandably berated, finally appear as the cutting edge of a serious intent: demystification, exposure, a railing at taboos. Or, indeed, simply swindling. *Rock 'n' Roll Swindle* is baffling, maddening and intensely provocative.

And, mostly, it's successful superpop: flashy, trashy, trivial, mythic.

So, what is 'Sex Pistols'?

Superficially, that's what the court case was about. Here, 'they' are seven vocalists, an uncredited orchestra, and various Cook/Jones permutations. The concern of Lydon is fully understandable. But beyond that, *Swindle* appears in its rush release as a victim of the conflict it epitomizes.

The frantic manoeuvring of Glitterbest and Virgin to outflank each other – one seeking, if not to destroy, to rob rock 'n' roll outright, the other to rob in a different, more subtle way – package and sell – has been one of the funnier and more serious pantomimes of the last few years. Virgin win the court case, and rush release a film soundtrack, intended as such, without the film, forcing it to be treated as a 'rock album'. Now, who's swindling who?

The 24 tracks fall into three rough categories: firstly, those by the 'proper' Sex Pistols. The bulk – six cuts – come from the August 1976 Dave Goodman-produced sessions, most of which found their way on to the *Spunk* (a.k.a. *No Future*) bootleg, itself rumoured to have been produced with the collusion of Glitterbest. Certainly, at this stage, Rotten's voice was rough in the extreme, but for that all the more effective; the Matlock band were simply very tight.

'I Wanna Be Me', with its brilliant media-obsessed lyrics ('This is brainwash/This is a clue') was the flip of 'Anarchy in the UK'; the version of 'Anarchy' here has already appeared on *Spunk* (beefed up by Cook and Jones in the studio), Rotten's vocal screwed tight into an explosion of rage – if anything, superior to the single. The three that open side two, 'Substitute', 'Don't Give Me No Lip Child' and 'Stepping Stone' staples of the early set (and again, only available on bootlegs), are massive, confident – equal to anything the Sex Pistols ever recorded.

'Johnny B. Goode' and 'Roadrunner' are hilarious; the band run through both effortlessly but Rotten can't remember the words, bleats, lets out a string of expletives – hearing him attempt Richman's American accent and break into broadest Finsbury Park is worth the price of the album alone.

And 'Belsen was a Gas'. Discography first; the SP's version is recorded in San Francisco, Winterland, their last ever concert, On the *Gun Control* album, it was the only presentable track: here it's cleaned up and is a stunning performance, a summit of desperate disgust, an ultimate statement. The band stop for a few seconds of stunned silence, a lifetime long, and then the cheers.

Fine, marvellous. But … is making a sick joke out of those horrors (which have scarred Western man's psyche for the future) in *any way* defensible? Is it not sensationalism of the worst order? It's never possible to be certain, but, I think, the answers are yes and no respectively.

It's very easy (and, no doubt, a corrective) to rear up on your hind legs and say: 'Belsen! Bunch of fascists!' – but think for a moment. Although admittedly ambiguous, 'Belsen' is here not only the epitome of the cry from the abyss that punk always was, but one of the most severe attacks ever on a whole strain of human behaviour: 'Be a man, kill someone, be a man, kill yourself!' Not an exhortation, but a terrifying description and critique of the processes that led to the concentration camps.

In screaming the unspeakable in an impossibly raw, painful way, the Pistols confront and warn, reopen wounds which, once healed, may be forgotten, and so may happen again. After 'Belsen', the Sex Pistols could only split up.

Next. Those recorded by Cook and Jones together, and with Ronald Biggs. This last episode seemed to confirm the views of those who thought McLaren was a mindless *Sun*-oriented sensationalist, and divided for ever those who had stretched to assimilate previous Pistol antics.

It's still very confusing: any attempts at analysis founder on the double-think that McLaren has always, and so successfully, erected. However seen, it kept open that edge of uncertainty, the cutting edge, that has always characterized and made effective the whole Sex Pistols project.

Always, the approach to media was hard, realistic, not liberal (i.e. thinking what you wish to happen is happening); the trashiness and superficiality inherent in most forms of media were understood and worked on: exploitation and sensationalism (often a reduction of events into the simplest possible terms) were not spurned. Both the Pistols and Glitterbest realized that to use any medium to its fullest extent for your own ends you have to accept the medium and confront it on *its own terms*. As here, a sure-fire hit album is used to deliver a swingeing attack on the medium through which it will be a hit: what could be more perfect?

The reverse side of this hard approach, this prodding at conventional methods of media use, is the ambiguity, the apparently careless trampling of sensibility, the toying with violence which can be seen as, and often is, irresponsible.

Put it in simplistic, sensational terms: the Sex Pistols and Glitterbest cut as broad a swathe as anyone can, really shook things up; the reverse side of the coin was Vicious's death. With hindsight, it looks as though one couldn't have happened without the other.

Biggs covers 'Belsen was a Gas': the message seems to be – 'If you can stretch to accept the Pistols' version, then this'll totally confuse you'; it does.

Both sides of the last Sex Pistols single, 'No One is Innocent' and Vicious's 'My Way', are also included. Remember that the single entered the Top Ten, as high as 'Public Image'.

Vicious also runs through competent, slightly anonymous versions of 'Something Else' and 'C'Mon Everybody': two Eddie Cochran songs whose lyrics – '*Who cares?* C'mon Everybody!' – and flash merely reinforce the hint that the Pistols are very much an update, 20 complex, accelerated years on, of the primal two-fingered rebellion that rock 'n' roll ever was.

The Cook/Jones songs add to this view: both 'Silly Thing' and 'Lonely Boy' are trivial, slightly perverted, thoroughly enjoyable updates of pristine teen-beat love themes; only here does the persistently excellent production falter slightly. As a final piece of off-the-wall puerility. Jones adds 'Friggin' in the Riggin'' – a punked-up rugby song. Harmless smut, it's bound to be the album's bid for popular identification: Singalonga Pistols.

There's no business like … The last batch – eight in all – are the sound track songs: this is where the album (and, no doubt, the film) is tied together – the stroke, the scheme, the swindle. In all, they comprise a consistent and savage attack on both the Sex Pistols myth and the leisure industry.

The album opens with an orchestral version of 'God Save the Queen' with narrative voice-over by McLaren: in best Fagan voice, dripping with 'Jewish' greasy greed, he exposes his Dickensian fetish, with relish running through the media-fossilized version of the Pistols myth and playing his own role of Stage Villain, seedy Svengali up to the hilt. Later, he sings in a deadpan gluti-nous croon the old schlock standard 'You Need Hands' – ripping into shreds trad biz hypocrisies and in particular the Jewish showbiz Mafia (note the Hebraic lettering on the sleeve), and thus by direct implication himself. It's very funny.

Elsewhere Bernie Rhodes's protégés, the Black Arabs, pull off a disco med-ley of 'Pretty Vacant', 'Anarchy in the UK' and 'No One is Innocent'. I don't know whether it's good disco, but it's audacious and silly and sounds great.

The last nail in the coffin is Steve Jones's 'EMI': over a luscious swelling orchestral backing he intones, in the clipped voice of the stage upper middle classes, lyrics hitherto spat out with such venom:

> And you thought that we was faking
> That we all was just money-making
> You do not believe that we are for real …

Finally, Ten Pole Tudor's two: only here do you miss the film as visual com-plement. The suitably ludicrous version of 'Rock Around the Clock,' with shrieks substituted whenever the word 'rock' occurs in the song, makes the point clearly enough but sags in the middle of side three. 'Who Killed Bambi' (the working title of the abandoned Russ Meyer project) is a pastiche of soundtrack tunes, all jaunty sprightliness – only Tudor's distorted and ridiculous vocals give the game away.

Again, as ever, McLaren has his cake and eats it. If you baulk at some of the material on *Swindle* that seeks to shock, or if you get angry, you fall into one trap – that's what he wants; if you ignore it, you cop out. It's the double-think that has run through the whole Sex Pistols project. How deliberate is it? Where does the swindling stop? Is it not a cloak to get away with anything?

By setting himself up as a grubby, sordid sensationalist, he avoids preaching and platitudes, and gives a harder edge to the wholesale assault on 'rock music' relationships – company to performer to audience – that this hugely enjoyable album represents. Gross sensationalism and dangerous propaganda are an unusual and unbeatable combination.

Especially now, when the threat that the Sex Pistols originally posed has been institutionalized, that assault is valuable in continuing to demystify, annoy and make trouble.

The 1960s generation turned the anarchy that was pop into rock, respectable and corporate; this album attempts to redress the balance – instead of being frightened by the accelerating disposability of pop it celebrates and wallows in it, undercutting the frantic desire for stability, imposed over the desperate attempts to gauge the market that define the operation of most record companies these days. And beyond: if it can be thrown together as easily as this and be a hit, the hint is, is pop not totally useless, the ultimate con?

You'd never guess the Sex Pistols were a supermarket. There's something here for everyone. *Swindle* is a flawed, brilliant, important record.

Melody Maker, 24 February 1979

Joy Division: *Unknown Pleasures*

'To talk of life today is like talking of rope in the house of a hanged man.' Where will it end?

The point is so obvious. It's been made time and time again. So often that it's a truism, if not a cliché. Cry wolf, yet again. At the time of writing, our very own mode of (Western, advanced, techno-) capitalism is slipping down the slope to its terminal phase: critical mass. Things fall apart. The cracks get wider: more paper is used, with increasing ingenuity, to cover them. Madness implodes, as people are slowly crushed, or, perhaps worse, help in crushing others. The abyss beckons: nevertheless, a febrile momentum keeps the train on the tracks. The question that lies behind the analysis (should, of course, you agree) is, what action can anyone take?

One particular and vigorous product of capitalism's excess has been pop music, not so much because of the form's intrinsic merit (if any) but because, for many, bar football, it's the only arena going in *this* country, at least. So vigorous because so much has to be channelled into so small a space: rebellion, creation, dance, sex energy. And this space, small as it is, is a market ruled by commerce, and excess of money. It's as much as anyone can do, it seems, to accept the process and carefully construct their theatre for performance and sale in halls in the flesh, in rooms and on radios (if you're very lucky) in the plastic. The limits imposed (especially as far as effective action goes) by this iron cycle of creation to consumption are as hard to break as they are suffocating.

'Trying to find a clue/trying to find a way/trying to *get out!*' *Unknown Pleasures* is a brave bulletin, a danceable dream; brilliantly, a record of place. Of one particular city: Manchester. Your reviewer might very well be biased (after all, he lives there), but it is contended that *Unknown Pleasures*, in defining reaction and adjustment to place so accurately, makes the specific general, the particular a paradigm.

'To the centre of the city in the night waiting for you … ' Joy Division's spatial, circular themes and Martin Hannett's shiny, waking-dream production gloss are one perfect reflection of Manchester's dark spaces and empty places, endless sodium lights and hidden semis seen from a speeding car, vacant industrial sites – the endless detritus of the nineteenth century – seen gaping like rotten teeth from an orange bus. Hulme seen from the fifth floor on a threatening, rainy day … This is not, specifically, to glamorize: it could be anywhere. Manchester, as a (if not *the*) city of the Industrial Revolution, happens only to be a more obvious example of decay and malaise.

That Joy Division's vision is so accurate is a matter of accident as much as of design: *Unknown Pleasures*, which together with recent gigs captures the group at some kind of peak, is a more precise, mature version of the confused anger and dark premonitions to be found (in their incarnation as Warsaw) on the skimpy Electric circus blue thing, the inchoate *Ideal for Living* EP, and their unreleased LP from last year. As rarely happens, the timing is just right.

The song titles read an opaque manifesto: 'Disorder', 'Day of the Lords', 'Candidate', 'Insight', 'New Dawn Fades' – to recite the first, aptly named, 'Outside'. Loosely, they restate outsider themes (from Celine on in): the preoccupations and reactions of individuals caught in a trap they dimly perceive – anger, paranoia, alienation, feelings of thwarted power, and so on. Hardly pretty, but compulsive.

Again, these themes have been stated so often as to be clichés: what gives

Joy Division their edge is the consistency of their vision – translated into crude musical terms, the taut danceability of their faster songs, and the dreamlike spell of their slower explorations. Both rely on the tense, careful counterpoint of bass (Peter Hook), drums (Stephen Morris) and guitar (Bernard Sumner): Ian Curtis's expressive, confused vocals croon deeply over recurring musical patterns which themselves mock any idea of escape.

Live, he appears possessed by demons, dancing spastically and with lightning speed, unwinding and winding as the rigid metal music folds and unfolds over him. Recording, as ever, demands a different context: Hannett imposes a colder, more controlled hysteria together with an ebb and flow – songs merge in and out with one another in a brittle, metallic atmosphere.

The album begins unequivocally with 'Disorder'; 'I've been waiting for a guide to come and take me by the hand'; the track races briskly, with ominous organ swirls – at the end, Curtis intones 'Feeling feeling feeling' in the exact tones of someone who's not sure he has any left.

Two slower songs follow, both based on massively accented drumming and rumbling bass – in their slow, relentless sucking tension, they pursue confusion to a dreamlike state: 'Day of the Lords' is built around a wrenching chorus of 'Where will it end?', while the even sparser 'Candidate' fleshes out the bare rhythm section with chance guitar ambience. In a story of failed connection and obscure madness, Curtis intones, 'I tried to get to you', ending with the pertinent:

> It's just second nature
> It's what we've been shown
> We're living by your rules
> That's all that we've known.

The album's two aces are 'Insight' and 'She's Lost Control': here, finally, Gary Glitter meets the Velvet Underground. Both rely on rock-hard echoed drumming and bass recorded well up to take the melody – the guitar provides textural icing and thrust over the top.

'Insight' leads out of 'Candidate' with a suitable hesitation: whirring Leslie ambience leads to a door slamming, then a slow bass/drum fade into the song. The attractive, bouncing melody belies the lyrics: 'But I don't care any more/I've lost the will to want more'. At the end Curtis croons, his voice treated, ghostly: 'I'm not afraid any more' to drown in a flurry of electronic noise from the synthesized snare.

'She's Lost Control', remixed to emphasize guitar and percussion, is a possible hit single: it's certainly the obvious track for radio play. Deep and dark vocals ride over an irresistible, circular backing that threatens to break loose but never does; the tension ends in a crescendo of synthesized noise.

On the 'Inside', three faster tracks follow – mutated heavy pop, all built around punishing rhythms and riffs it'd be tempting to call metal, except control is everywhere. 'Shadowplay' is a metallic travelogue – the city at night – with Curtis fleeing internal demons; the following couple, 'Wilderness' and 'Interzone', wind the mesh even tighter.

'Wilderness' externalizes things into Lovecraftian fantasy, all echoed drumming and sickening guitar slides, while 'Interzone' moves through a clipped, perfect introduction to guitar shrills and 'Murder Mystery' mumbles:

Down the dark street the houses look the same
Trying to find a way trying to find a clue
Trying to get out!

Both sides, finally, end with tracks – 'New Dawn Fades' and 'Remember Nothing' – so slow and atmospheric that alienation becomes a waking dream upon which nothing impinges: 'Me in my own world ... '

Leaving the twentieth century is difficult; most people prefer to go back and nostalgize, Oh boy. Joy Division at least set a course in the present with contrails for the future – perhaps you can't ask for much more. Indeed, *Unknown Pleasures* may very well be one of the best, white, English, debut LPs of the year.

Problems remain: in recording place so accurately, Joy Division are vulnerable to any success the album may bring – once the delicate relationship with environment is altered or tampered with, they *may* never produce anything as good again. And, ultimately, in their desperation and confusion about decay, there's somewhere a premise that what has decayed is more valuable than what is to follow. The strengths of the album, however, belie this.

Perhaps, it's time we all stand facing the future. How soon will it end?

Melody Maker, 21 July 1979

Gary Numan: In Every Dream Car, a Heart-throb

… thinking of it now, a television documentary would be preferable. On a commercial channel, so it's nicely broken up. A working title could be 'Some People Need the Heroes': it's a line from one of the subject's songs which seems to fit.

After the station ident, there'd be a short piece of film before the titles start running. This section would consist of two short, contrasting clips.

The first would show the subject (Numan) on stage in mid-performance, just out of reach of scores of hands which wave like sea-anemones (nice simile – beguiling but deadly): he darts, suddenly, to brush a few of them lightly. There are a few sharp, clipped screams.

The second clip would show Numan talking, white T-shirt, wary eyes, soft voice, in a hotel room: he doesn't really understand human nature, or even like people very much – he worries about not getting close to people, becoming a product.

The clips (perhaps melodramatically, but That's Entertainment!) would take Numan's pop stardom as granted, and go further to pose a series of questions – Did he choose to become a star? What kind of star is he? What does 'star' mean? And what is he going to do with this peculiar but pressured position? In short, to try and sort out, a little, what is this thing called Numan.

Which, it not being an ideal world or even America, will have to comprise the parameters of this article.

At the time of writing, Gary Numan's first tour since becoming a major 'pop' attraction has just finished. He has played to – roughly – 40,000 people in 16 capacity venues throughout the country: by the last date, at Sheffield's City Hall, he has lost about £30,000 on the tour (including the £3,000 or so given to Save the Whales from one of the Hammersmith nights) which has, nevertheless, confirmed Numan in many people's eyes (including pop columnists on the national tabloids) as a new pop star. It can be said to have been very successful.

The writer is to join the tour for the last two dates – Wolverhampton and Sheffield. In adapting to the different time-scale of the touring party, he will become a different person: the Journalist. In writing about it, he will swap lived experience for self-conscious experience. As the Journalist, he will interview the Pop Star: perhaps they'll both be able to crash through their roles a little.

Jill Mumford: *Gary Numan collage*, Autumn 1979

The writer's motives: he's interested by the fact that the music press have so consistently got Numan completely wrong; he likes the last two singles enormously (but worries about the albums) and is interested enough to have quite a lot of theories about the phenomenon.

The facts about Numan's success are obvious if you flick to the charts pages. The don't need reiterating. Like other recent developments – the mod thingy, the recurrence of 'teen-style' following for figures as diverse as Geldof and Sting, and the emergence of political groups far more explicit than the TRB ever were – this success illustrates a concerted stylistic shift further away from the last and current stylistic evolution (punk) and its more directly related products. As far as the mass market is concerned, passion – being an uneasy fashion – is being hesitantly or otherwise discarded as a new breed of heroes (as opposed to anti-) emerges.

The strands contained in the 'original' groups and followers (which compression enabled them to make such an impact) are now being unravelled further and further and turned into careers by groups who are trying to extend the ideas even further than imagined. The context is lost. Whether you approve or not, it's inevitable. In this confusion, the writer, more by accident than design, becomes involved with documenting this particular strand: a post-punk mutation (rather than plagiarism) of certain Bowie and Roxy

Music elements, made new, different, attractive and successful.

To the narrative. Armed with all this speculation and more, the journalist arrives at the Walsall Crest Hotel in time to meet the mediator between him and the star, Su Wathan, and to be transported to the concert. He tentatively meets Numan, helps himself to some food, and watches the event, attentively.

It reinforces his earlier impression: it is a Spectacle. Numan takes his environment around with him: tonight it's set up in an ugly, functional 1930s hall, more suited to school speech-days than to any kind of celebration. Conceived with an architect's megalomania, the stage set dominates and frames both audience and performers alike: impossible to escape or ignore, exactly contemporary in its practical futurism. Measuring time by decades, we look spuriously to the new to bring up new things: to a considerable section of the pop audience, Numan is it, a successful percentage of new information grafted on to old.

From inside the towers, fronted by banks of light, the five-piece band – clockwise from bottom left: Paul Gardiner on bass, Chris Payne on keyboards and violin, Ced Sharpley on drums, Billy Currie (Ultravox!) on Moog, and Russell Bell on guitar and noises – and Numan, stage-centre, run through 20 songs to tumultuous applause.

The audience is a mix of archetypes and individuals: some 'punks', some electric friends, some who've paid Numan the ultimate accolade of exact imitation (the 'Replicas' phase) and quite a few who are non-aligned but who've simply come to see what this thing called Numan is about.

A common factor is their youth: a large number are under 21 – a generation too young to have seen Bowie or Roxy Music, or even later heroes or anti-heroes from the Sex Pistols onwards. Some people need the heroes, and Numan is the one they seem to need: they don't dance, much, but watch intently. The ones at the front stretch their arms out with palms open, like drowning men: these gestures of innocence turn to one of danger when the star gets closer – they'd tear him apart if they got the chance.

The quality of the material is variable, but they don't seem to notice. The performance suffers from a sag in energy after the initial impact of the opening: some of the songs, too, sound remarkably similar to one another. The band play the material better than they did at Glasgow, and Numan's performance is more confident, able to encompass a relaxed spontaneity.

But what's missing in Wolverhampton is a sense of occasion. It's special enough for the audience, but not for the band: the same old story of what the audience sees once, group plays scores of times. This time they're professional enough – remarkable, considering their experience as a unit – to maintain standard, but there's little extra.

The journalist gets a bit bored and concentrates more on the audience.

What is strange about the event is their relationship to Numan: it's peculiarly passive, bereft of direct contact (bar the hand-to-hand brushes) – as if there's an invisible wall between them. He's up there, they're down there: it's a more formal relationship than the journalist is used to, made perhaps more obvious because of Numan's comparative inexperience and the particular performer-mode he's chosen.

'I think … no, I'm not really interested in a new relationship: I'm sure what can be done. I've really no idea … apart from the fact that you talk to the audience and claim to be one of them, or admit that you're not one of them, which is why you're singing and they're not, and get on with it, which is what I've done … I've very little to say to them. They know what the songs are, I'd imagine. I really wouldn't want to tell them what the songs are about before each number: there's no need to tell them what they are because they already know. There really isn't much more to say – you can't have a conversation – it's very false with between and four thousand people …'

Captured on CRO2, Numan's voice plays back to the writer a few days later. At the time he remembers being impressed by Numan's candour and directness, whether or not agreeing with all of what he said. As the journalist, he'd chaperoned helpful, courteous Numan to the tape in a hotel room in Sheffield, the day after Wolverhampton.

By this time he'd met the rest of the band, the support band (Orchestral Manoeuvres in the Dark) – the tour organizer, the security people, Numan's parents and others whose presence is necessary for the tour, and had been impressed by the subtle sanity of this operation under strain.

Although by now locked with each other into the different time schedule that constant travelling and the long, peculiar hours induce, those on the bus retained an air of innocence even when winding down from the adrenalin of the performance: the more traditional 'rock 'n' roll methods of alleviating the strain – both chemical and attitudinal – were, mostly, conspicuous by their absence, much to the journalist's relief.

Continuing: 'That kind of performance suits me. I don't like being very close to them. I don't like standing gigs at all: I get very worried when they get near the front. I know it's very flattering that they rush the stage, but it does worry me when they get that close. I like there to be a gap, I like theatres where there's an orchestra pit. It isn't that I don't want to talk to them, meet them, or anything like that, it's just that I don't feel safe with all those people trying to get there …'

We're eavesdropping on the tape again. At this point, the journalist is easing the interview in by discussing the previous night's performance, commenting on the gap he'd noticed between performer and audience: as he suspected, it has something to do with the quick, massive pressure of the situation in which the 'star' finds himself. Numan is sitting, concentrating on the replies as hard as the journalist is on the questions, at the same time watching his reflection in a mirror as he nervously teases his recently washed hair.

They've both established enough of a rapport to attempt communication by prior arrangement, both have driven up from Wolverhampton to Sheffield in Numan's shiny, white, expensive American sports car – a present from WEA Records to induce Numan to enter into a fresh five-year deal.

The drive is uneventful, pleasant through the West Midlands to the M1, then north – as the journalist gradually forgets the incongruity of his position in the careful, respectful attempts to find common ground. Naugahyde, red on white, Neu, Bowie and Roxy Music tapes, stunning acceleration – a complete sense of unreality in the Indian summer, amplified by lack of sleep. Here, in his car, Numan appears safest of all.

By his own admission not a natural performer, he decided to do the show as it was, and lose money, because 'I thought there was no point in going out unless you were going to give people something to remember and to make it worthwhile. There's no point in being top of the pile unless your show's going to be top of the pile as well.'

Was the lavishness not also to cloak his inexperience?

'You mean to take the limelight away from me a bit? No, it wasn't really. To be honest, the show was put together to be something to look at. I merely thought that being new at it, I wouldn't be very interesting to look at for one and a quarter hours.'

'I don't think I am: I can't do enough different things or look in enough different ways to keep people interested for that time – apart from the real diehards who'll gaze at me for hours. Obviously the majority of the audience isn't like that – especially at this early stage, a lot of them – half of them are just going to see what the fuss is all about.'

What was the point of the pyramids?

'On the cover? It was an image. On stage, the robots are pyramid-shaped – that's to tie in with the cover, and also because I thought that robots … you say a robot and people think of something that does this (*gesticulates mechanically*) and clinks about, and really that's the most unpractical shape you can think of because it's so unstable, it keeps falling over all the time. A pyramid is, I think, the most stable shape you can have: it really is hard to budge. Talking about a straight-thinking machine, it'd have to be that shape where it didn't fall over and damage itself.

'So I thought – well, if I'm going to do it, let's do it realistically, in the proper shape of what they'll be and not go for the image. We had enough impact in the show itself – it'd be nice to put some realism into it.

'I also think that the panels, the walls, it looks like they just light up. I think that one day probably we'll have buildings like that where you don't have streetlights, but the walls of the buildings themselves light up outwards, so it's not like streetlights and shadows and little quiet corners that you could get mugged in …'

The journalist demurs. It transpires that Numan was actually beaten up in a 'quiet corner': that would account for it.

The journalist then mentions that the show lacks an edge of provocation that he's become used to, and likes in his entertainment.

'I think it's just taking it back to cabaret – showbiz for showbiz's more than anything. That's trying to explain what we're on about, and use this as a visual expansion of our songs … To be honest I used to hate all that stuff (*cabaret*), but fairly recently I've got to really like Bing Crosby and now I like Frank Sinatra. I never did before, but the way he just breezes among his crowd as if they're in a circle and not on stage, and he's so relaxed.'

To compound the impression, Numan performs a reasonable version of 'On Broadway' in his current set: it's a song he's always liked, and featured in the lyric of 'You are in My Vision'. The journalist finds this conservative aspect a little hard to take, and wonders how Numan felt about the atmosphere in which he moved in late 1976 and early 1977, seeing the Sex Pistols early on (at Notre Dame Hall) and playing the Roxy in June 1977.

'I always thought it was a movement, especially so in the early days before it became fashionable. I don't feel part of it, no. I don't think I'm doing what I'm doing now because of it, I think I'm doing what I do now *quicker* because of it, if you know what I mean. The business side of it changed: people got signed up, so I went out, crashed away for a few months, got a contract, and then away we went on our own tack. Which is because of punk, but I didn't get into it and evolve into what I'm doing now, I just simply used it.'

It fits. In moving the audience away from the last fashion, Numan mustn't appear to identify too much with it. For good or ill, his preoccupations are different.

'I did use it solely as a means of getting a contract. I didn't see it as going anywhere, I don't think it has gone anywhere. I was excited by the thing as a whole, that all of a sudden there was something that was completely new – new fashion, new music. Hopefully when it got started, something really great would come out of it, but it sort of got destroyed by its own ideas. The anti-hero thing could never happen because this country has always had the heroes, it always will do – I think it's a very English thing to make heroes.'

He replies to another question: 'I never agreed with coming on and being the same as the audience, I never liked that side of it. I never liked it as a personal taste. Also I thought it was very false: you'd see bands as they got more and more well known get more and more distant. Half of it is necessity, really – you have to …'

The phone rings, interrupting the journalist's question. After it's finished, he suggests that Numan's lyrics project a nightmare, depressing view of the future …

'It's an extreme view of the future … from what's happening now, but only one view. It's not necessarily the only one I have, the only view I think there could be – it's possibly the most interesting to write about. It's what I see around me. I'm obviously very affected by things – the violent side of human nature. Human nature itself is quite interesting to write about, if you take it to its extremes.'

The journalist agrees. Does he think they'll get more extreme over the next decade? And if so, what can he see himself doing in that case?

'Hopefully I'll have enough money, whatever, to get away from it …'

Wouldn't he try to do something about it, even if it was merely banging his head against a brick wall …

'I haven't got the interest to want to prevent it or stop it – I tend to be much more selfish and think how I can get out of it, rather than help other people out of it – that may change as I grow older and hopefully grow up a bit more. I know a lot of things I do are very selfish – there must be a word stronger than selfish to cover it – something I'd imagine really isn't quite right.'

Numan's honesty, at least, is refreshing. Like many who are in control of their own destiny, who've struck lucky at an early age, he often seems impatient of those weaker than him, heartless perhaps. He's done it *himself*: why shouldn't others?

His business instincts are acute: apart from having as much control over his career as is possible – he writes the songs, produces them, designed the stage set, hired the band – he's recently entered a favourable deal with WEA, through licensees Beggars Banquet. The deal is for a five-year period, with an advance of £17,000 for the first year.

The deal, further, binds him to seven albums over the five years: 'But we did write into it that if any time I wanted to, I could stop – we talked a lot about this, and it's quite important – that any time I wanted to, I could stop and they can't withhold the royalties, but the agreement is that if I ever do go back to writing I go back to them and finish off that deal. I've signed up for albums, but not for any special time – I don't have time schedules to meet. It's quite nice.'

Earlier he'd mentioned his appreciation for the down-to-earth aspect of WEA, their honesty about the nature of their involvement with the music business (cash). It is suggested that he's extremely fortunate in his relationship with them – in being in a position to dictate his own terms: they need him more than he needs them.

He ponders, and replies: 'I think on a straight 50-50 basis, that's true …

because we could go elsewhere, and they couldn't go and just nip out another number one.'

His position is even stronger because of the current comparative weakness of the music business: 'No, no. They need everyone they can get (*laughs*). WEA have got Rickie Lee Jones and … God knows … Fleetwood Mac, the Eagles … they're all right. They're doing quite well at the moment, but these people may not be going for another five years, and they think I will.'

The white sports car with red naughahyde is an illustration of that strength, that bargaining power. The journalist enquires whether Numan thinks it's a toy.

'It's a toy for a little boy, to keep him happy, that's the feeling I get … It doesn't bother me. It appeals to that side of me … "Oh wow!" that sort of thing. Also, it's a good move on their part to make sure I didn't bugger off to CBS: it's a good little lever they had to make sure negotiations went on in a semi-friendly fashion. I found it flattering that they gave it to me, I don't think they've done anything like that for quite a while.

'I was told that when they heard that I'd been talking to CBS – it was only really to find out what the going rate was – it got back to WEA and they thought, "Oh my God, he's signed to CBS," and – this is what I've been told, mind you, it's not me shooting off my own trumpet – I was told that they said, 'Get him whatever he wants, just make sure he doesn't go' – which is fair enough, cos I suppose I make a lot of money for them.'

All isn't entirely roses, however: 'It's easier for me to dictate now than it was before … to say we want to do this, we want to do that. Before we used to have big arguments about it, now I sort of say, 'That's what we're doing,' and then make sure that they do it – but that's half the problem, to get them to do what they say they'll do, because they say OK to keep you happy and then worm their way around it. They need constant watching.

'But … I don't know if this is true, but they seem to have a respect for my intuition, if you like, in what's to be done. I think they're realizing I know more of where I'm going than anyone else does, because they've no idea of what I'm going to do until I have. Problems haven't arisen.'

Numan's known where he's going for a long time. Is 'being a star' what he'd always wanted?

'Yes, very much so, that's why I went into it. It's the only thing I've wanted, you know, for such a long time. I … always thought I'd do it – looking back on the material I based that opinion on, I'm very surprised. It really is awful. That was when I was 15, 16, hadn't even written a song, still thinking I was going to be a star.

'There was something about the atmosphere of the business that interested me – I can't really give a definite thing that gave me that atmosphere, I just remember reading about it. I had a cousin about seven years older than me who was really into it – and an uncle, Jess (Lidyard) on drums, who I ended up working with – and obviously being around this side of the family, I'd read their papers, I'd watch *Ready Steady Go*, and I just got the atmosphere from it, really from when I was quite young. I thought: Well, that's for me, even though I didn't really do anything constructive about it until I was 16.'

Often becoming a 'star' is as much an act of will as of talent: some have the talent, few the *will* to enter the marketplace in the manner that being a 'star' demands, and then to hold on to what you've got; the consolidation is even more of an effort than the initial rise.

At one stage, Numan admits that he's sometimes at present holding on by his fingertips. The pressure is intense, and increasing: 'I didn't play on it being this early.'

Gary Numan is 21. He still lives with his parents in Wraysbury, near Slough. His father, who works for British Airways, sank all his savings to enable Gary to continue without a record contract. The family are obviously very close.

The journalist asks Numan about his adolescence, seeking clues. Numan admits he was a very isolated child, by choice: 'I wasn't bothered trying to meet other kids. It didn't bother me … to meet other people and talk to them.'

He left school at 16: 'I went to a grammar school: that's the one that sent me to a psychologist. I was expelled from there, eventually … I was a disturb-ing influence. They did try to help me, they were quite nice: they let me stay an extra year – they didn't expel me – even though I should have been. I went into the top class, the A stream. The next year, I was demoted to the bottom class.'

Numan's 'problem' was irrational fits of violence, which came on quite sud-denly, without warning. In the next couple of years, he went from school to school.

'I then went to – I had a talk with my dad who said (in 'Summertime Blues' voice) 'You really need an education, son,' (*laughs*) so off I went to technical college and … it happened again. I had some really weird experiences there … I was just sitting down, and all of a sudden you feel like a bubble forms and people's voices stop making sense.

'I couldn't understand what people were saying, and I could feel myself actually moving back into it, and my head became the bubble and I was going inside that. It only happened about three times: it made me feel quite strange, occasionally – it really did affect me quite a bit.'

On playback (and at the time) the writer wonders whether Numan hasn't swapped that bubble for the society-approved bubble of stardom.

Numan continues: 'I left college and went straight into work. In the day-time I put air-conditioning into buildings, I was a driver, a clerk, really just everyday jobs, all the time planning as well as writing. I was always intending that it was just for now, so it was about bearable. It was enjoyable if I enjoyed the people I was with – one job I had, I was there exactly for a year from birthday to birthday, and it was the worst time of my life – horrible. Horrible people. They hated me because I dyed my hair: they used to call me 'Wally Wanker' – that was my nickname.'

It is guessed that this might be one of the reasons why he doesn't like people very much …

'Obviously this is where … at various times I met people, particularly when I was younger, and I've taken it all in. Another thing that affected me a lot was when I split up with the girl – that was quite some time ago, about two years ago in September, or three years ago. It was very painful. That possibly affected me more than anything, particularly in terms of me getting close to people. It was the one and only time I've ever loved someone outside the family.'

Still, it happens to a lot of people and they survive.

The journalist asks whether he has any of his old friends left. Numan comments with a certain objective humour: 'No, they dropped me long before I became famous. Quite some time ago. They got rid of me because … I was singing in a group and they didn't want me writing the songs any more, so I said, 'It really doesn't bother me' – and I didn't intend at that time to become a big front-man pop star anyway – I was just doing it to gain experience, but they weren't writing any songs.

'So I said, 'Well, write them then, I don't mind,' but that wasn't very good. And so they got rid of me, then went out and did their own set – it took them about six months to write their set, and they had a couple of my songs in it anyway – there was only about a 30-minute set, and it really was awful. The group? It was Mean Street. They were on the *Vortex* live album.'

The journalist dimly remembers hating it.

'I was disgusted. And all my so-called friends at that time would follow them around religiously, and pogo at every gig, it was like rent-a-crowd. And they dropped me completely from their parties, from anything at all … I had one other one, called Gary Robson, who's the only friend I've got … the only friend I trust completely.'

Presumably that's where 'Are "Friends" Electric' came from?

'Yes, it was based on that. And then obviously being deserted made me very paranoid in my attitudes towards friends. I often tended to write "friends" in inverted commas in a lot of the songs.'

The journalist mentions that he seems to have had a lot of unpleasant experiences …

'I don't think any more than most people, I just think that I take them … badly (*laughs*). I find it hard to accept that and understand it – I find it very hard to understand human nature a lot of the time, which is part of the problem.'

The same day, in the *Daily Star*, Numan is given the honour of a centrespread: 'Gary – We Love You! But Numan is So Alone'. The image machine revs into higher gear; Numan is typecast as 'aloof and arrogant' …

'Most of what I said had been blanded out – I didn't actually say what he wrote down, he took the gist of it only. It wasn't done in a nasty way, so I didn't mind it – it was a bit sweet and sickly. I'm not like that.

'The image doesn't worry me. From an outsider's view, it's probably accurate. I think I'm quite strong-willed and know exactly what I'm doing – which is mistaken for arrogance. The "aloof" bit is my wish not to get too close to the audience … which isn't being aloof. It's more survival, really.'

His audience?

'It's very awkward – to be honest about it without giving the wrong impression. I don't feel any … I won't say loyalty, I don't feel that I owe them anything. I made the records and they bought them. They owe me as much as I owe them, so they cancel each other out, really. I don't now have to make another album; I get very annoyed when I hear these things like, oh, people saying, "We made you." They really didn't, they really didn't make anybody at all. We made ourselves, they simply bought the records.'

What kind of hero does he think he is to them?

'It's a bit difficult to answer. I think possibly to a lot of people I'm a symbol of something new – I wouldn't venture any more than that. The … *pose* element is an image: they'll see that, and then they'll go home and imitate it in a mirror and do G. Numan handclaps. That's thought out the same as the image is thought out, to give people something to latch on to. It's taking everything I did when I was young and when I was a fan – and using that, knowing that other people somewhere must be similar to me: I'd like them to do what I did to my heroes.'

As the pressure mounts, and the demands increase, Numan's attitude to his audience, like most stars before him, becomes all the more equivocal. But then, this attitude is, as well, an extension of his attitude towards people in general – wary, mistrustful.

Most of Numan's songs are about alienation, distances between people, failures to communicate. By becoming a star, he's given societal approval to live out those states without attracting attention, as he used to in school …

'I used to live it out quite a lot before, really, because I didn't go out much, I've never gone to parties. If I go out, I normally go out on my own, in the car, driving …'

Is he treated as an object?

'Completely. As a product, yes.'

Would this encourage him to treat others in the same way?

'I think it would do. I think it's a bit early yet for me to change my person-ality to that extent. I find it difficult when the audience meets you, because when they do they're obviously nervous or edgy because they're not sure how you're going to be. Most of them you meet are completely unnatural: they're not giving you their real personalities at all, and you have to accept that. Obviously a lot of them are impolite because of that, and a lot of them try to give the impression that they're not bothered a bit about meeting you and put on this big air of indifference. That upsets me a bit, because it's unnecessary, as well.'

Does he *feel* under pressure?

'Not consciously, but all of a sudden things get on top of me for no reason whatsoever, and really it can come on within minutes. I feel as though I have to do something, but I don't know what it is, and nothing you do seems to be it. I'd imagine it's like getting stuck in a lift, it's the same sort of helplessness. It's very frightening, sometimes.'

The journalist empathizes. He's noticed previously, as one by-product of this tour, numbers of stray females making their way in a very determined fashion into the hotels where the group are staying. Numan finds this hard to cope with.

'I find it very unnerving when people come back to the hotel, because being a "rock 'n' roll star" you're obviously expected to pull the lot – so you've got to come out with the smooth talk, and I'm just not like that. I don't chat people up at all. I find that, trying to get the role right – my 'position' – what they expect me to do and what I want to do aren't the same. I find *that* possi-bly the most difficult part of it.'

Numan has to go to a sound-check. The journalist packs up the tape and goes off to visit some friends. During the sound-check, the tension rises: Sheffield is the last night, with all that entails.

The gig is superb – the journalist is, wittingly, quite riveted by the specta-cle: on top of the edge given by the knowledge that this is the last time that these songs will be played so, the road crew are dicing with the robots and dry ice, upstaging Numan whenever possible. Numan enjoys it, perhaps relieved of the burden of his image, and rushes round the stage like a boy let out of school.

While it's the best performance out of the three that the journalist has seen, out of a tour whose staging is such that all others will be measured against it for some time, he still worries (he likes to, really) about the saminess of much of the material and the spectacle's ultimate reinforcement of passiv-ity, of one-way experience. There's no doubting, however, its objective suc-cess; the crowd go tapioca.

Back at the hotel, there's an end-of-term party, featuring band members, WEA functionaries (including a recognizable but definitely sleeker Dave Dee: the irreverent journalist thinks of 'Bend It' and smiles) and assorted liggers, with a chastened Human League in tow.

It's brought home here to the journalist what a cruel game the music business can be, for both those at the top and the bottom of the pile: one is fawned over, the other is ignored, excluded. *Both* involve a dehumanization. The journalist, as voyeur/vampire, submits and watches. He argues amicably with the classically trained Chris Payne (about *noise*), talks to the careful Paul Gardiner, and discreetly slips out of the circus, early.

The next day, he ties up some loose ends with Numan. Gradually he finds out that, last night, Numan finally buckled under the strain.

'I like it exactly as I am at the moment, except that there's still too many people getting to me after gigs and before gigs. Last night there were too many people that I didn't know and that I didn't want there, but I was just too polite to say "Get them out," so I ended up having rows again.

'There was another scene last night: during the last two or three days I finally decided that I'm fed up with trying to be nice to people, because it isn't respected or appreciated for what it is. I don't think they realize the effort that's needed for somebody in my position to constantly be *nice all the time* and sign every little bit of scrap paper they stick under your nose ... I think there should be quite a change of attitude from now on.

'If people are rude to me because of nerves, or anything else, then I'm really getting into the frame of mind that I really won't put up with it, which isn't getting ... tired, already, of their attitude. I don't want to be treated like a product any more. I think it's about time I put my foot down, make people realize that they can only get so much before you get fed up *like any normal person.*'

The journalist supposes that it's the pressure.

'I think, finally, yes. I was really surprised – I was completely calm, and then something happened and then I went off ...'

But surely that's a direct function of his wishing to become a star, which is what he always wanted ...

'Yeah ... but then I don't have to take it when I get it, I can adapt.'

It transpires that the straw which broke the camel's back was a pair of professionally stray females, shrink-wrapped in lurid turquoise, who gained admittance to the hotel and then settled in. Apart from harassing Numan at a particularly vulnerable point, they attacked his assistant, Su. These irritations were compounded by further annoyances, until ...

'I finally said – this is the first time in six months I've done that – I just said, very loud, "Get off my back," and just shrugged them off, walked away. And then I walked past this woman and she goes: "Temper, temper" – and that was the last straw. I started slamming doors and throwing things. Then I calmed down a bit, but it then flared up again later on and I wrecked a radiator – threw a fire extinguisher and smashed a radiator – and a phone and a chair … then I was all right. I was completely calm …'

It reminds the journalist of Numan's description of his schooldays, a little unnervingly.

If the pressures of touring are now over for a while, others begin: the next day, Numan and the band are going into the studio to begin demoing the new album, *Telekon*, which is about:

' … a man who can finally harness the power of telekinesis, who can move things by thinking about it. He realizes he can do it, and it just increases and snowballs: because of his power he ends up destroying everything, including himself. That's planned, but it's not definite yet.'

… and mixing live tapes recorded at Hammersmith for an EP: one of the tracks included will be the live version of 'Bombers'. Then it's Europe, America, Japan, the world. Ad infinitum.

The journalist looks at the fragile, incredibly determined youth and wonders. He's impressed by Numan's candour, humour, directness and willingness to communicate, and respects the sweep of his vision and his nerve in carrying it out. He thinks, later, that Numan might still be seeing only part of the overall picture and that, although young, fast, bright and learning on his feet, the pressure of his chosen position is going to give him severe problems. Numan is controlled to a 't': it makes the break all the harsher when it occurs.

The writer later flicks lazily through a Burroughs/Gysin volume, *The Third Man*, for lateral illumination: he reads a piece about coincidences, synchronicity. On transcribing the tape he finds it's broken off in mid-sentence, with Numan finally saying, hesitantly: 'I … don't know …'

Some people need to *be* the heroes: but what does it do to them?

Melody Maker, 20 October 1979

In time it could have been so much more
But time has nothing to show

Style

1980–88

David Bowie:
The Gender Bender

'What do you want to be when you grow up, David?'

'Mum, I want to be an Artist, even a Star.'

And so it was. Most pop stars are transient moths to the flame; David Bowie has lasted. To the public he's beyond Pop Star – he's Star Artist. On a plateau: untouchable and mysterious.

His influence has been huge, and not always healthy. You can see it on any High Street, in any disco, in any gay club, on any record sleeve, in every kid who works at Woolies and wants to be a star. Worst, you can see it in the current crop of 'art' bands. Mary Harron remarked disparagingly about the recent Futurama Festival that Siouxsie and the Banshees 'taught a whole generation to pose without humour'. But who taught *them*?

'They shared lovers of both sexes, freely and openly … they shaved their eyebrows and dyed their hair outlandish colours …' So ran the intro to Angie Bowie's kiss 'n' tell memoirs from Two Goose Ranch. This is exactly what most people want to know – you can always rely on the *Sun* to give the public what it wants – about David Bowie, Star: a bit of futurism, a bit of make-up, but best, lots of Gender Confusion.

This, just as much as the 'art', is the key component to the Bowie monolith: Ziggy, you see, hit home. The great public only ever likes to keep one thing in its head about people: just as Bowie's massive contribution to fashion was in the fact that you can *still* see the glam uniform of baggies, tank-top and platforms on provincial streets, so the spice in his image was gayness. Ultimately, if Bowie has invented a whole language of 'art' posing, he's invented more specifically *the* language to express gender confusion. It still hasn't been superseded.

Only last summer a group was to be seen on the stage of a more liberal Manchester club; called Spurtz, they featured two girls who knew what they were doing and one chap who didn't really. They weren't much – noisy and atonal – but what struck me was that the lead singer, banging around in a lurex mini-dress, was drawing entirely from a vocabulary invented by Bowie. And people stood and took it.

These days Bowie is beyond such open campery, as he's beyond open doom-peddling and glam: that's just to say that he's more subtle about it. Angie panders to the great public's view, whereas 'Ashes to Ashes' – his first number one for ages – pandered perfectly to the pop public's view of Bowie: now taking his place as a mature artist.

David Mallet's brilliant video packed in the usual elements – space, madness and lots of blurred gender – in three minutes that seemed pregnant with meaning; once examined, of course, it was at best expressionist and haunting, at worst hooey, but that's irrelevant. It dazzled, it sold. For that one time you saw it on *TOTP*, it was everything.

Ultimately Bowie is still worth watching, if not for the actual performances, many of which have been perfunctory over the last three years, because of the dreams that reside in him; not as a stylist now, but as a totem. If all his costumes were sold, like Judy Garland's in the MGM lot, what kink would wear them?

> What the hell is wrong with me?
> I'm not what I want to be!

Bowie's real success (after all, any tart can bang around the stage in a skirt) is in tapping that perennial teen equation – which, so protected have we been, has carried over into the twenties. Teenage isn't a very nice age – most people who tell you so are lying – and it's one of the immutable facts that the pop biz is staffed by 20- and 30-year-olds reliving the teenhood they never had in re-creating it for consumption by teens.

David Bowie came out properly in a blaze of obvious self re-creation – from Terry Nelhams through Andy Warhol – and it touched every suburban heart. You don't need to be you: you can change your clothes and your name and your hair and be an entirely new person!

Where Bowie upped and out was in changing this on *every* album. That is the promise, the premise of pop and teen fashion: overnight, you can be transformed into something superhuman. Not very pretty, but, to date, necessary. Bowie is the agent of that transformation made manifest and perennial: 'Every man and woman is a star.'

David Bowie started as Davy Jones, a dedicated follower of fashion: from sharp mod to fuzzy hippie. 'Space Oddity' caught the end of hippie idealism perfectly. It was too fragile, but it crystallized one aspect of the myth – futurism. The breakthrough came later. Along with Mark Feld – another mod re-creation – Bowie reacted sharply against the supposed 'authenticity' of hippie.

The sleeves of *The Man Who Sold the World* and *Hunky Dory* set the correct note of ambiguity (at the time, so shocking that the sleeve to the former was changed so as not to sully clean-cut American youth) and *Ziggy*, a truly plastic piece of posing worthy of Paul Gadd, cracked it. All Bowie did was release a catchy 'Over the Rainbow' steal in 'Starman', say 'Hi! I'm bi!' in the MM and hey presto! – Glam!

This really was news. Homosexuality, if not bisexuality, had always been part of pop, both in the process (managers picking up potential singers) and the appeal. Film stars like Montgomery Clift had killed themselves in a previous age trying to deny it, but this was the first time, five years after legalization, that any star came right out and *said it.*

People might have nudged each other about Cliff, would have made jokes about the Beatles if Lennon hadn't been such an obvious thug, might have suspected Ray Davies, but the only heavy ones were the Stones and the Velvet Underground. Even in their case, it was only the icing on the cake of outrage; for the Stones (in particular Jagger and Jones) it was part of their package of anarchic assault, while the Velvets just used it as sleaze – homo mixed with heroin, drag with S&M.

Bowie acknowledged the debt by recording 'White Light' and 'Queen Bitch' – the one that got everybody going with the glitter eyeliner – finally paying tribute on 'Let's Spend the Night Together', a version that made full use of any possible ambiguities. John, I'm only *dancing.*

What he did, then, was to open Pandora's box: by making homosex *attractive* (rather than a snigger) he liberated and brought into the mainstream a whole range of fantasies which had hitherto been repressed. Naturally they came out with great force.

Make-up definitely beat dope as the thing to shock your parents with. It must have taken a degree of courage – although Bowie had the distancing effect of the Ziggy superstructure – but in a way it was inevitable. Some of the results were OK, others awful. At best it was a healthy reaction against gender stereotypes, puritanism, and gave people a chance to move out of whatever closet, but at worst it was Jobriath.

Naturally the puritan hangover still bit; homosexuality had to be perceived as part of some greater decadence. Bowie's gently pessimistic futurism had gradually been replaced by a harsher, apocalyptic view. Distorted as all pop messages are, it read: if it's all ending, anything goes. As such, it was rationalized and easier to cope with: homosex became so pop that it even entered the puritan world of Richard Allen's *Skinhead* series, in his peerless *Glam* – whose cover is an exactly inept copy of *Aladdin Sane.* In the book, his descriptions of faggy glam rockers conform absolutely to public stereotype.

If he wanted to avoid becoming a sharper, futuristic John Inman, Bowie had to move fast. He was already outgrowing Glam and its restrictions, while the public was celebrating Slade and Sweet – brickies dressed up as rent-boys. It'd been good to him: the homosex angle had provided the scandal on which any sound teen career is based and glam had handed him a generation on a plate. Increasingly he pushed at the limits, offering himself as Artist, a generalist adopting different roles over a series of brilliant, yet reactive and reflexive albums.

Each had a high-profile, visual identity that went with the product: these identities Bowie would live out for the duration of the product's life, to such an extent that the albums seemed to have more life than he did – possession in reverse. However dangerous, it was a shield, and gave the advantages of being an Artist rather than an Entertainer: a more permissibly aloof stance, a greater deal of privacy, a greater ability to chop and change. Ultimately, it was that distance involved with Ziggy – avoiding the effects of being a pop star by adopting it as a *role* – amplified.

All this time he was being watched. And copied. Having taught a whole generation to pose, they weren't about to give up: they just got arty as well.

To fit the artist image he began to steal from, and model himself on, specifically literary sources: particularly from William Burroughs and Christopher Isherwood. Both are more or less specifically homosex authors, both deal with smut and totalitarian control as *part* of a slightly more complicated world view that Bowie seemed to transmit. Pop will never leave anything alone, and automatically trashes literary models; either way, Bowie's dilettante fancy flashed on the most obvious elements of both.

From Burroughs he took cut-ups (now translated as random or planned accidents) and control paranoia, *1984* style. From Isherwood the physical spirit of society in decay, pushing the Weimar parallels – the dyed blond, ambiguous dandy, the Thin White Duke. All this was at best stylish and illustrative, particularly when coupled with some good torch songs like 'Win' and 'Golden Years', at worst silly and dangerous, not only generally – in reinforcing the role of the Artist as divine, separate from society and responsible only to himself – but personally. Role assumption became loss of identity: crack baby crack.

By this time Bowie was reaping the seeds of what he'd sown. Punk professed to take *Diamond Dogs* seriously. In a fit of cultural Stalinism, it ignored homosex and concentrated on decay and 1984: true to its roots as an Art Movement (rather than a True Expression of Working Class Revolt) it was fascinated by Bowie as Artist, as Art object, in any orgy of self-re-creation. Remember those wacky pseudonyms?

The man himself had meantime Come Down to Earth and sidestepped into withdrawal, away from any sharply defined image: always at his best when writing for kids ('Kooks'), Bowie introduced a note of autism into the next two albums, *Low* and *Heroes*, as well as several electronic pieces that had the power to disturb dreams. True to form this mode was discontinued.

As Bowie marked time for two unproductive years, his electronic pieces were copied endlessly and became Moderne. He spent the rest of his time appearing in a silly 'decadent' movie and endorsing various people on the sillier end of arty pop: with the exception of Talking Heads, he was to give his direct or indirect approval to Eno, the Human League, Devo and Siouxsie and the Banshees. His approval meant they were all taken *very* seriously.

Most revealingly, he made heavy use of Steve Strange in the 'Ashes to Ashes' video – the most recent, the most absurd, yet *the* most magnificent, exponent of the Suburban Pose which never dies.

At the time of writing, what was once irritating and daft is fast becoming a bleat of defiance; *Scary Monsters* arrives in a climate which is hard on both Bowie and his chosen children. In the face of increasing hardship and political polarization, arty posing and homosex – inextricably linked too often thanks to Bowie's example – are definitely seen to be *out*: the former as a childish luxury, the latter as a definite social disadvantage as dog eats dog.

The new album attempts to come to terms with this: angry, disturbed singing over harsh, distorted noises laid on a familiar Bowie beat. A careful mix of the familiar and the novel, some of it is impressive, all of it beautifully crafted, while two tracks – 'Scary Monsters' and 'Scream Like a Baby' – are equal to anything he's done.

Bowie would like to sound worried and he no doubt is: at least three of the tracks contain a futurism that is only a small projection from present trends, call it Alternative Present if you will. 'Fashion', always an acrostic of fascism and passion, accurately catches the soldier talk coming from the middle classes, while 'Scream Like a Baby' makes it quite clear that if you're gay or socially at all divergent then they'll come for *you*. He even appears to give the lie to his own position on 'Teenage Wildlife': 'I feel like a group of one.' AAAaaah.

And yet the record arrives in a typically frittery sleeve. In the same week the press is full of glowing reports of *The Elephant Man* as Bowie keeps his options out of pop, out of Europe. You can't teach an old dilettante new tricks.

Each new album raises the question of his relevance: this one even more so, as Bowie backs up against the wall of his futurism. What was once projection is now fact. But each time, his relevance is reaffirmed: not necessarily because of musical or pop merit, or even brand loyalty, but because it touches on the concerns that are expected.

No one has surpassed Bowie as gender bender, not even the Village People, although macho styles have taken over in the gay discos, parodying the world outside. Tom Robinson's committed model didn't last either. Nor has anyone surpassed him as Artist model, not even Eno. Even though he's due for supersession on both counts, the mystique is still deadly strong. Why? Because David Bowie has entered British life as the model for every kid who says 'I wish I was …' He's the creation of that need, and as long as it remains, so willhe. Will it be for ever?

The Face, November 1980

Vivienne Westwood: Rich Pickings at the World's End

At the time of writing, clothes – that fleeting yet sharp mirror of the soul – reflect depression, poverty, a fear of the present and – worst – militarism. Music isn't even worth talking about. If clothes carry dreams, most are stillborn. Oxfam food goes to nations below the breadline. Oxfam clothes to kids on the breadline. In this climate anyone brave enough or bright enough to suggest you can raise your head above the shit is worth a hand.

'To look rich is *great*.'

Will you wear your heart on your sleeve? Can you?

'One of the reasons why I'm not sure about doing interviews is that I don't want to put people off and make out there's some big heavy number about getting these clothes. They're colourful and they're exciting: I would really prefer the clothes to speak for themselves.

'And if they're worn on those – I don't know what you call those golden girls with wonderful bodies on the beach – if they're presented in that way I think it's really great. That's who I want to wear them. I don't want people to think 'Gosh … if I wear these clothes I've got to be part of the underground.' These are definitely overground clothes. They're for chatting in aeroplanes in, not tunnels.'

There is a paradox in creating your own dreams out of someone else's visions: each time, from Let It Rock onwards, Vivienne Westwood's clothes confound it. Partly it's the fact that they're so extreme: no skulking in doorways, you have to *wear* them.

This winter Vivienne is trying again. After Let It Rock, Sex and Seditionaries – the straps that changed the world – she's designed a new collection called World's End. You've seen the clothes on Bow Wow Wow, but this time it's only ancillary; someone could wear the clothes who'd never heard of the group. The important thing is pride.

'I think that once you put my clothes on they make you stand in a different way for a start, you don't adopt the same postures, you can't be anonymous, you have to sort of … strut around. They just give you a great lift.'

In 1977 Vivienne and Malcolm McLaren unveiled the only modern look of the 1970s – Seditionaries. If the first punks had thrown up every youth style since the war and then stuck it together with pins and panache, Seditionaries avoided retro yet caught the confusion: the look – both in the original and the imitations – spread throughout the world.

'With Seditionaries, I didn't bother to sell the clothes in any other venue at all. I didn't have any outlet for the clothes other than that shop because the extension of my business was the Sex Pistols. This time I'm doing it on my own: you see, I am the best fashion designer in the world but I'm also, by that definition, an enormous fool because I've never got the juice out of what I do.

'I never thought I'd do another lot of clothes: I was doing it for the purpose that I wanted to establish myself in that way, really, in the fashion world. Because being so good, it's just a shame. Looking at it objectively, it's not a question of one's own personal ego, I just don't want to see myself as a fool – in this commercial world, anyway.'

World's End is Westwood without McLaren, working with a designer, aiming upmarket. The clothes are much more extravagant, richer, than previously – perhaps too much so, veering dangerously close to the 1966 Chelsea preciousness hinted at by the name and the locale. But then times have changed: what may seem precious is actually a blast of defiance, colour splashed on to a canvas of drab.

'The shop will be called World's End, and there's the idea of time in there – I may have one wall all broken glass or something like that and I have this clock that's got 13 hours on it going backwards.'

Vivienne isn't nuts about being interviewed; I can see her point. One picture is, after all, worth a thousand words. Yet what she has to say is worth repeating: she says it quickly, jerkily, as the ideas pour out, changing mid-sentence in a twist of logic that is hard to follow but which on playback makes sense.

'I'm an anarchist, and your position is forever changing really. I mean, for instance, if Wedgwood Benn and his lot were very strong I'd vote for them. For the first time in my life I'd vote for somebody. One reason that I didn't vote is that I don't believe that anyone is as clever as I am or whatever … I think Margaret Thatcher's an idiot. I mean, she's clever on one level, but she's an idiot because she's leading a life that's just a waste of a life. She's clever at destroying other people's lives.'

McLaren and Westwood have always understood the crux of pop politics – should there be such a thing: pop is only ever good at communicating attitudes and moods, never specifics. That's where the traditional left approach falls down, and where the two of them come into their own, in their different ways.

'My job is always to confront the establishment, to try and find out where freedom lies and what you can do: the most obvious way I did that was through the porn T-shirt (a series of them, including the cowboys with touching dicks) and so you find out what's going on.

'I love Malcolm's idea of the *gold* – the business of the kid at his art school when they had that sit-in at that time, and they were moaning about how they couldn't get the right materials or facilities in the college or whatever, and he stood up and said he's always wanted to sculpt in gold and didn't see any reason why he shouldn't … that was a great statement. The children shall inherit the earth: they should be allowed to cover themselves in gold dust if they wish.

'To look rich is great. Malcolm's opinion of why the Two-Tone thing did so well is because of the clothing. The clothing is always so important. I mean, you couldn't imagine the punk rock thing without clothing – it was so easy for some bank clerk to wear that all of a sudden you've got loads of people who've never stuck their neck out before suddenly thinking they can wear what's fashionable and that's why it caught on.

'I don't really want to talk that much about fashion. It's only interesting to me if it's subversive: that's the only reason I'm in fashion, to destroy the world 'conformity'. Nothing's interesting to me unless it's got that element. Either it'll have to go out of the dictionary, or it'll change into something … what I'm saying is that if everybody wore these clothes, that word conformist would go because a conformist would then be a great thing to be.'

Whether that's the way it'll turn out is another matter. The clothes are superb: inverting style codes in a way that's both subtle and shocking. A jacket uses a traditional sports jacket cloth, yet is cut medievally, with the sleeves slashed to reveal a dazzling flash of orange patterned satin. Beautiful. But will people *wear* them? Vivienne is confident, but then she has to be.

'I think the clothes are better than they've ever been and I do hope, if they communicate to more people, they must be better than clothes that communicated to a minority, by definition.'

Although they're not completely tied to Bow Wow Wow – part of the collection is in the West End, in Joseph shops – the clothes still need the band for lift off in the youth market. And they're obviously closely linked: both use a burst of colour to highlight the surrounding drabness.

Can they both realistically expect that buying their product will cause a change in people's lives? Vivienne insists that clothes can change your life.

'Malcolm has always been totally fascinated by clothes. They're the most important thing in his life really. And when he came out of college, I'd been teaching and he was looking for a job and I could sew and I helped him. And then he went off to do the New York Dolls and the Sex Pistols and I got left with it. And I, somehow or other, couldn't put it down because I'd begun something, and as I say, the culmination of that is communicating with a big audience. But having said all that, I can't think of a better way of putting yourself on the spot. I think fashion is the strongest form of communication there is.

'With these clothes, you want to look rakish, you want to look like you can walk down the street feeling like you can *own* the street and you're Jack the Lad or whatever, you need to do that – you know fashion's just *life*, and I do believe that appearances are everything.'

Rhetoric aside I *am* interested in how she works.

'I've always worked through a process of research. The punk rock thing came out of the fact that I got so intrigued, when I started to make clothes in rubberwear in Sex, by all those fetish people and the motives behind what they did, that I really went into the whole research of it. I wasn't content with thinking: Oh, I'll just do something that looks a bit like what they wear. I wanted to make exactly what they wore, or get them to make it, understand all the things there.

'Out of that, that's where all the straps and things came from. You can never create anything in a vacuum. It comes out of chaos really, but that chaos is something you're continually piecing together and discovering.

'When I did the clothes this time, I started to try and do something I had a feeling for, and I couldn't get anywhere. Three months' work I threw away and what I did, what I had to do, was to go to museums and find old patterns and things like that, and scale them up, and look at how people made patterns in those days.

'And I really then got a breakthrough, because their priorities were totally different from what our priorities are: they didn't want to cut a trouser that neatly defined the two cheeks of your bum. They weren't interested in that. They were interested in sexuality in a totally different way. The dynamics were different but the sex was there. And I only find it out by research really: my stimulus is always academic.

'This time, what I've used is anything that personally excited me, and so I took things like Apache Indians, who were just great, those people after the French Revolution who were called the Incroyables, and pirates. Those three things are basically the things that this whole collection's been based around.'

If she's moving upmarket, isn't she going to price herself out of reach of most kids? She denies this, saying that the cost of the clothes will be kept low enough and that kids will scam the money somehow if they really want them.

'If I don't manage to keep the prices down, then what I'll do is make some more clothes in cheaper materials, the same things, so that people can buy them, you know. That's what I'll work on next … because I love selling to some of those kids. You can't always put yourself into the role of performing a social service, but I do nevertheless care that a kid who wants to buy my stuff can get something together.'

Are clothes really worth a candle? Oh yes. They are trivial in themselves, but they're dream carriers – perhaps the most expressive and accurate one now that music is useless and papers even worse.

Are Vivienne's worth a candle? Yes again, provided you've got the guts to wear them. You *can* go out dressed like that but the effect is shattering. I hope they're not too tied to Bow Wow Wow: they'll escape then the slight sense of anticlimax. And I hope a lot of people wear them: at least the world will be a little livelier. Because silly though they are, clothes are one of the few things left.

The last word lies with Eric Joy: 'At the beginning of empires, when a nation is healthy, no one can be bothered with fashion as such. They're much too busy. But when empires crack up, like the decline and fall of Rome, you get sartorial spendour, homosexuality, the equivalent of drug addiction and Mickey Mouse T-shirts ...'

The Face, January 1981

An Enclosed World: The New Romantics

At the end of 1980 Spandau Ballet and Adam and the Ants played *Top of the Pops*. They dressed funny, sang songs that weren't about much, performed quite badly. It didn't matter. They looked young, confident, *new*. Together with Malcolm McLaren's protégés Bow Wow Wow and others, they are the reaction to the cardboard cut-out that punk became. Glamour replaces grubbiness, naked élitism – inverse élitism, withdrawal – commitment, dance – thought, gold – grey.

They share an exaggeration of dress and a bright, attractive dance music. Adam and the Ants and Bow Wow Wow are closely linked through shared players, McLaren's moulding and the Burundi beat used throughout. Their emphasis is on richness, sexiness and a certain humour. Spandau Ballet are colder, blander. The music and the pose is full of mid-1970s ironies – Roxy Music and Bowie filtered through the German synthesizer disco that had become mandatory for the 'new musick' in 1978.

The lessons of the 1970s have been assimilated. The truism about the decade was its grimness. Yet for many it marked an explosion of fantasy – from the sophisticated statements of early punk to the rack of stereotypes they can now choose from, ironically or otherwise.

This assimilation finally ran the stake through the 1960s ideal of a rock 'n' roll community, pop as the binding gel in the youth culture kit. That community has now irretrievably fragmented into myriad markets. Now comes the first proper sub-generation since punk, and a further and complete loss of innocence. Most responsible for both is Malcolm McLaren.

Always a stylist to whom clothes were everything, he injected into the Sex Pistols fashion and political theory. The band started as hangers for clothes. These clothes, designed by McLaren and Vivienne Westwood, mixed sex, shock tactics and stylistic cut-ups of most post-1945 youth styles.

Johnny Rotten would wear a velvet-collared drape (Ted) festooned with pins (punk), massive pin-stripe pegs (modernist), a pin-collar Wemblex (mod) customized into an 'Anarchy' shirt (punk) and brothel creepers (Ted).

This compressed, brilliant look unravelled during the next three years, to become individual styles. There have been Ted, mod, even *punk* revivals.

McLaren pulled his last stroke early in 1977. He turned EMI and A&M's sackings of the Sex Pistols into parables. *This is the way the music business works*, the story ran, *it's frightened and vulnerable, and you can con it to make lots of money*. Pop was never quite the same again.

The new big three have learned, and opted for multinationals: CBS for Adam, EMI for Bow Wow Wow and Chrysalis for Spandau Ballet. This is lucrative in the short term as company pressure gets you on radio and TV but dangerous later when you need manoeuvrability.

Stylistically, they've taken note of the cut-ups that early punks made explicit. Those who participated in punk and then the 'Blitz' culture, specialize in different looks – the pierrot, the toy soldier, the ghoul, the squire, all done perfectly for a month at a time.

They have also done one very bright thing: they've left the twentieth century.

If all the other revivals have come up short against time the New Movement has made it irrelevant. They are going back to a different age, and a pre-industrial epoch. Most other youth styles have just tinkered with a nine-to-five structure born of Victorian society, where you knew your place. They have stepped outside it.

Adam is a sixteenth/eighteenth-century pirate; Spandau Ballet opt for Culloden chic, the skirts of 1745, while Bow Wow Wow's designer Vivienne Westwood harks back to the French Revolution and the eighteenth century.

They are hinting at pre- and post-industrial attitudes. So the past is looted. What they are saying is, our society is obsolete, and unconsciously they hint at a new world.

In making no statement, they make one: the ultimate extension of the 'Me Decade'. Self is now turned into an Art Object, while relations with the outside world are carried out from within self's constructed cocoon. Self finally retreats into a fantasy vacuum, with its micro-cassette, video and replacement of 'nine-to-five' by the micro-chip and, of course, the dole queue.

Time Out, 30 January–5 February 1981

Mutations on a Theme

It's the haircut. They all have it. The Thin Boys practising their Anthony Perkins, razor-eyed outside the tube station. The clothes vary but the haircut is universal. It's rocker but it's obscure: the scores of different yet similar quiffs all reproduce a face from some arcane 1950s rockabilly album. Pinched, pouting, crowned by a flat-top long at the sides, fronted by a greased lick … the Mac Curtis.

A couple of eternal style laws: when things get complex and flaccid, like now, there's always a return to basics. Each time a style goes overground – like the Blitz crowd have recently – there are always several others lurking to take its place.

The latest is a new rocker – the Cat. Not Teds, not rockabilly rebels, neither swinging nor stray, just Cats.

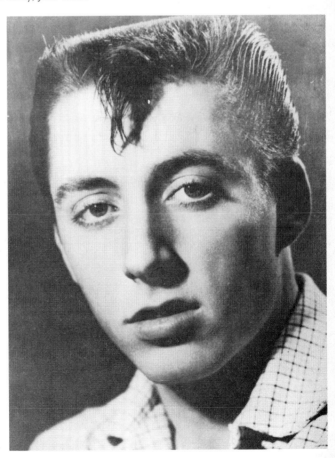

Mac Curtis,
mid 1950s

Fuelled by cheap and accessible American imports, the strategy is simple. Concentrate not on the handed-down English mass Ted uniform – all those drapes, too chintzy – neither on the rockabilly look, but on the American look *as it was*: the College Boy.

Flecks, black pegs, white sox, simple white or plaid shirts, loafers. If you're pushing it, two-tone shoes and white ties, mohair box jackets but that's … flashy. For everyday tubewear, the windcheaters, reindeer sweater, and drainies. The 1950s, stripped down, for 1980s style. Finally, for those who can afford it, the ultimate accessory – a 1950s or early 1960s Vauxhall.

Saturday night inside and outside Shades – a small disco above the Manor House pub in north London – the atmosphere is up, fast, friendly and fanatic. And innocent. Minute detail of haircut and cloth, of behaviour, are clocked and traded. Several are new converts: the Mac Curtis tops Madness badges. The crush on the door is like that outside a Sex Pistols concert in 1977.

Inside the emphasis is on the dance: the guys dance the bunny hop, while the girls swirl the rock 'n' roll by themselves. One stunning pair jive with precision and accuracy, perfect 1950s re-creations. Elsewhere it's gang gossip and Boy meets Girl – somehow, through a trick of time, the untainted vision of a teen paradise that those rockers of the 1950s dreamed of but could never ever reach.

But now they've got it, the celebrants are besieged by complexities and forces that their 1950s counterparts could never have dreamed of. Hence the time-warp. The past is packageable – into a lifestyle that contains certainties lacking elsewhere. It seems that certainties don't exist beyond leisure packaging, yet. Faces on the spot – like Jay, partner in the Rock-a-Cha clothes shop in the Kensington Market – insist that these rockers avoid the racialist, conservative overtones of recent rocker-derived styles, the 1977 Teds and the rockabilly rebels, and it certainly isn't in the air. But, when the complex is reduced to the certain, who knows?

Meantime who cares? It's that cliché – true teen heaven. A place to go where the cops don't go and the trendies don't show. Their very own cloud.

The Face, February 1981

Punk Five Years On

I've been waiting for this one ever since the Black Arabs: a classic rip-off single to finally mop up any punk detritus and put it where it belongs – in the dustbin of Embassy pop. Now that everything is Woolies, it's happened, showing that the old pop art of acute exploitation is far from dead. Put together by Pistols producer and bootlegger Dave Goodman, packaged in pseudo-Jamie Reid ransom-note graphics, performed by bits of archetypal loser punk group Eater. A *Tribute on Red Vinyl*, disco-segues six 'classic punk rock songs' – 'Boredom', 'Outside View', 'New Rose', 'In the City', 'Career Opportunities' and 'Liar' – into the usual metronomic mess after *this* hilarious introduction:

Hey hey
Remember that day
In nineteen seventy-siiiiiiix
When the music scene was boooooring
And then along came THIS!

The boy at Virgin Megastore thought I was mad, but I haven't laughed so much in years – for where else outside the *Rock 'n' Roll Swindle* could you find a disc so frankly paradigmatic of the cycle that turns revolt into trash?

And where, indeed, better to start than here, at the heart of a generation's shibboleths? For the sacred cow of punk rock – its bones picked and scavenged white – is still invoked in the spirit of 'change' and 'youth rebellion', in the ceaseless search for the shiny and new that *still* permeates pop theory and commentary, or at least the dumber end.

I mean, it doesn't really matter either way as long as you're enjoying yourself, but at the time of writing, pop and its satellite industries are permeated with a spirit that can only be described as conservative, self-important and escapist.

To pretend otherwise is to ignore the facts; to pretend otherwise is also futile and fruitless, the spasms of people who really should be old enough to know better. I still like records and things, but I don't make the mistake of thinking that they'll fuel the barricades; for in pop there is no Youth Rebellion, only Youth Consumption. And *that's* built on foundations of sand as well.

So I listen to, laugh at, and love *Tribute on Red Vinyl*. Because it's trash, and trash of the best kind, made all the more piquant by the wrecked ideals and cancelled dreams that inform it. Because it makes a few things very clear: the way that things really are in youth culture – exploitative and trivial, with rare flashes of insight and illumination; the way that our sophisticated, con-sumerist, capitalist economy diverts any possible threat (however silly) – not by force, but in collusion; and finally that laughing with this trash is the only possible reaction to punk's 'sacred' memory and, in laughing, to reflect that believing in itself isn't ignoble but that clinging on is.

> *October 29 1976. I go to see my first proper punk group. I know what it's going to be like. I've been waiting for it for years, and this year most of all: something to match the explosions in my head – stacks of smoke rising from the parched country-side, like cities burning in the Blitz. The group are called the Clash; everybody I talk to says they're the best. Into a Victorian hall, half empty with people standing in bunches. Hostile, insecure. Suddenly, four men with brutally cut hair come on stage, tigers let out of a cage. Bark into a microphone, start making a pummelling, industrial noise. The noise – terrorist slogans knives in West 11 spattered sta-prest Red Guard stripped 1960s riffs apocalyptic drumming – coalesces with the speed and my internal explosions into a perfect chaos. One song: a genuine cry, a child screaming in fear: 'Waa waa wanna waa waa'. Within ten seconds, I'm transfixed; within thirty, changed for ever. That night, I go home and spray a shirt. I do it wrong: the paint congeals in lumps. I don't care. It all fits. Somewhere new to belong …*

Punk's antecedents, its effects and resonances, have been and are still being well documented. That makes it very difficult, and perhaps irrelevant, to try and recapture its initial impact. No doubt it's a sensation that happens to mil-lions of kids in different ways and at different times. This time, there were res-onances built into, left out of and fossilized in punk rock that have actually run quite deep – not only into pop (which isn't really that important) but into our society.

At the time, punk rock wasn't just music or going to the right clubs or wearing the right clothes (which is not to say that these matters were not absorbing or important). It was or *seemed* more: an all-in critique of, an all-out attack on, things as they were and things as they were going to be. Apocalypse now. Serious stuff; playing with, and stoking up, real fire.

For a brief time energy replaced 'reality' and anything seemed possible. When the inevitable slackening off happened throughout the second half of 1977, the consequent disillusionment and cynicism was profound, and still affects us today. Idealism turned into the careerism that was always lurking and that was that – Chaos into Cash.

For that short time many people had taken note: how could they not when in Jubilee Year both the Queen and Johnny Rotten – Britain's mirror image, like Jordan and Thatcher – were *the* two English icons in the world's press?

A week later out of bed with the flu to see the Clash again. I know now that that song is called 'White Riot' and that it's 'about' the Carnival riots. I understand. They play it two or three times to an abusive audience, a few fans and my tape recorder. At the end of – not a 'set', more a set-to – the singer jumps off stage with a helper, who's been lurching with speed-brimmed eyes just a bit too close, and runs through the dispersing crowd to hurl himself at two drunk longhairs responsible for the heckling and flying glasses. The crowd clears and circles. A messy and inconclusive fight starts among the beer slops on the wooden floor. People watch, hollow eyes; the PA plays the Stooges' vicious, vacant 'No Fun'. Everything fuses together. Threatened yet fascinated, I don't sleep that night …

What is interesting now is not lapsed idealism – inevitable in pop and renewable in other areas and at other times – but the process by which punk rock spread from a sophisticated politico/art vanguard into a mass pop movement with the simplification that necessarily entails.

What really sticks is the way that punk rock's sophisticated yet catchy rhetoric – No Future No Fun No Time White Anarchy Antichrist Riot On Bondage Boredom Breakdown Up Yours – eventually reached the people it was 'supposed' to reach. Mark P's 'audience … waiting out there in the discos, on the football terraces and living in boring council estates'.

What happened to the early punks was predictable: for all the rhetoric, most did very well career-wise. Indeed the disparity between punk rhetoric and its then reality was no better illuminated than by the fact that earlier this year, while the two figureheads of 'Anarchy' and 'Riot' were in New York taking their place in the art and rock establishments (and, of course, who can *really* blame them?), that rhetoric was being acted out on the streets in Toxteth, Moss Side, Brixton, for real by the kids who really *did* have No Future.

Suddenly, all over the country, people weren't too thick to try it: in doing so, they reinforced pop's current irrelevance.

Fairly soon, I've seen the Damned and the Sex Pistols. I see the latter while they're being filmed for Janet Street-Porter's 20th Century Box. Punk rock prepares to go public. People are dressed up to the nines; ready to act, to explain their novelty to the cameras. The hall is drenched in a dazzling white light – one huge stage. It all seems so self-conscious that I think I'm in an English version of the Factory. The same speed; represented and real violence; self-re-creation; identity fuck; outrage; and another loud, mad, sado-masochistic pop group: a kid who looks like Steerpike, with burns on his arms, who murders you every time he looks through you. They sing 'No Fun', and a song called 'Anarchy in the UK'; people get up and dance on

stage, pour beer over each other, while the singer insults the audience and
goads them to jump on one another. Two girls are dressed in a bin-liner and
tied up to one another with a dog chain; the boy next to me has his face
obscured by a Cambridge Rapist mask while cameras hover, kids jump and
wrestle by the stage as the noise mounts. The singer, Johnny Rotten, wears
a dirty white shirt ripped and festooned with stickers saying 'I survived the
Texas Chainsaw Massacre' … All the time, the cameras: some of the kids
get their own back though, and spit on the director … On the way out, I see
some of the Clash; they're talking about record deals with Guy Stevens …

The violence, real or represented, was the crucial link in punk rock's going
public for real. Previous to the Bill Grundy incident, it had been *part* of a mixture that included a good deal of posing and sex-bondage, leather and rubber
– and situationist politics, which were bound to get everyone going.

What punks were doing was *representing* violence: because it meant they
got noticed, because it reflected apocalyptic politics and, frankly, because it
was a bit of a thrill, a bit of a giggle.

Pop hadn't seen anything like it for years: not since the mods, whose
iconography pervaded some of the more crucial graphics. Everybody had forgotten about the skinheads, which was a mistake. That violence was a crucial
part of punk's monomaniacal approach – having to 'destroy in order to create'.
Because it was placed in the music business and directed *at* it, it wasn't *that*
serious; more playing with fire.

Most punks understood this, tacitly. Many were art students, upwardly
mobile working class who were pretty wimpy: hence Sid Vicious' (it was he
who had helped Joe Strummer to 'sort out' the hippies at the Royal College of
Art) status and rarity as a gimmick, as a bit of real danger. Conversely, if
Strummer hadn't jumped into violence, then much of the undoubted force of
the Clash set-to would have been lost: he *had* to be seen to put his fists where
his mouth was.

For all the hysteria, this was understandable at a glance and understood:
what complicated things was Fleet Street. Every new youth movement starts
with some sort of moral panic: with Teds it was seat slashing; with hippies,
drugs. With punk, it was foul mouths, swastikas – worn by a few of the more
visible female punks – bondage and violence. The press sent to town over the
fairly pathetic Bill Grundy incident, took punk rhetoric at face value, and
plastered it across the country. Punk rock was hoist by its own petard: from
then on things did get quite hairy.

Later, in the lunch hour, I sit on the bog attacking cut-up bits of paper with
Pritt glue in a very real fever – got to do it now, now, now. 'It' is a fanzine.
It has a lot of very crude sub-Dada (so I find out later) graphics and a

LEFT Johnny Rotten at the Notre Dame Hall, Leicester Square, 15 November 1976 (Ian Dickson)

BELOW Jamie Reid, early idea for 'Anarchy in the UK' sleeve, November 1976

confused yet real article that talks about fascism and punk rock, the rise of Thatcher and the dangers of our total collapse. I need to give voice to those explosions in my head, so powerfully confirmed by punk rock. It is quite premonitory in a naive, spattered sort of way, and upsets quite a few people. I sell it at concerts and at Rough Trade. It gets me noticed: soon I start writing for Sounds. The first thing I help on is a piece called 'Images of the New Wave': it has pictures of tower blocks ...

Preceding the general rioting in Toxteth et al. – the rioting of those doubly disenfranchised from work (no jobs) and leisure (no money to fill the gap left by no jobs) – was another riot in Southall. Everyone was having a riot of their own: and now it was the turn of the Asians and the neo-fascists.

There seems little doubt, for all the writs, that it was triggered – rather than 'started' – by the incursion of the 4-Skins and a couple of coachloads of skins.

It is there, in Oi, or whatever it's called this week, that the tip of the fascist iceberg finds its expression in music.

For all the denials, for all the whimpering and mealy-mouthed double-talk, the neo-fascist *tendencies* of Oi et al. are obvious to anyone with a passing knowledge of twentieth-century history.

The links with punk rock are obvious and direct: 'authenticity', 'street credibility', the same represented violence and cultural brutalism. Punk played about with a lot of ambiguities because it made room, and because it was a thrill: it's not any surprise that those who weren't fortunate enough to be in the know took these ambiguities at face value. Granted, the skins who run around with the Oi bands may only be fascist in style, but then these days style is *all*, isn't it?

Punk rock as an art vanguard 'predicted' a great many things. As one kind of political vanguard – and the one, it is to be remembered, which floated the only *real* anti-Jubilee protest – it sought to help these predictions come true, to exacerbate the situation. And when I say 'punk' here, I mean the two groups and their managers who defined the style: the Clash and the Sex Pistols. In 1976, it, they, other people went on about unemployment, civil disturbance, increasing economic collapse, fascism et al. in suitably apocalyptic rhetoric; from here it looks as though they didn't have much to complain about – as *everything* has got worse.

Culturally, it cut up every youth style since the war, and threw them together sensationally; the legacy can be seen in the frantic resuscitation and adoption of each of these styles ever since – style itself having been severely discoded.

Business-wise, the Sex Pistols' public shenanigans with EMI and A&M didn't do the industry any good. To be sure, good old rock 'n' roll and pop consumption was revived in Britain, but it didn't go down a bomb in the Midwest, did it? Indeed, since 1976, the UK has become a backwater market, insular, ignored.

Better, the Sex Pistols ventured to suggest that you could enjoy what was pop without *buying* records – the experience, whatever that was, was enough. More generally, what punk rock represented in terms of youth culture has been reflected and amplified in other areas: the collapse of the liberal dream, political polarization, the election of a 'hard' conservative government (Mrs Thatcher as fantasy dominatrix) and a general minority hardening and isolation: from pop fragmentation and irrelevance to the aggressive, 'masculine' stance of the current gay style.

From here, I'm intrigued by punk's mixture of preciousness, utopianism, energy, brutality and above all ambivalence: did punks (if they thought about it), did I, really want what was being 'predicted' to happen? 'Prophets', fake or real, always have a vested interest in seeing their 'prophecies' come true –

that's why they're so often self-fulfilling. What somebody desires, they predict.

But now it's all history. All this, like *A Tribute on Red Vinyl*, is mopping up after the event: the punks were sick on the carpet and now it's been cleaned up. These days, I feel as though I know a bit too much about the pop process to participate with the naivety it demands: even trying to lay the process bare is interesting but pretty thankless.

I still buy records, but it's like it used to be: they fill in various gaps, rather than occupy the centre of my life – these days, I find sex, video and history *much* more interesting. As for punk rock, I don't regret it all, although I do wonder occasionally.

I wouldn't do it again.

If at all, it's remembered as a blurred, frantic, exciting period, which for all its paraded nihilism, negativity and stupidity actually held out hope as some reaffirmation of the human spirit in this collapsing society.

That it turned out the way it did really says it all.

Until the next time.

The Face, November 1981

Soft Cell: The Whip Hand

It happens less these days – the law of diminishing returns – but I still have occasional single obsessions that are also, thankfully, irrational. This summer's was 'Tainted Love'.

I was travelling around a good deal, in a variety of circumstances, and seemed to hear it everywhere I went: on the radio walking down Carnaby Street to work, on a record player near Nottingham, on the jukebox of a tiny pub in the Derbyshire hills, where its kitsch familiar blended perfectly with the Fablon, horse brasses and tightly gripped pints. I was fascinated, and reminded of one of the central attractions of pop music and, by implication, mass culture: that single moment of total access.

Later, when my obsession had cooled into familiarity, I started to find reasons for liking the record. It was light and made me laugh, it was simple to the point of folly, it annoyed all the right people, and it went beyond camp to be sexually ambiguous in a fairly direct way.

These impressions were confirmed by seeing Marc Almond gyrating on *Top of the Pops* when the record went to number one. The performance was hysterical: this wizened Duggie Fields lookalike appears, all in black, with

Marc Almond and Dave Ball of Soft Cell, late 1981 (Chalkie Davies)

bracelets half-way up his arm, and starts to do the frug and pout furiously, while his mate just stands there and plays with his toys. The director, thrown into what is obviously a state of revulsion, applies an excess of cosmetic Quantel to cover the whole mess.

Recovering, I was fascinated not so much by the humour and straight ineptitude of the performance but by the fact that, obviously, here were two kids who had gone away, thought about and gone through things, put them together in a way that reflected their experience and their lives (as opposed to a hypothetical concept or trend), here they were suddenly! Nuuuuuuumber one! *TOTP*! DJs! Go-go dancers! Weenies! Lights! Cameras!

In short, although hardly revolutionary, Soft Cell didn't quite fit this age of collusion, and I wondered how they'd wandered on to the set.

I wondered also how they fitted into the industry. For although that notion of the moment of total access is fine and romantic, it ignores the crucial fact that the moment must be paid for. The question is, the price.

The means of production whereby that total access is gained is controlled by various international and national organizations (both private and governmental; at present, largely reflecting the wider world of power politics outside, vested interests are very much in control. Certainly the threat posed on a multiplicity of levels by punk rock has been thoroughly assimilated and has had the convenient effect of training a new audience to be ripe for the picking.

At the time of writing we are, realistically, talking about simple bargaining power: in this the controlling companies have the whip hand. Although groups may *appear* to get what they want, they pay the heavy price of both being expected to provide an *instant* turnover on the company's investment – if no hit, *out* – and tailoring 'what it is they want' to meet these corporate pressures and expectations.

This spiral is fast, and vicious: Soft Cell have been plucked from cultish obscurity and flung into the whirlpool's centre.

Another thing I wondered about was their attitude to the sexuality they appeared to be displaying and exploiting. As befits a pop age which could either be called the New Hippie or the New Glam but which is a hybrid of both, most of its leading players are the time-honoured pop product, the Pretty Boy. Or, at least, the Wimp.

In theory, this is far preferable to that other time-honoured pop product, the Stud, but in practice it always dovetails neatly into received channels of expressing sexual ambiguity or divergence – which nobody has bettered since Bowie. Indeed only Bowie has the power to disturb on this level: it is a factoid that 'Boys Keep Swinging' actually went down in the charts after Bowie's dazzling, fragmented video was shown on *TOTP*.

As one who believes strongly that the human capacity for sexual experience is far wider than we are encouraged to admit, I'm always interested in manifestations of this sort: Soft Cell simply *seemed* to go a bit further, and, being mischievous, I wanted to see how far.

We start, Marc Almond and I, as we mean to go on.

MARC & FILTH: 1

JON SAVAGE: ... and there's the whole Playland thing ...
MARC ALMOND: I've done a solo single a 12-inch single. Both the numbers are about ... one's called 'Fun City', which is all about the Playland thing. I don't know ... it always fascinates me down there. I mean, I would probably find it very boring after going round Times Square [New York, where Soft

Cell recorded their Polygram LP]. The thing about Times Square that I thought was so heavy was that … like Soho you can just walk about and skip around … but walking through there you'd regularly get your pockets gone through, feel hands going down your trousers and your pockets.

JS: For your money? Or were they touching you up?

MA: Might be! I thought it was the money! It was just really horrible. Like gangs … when you're walking along they sort of crowd you, you feel hands going all over you, which is really like a horrible creepy feeling. I did two numbers about that sort of scene, one was the filthy side of it and the other was the sort of sad side of it.

JS: Are those the things you find interesting?

MA: I don't know … I've always loved the idea of glamour in squalour – filth and squalour and sleaziness and seediness – there's always something really glamorous and sort of really sparkling. That's what I … I find it curious, I have a fetish about those places. It's a bit like … I always feel like a spectator, which is wrong but I don't give a shit, I like it anyway.

MARK AS HUNK: 1

MA: (*Picks up copy of Flexipop, a pop magazine aimed at a young market, featuring diseased picture of Soft Cell, in colour, on cover*) Do you think I look hideously ugly in that picture? Really, tell me, please, I've got a complex about it, I really have.

JS: You do actually.

MA: I do look bloody horrible.

JS: You look 35.

MA: I look so Jewish, I'm not even Jewish.

HUW: (*Childhood friend and set designer who has wandered in*) No, I can tell it's you with a false smile.

MA: Really false. Really false.

JS: What happened that day?

MA: God … just so bloody awful.

JS: I mean did they just say 'Cheese'?

MA: Yes … they thrust all these things in your hand and squirt all this stuff over you and say pretend like you're having a good time.

JS: Do people expect you to do all sorts of ridiculous things now?

MA: Honestly, you just don't believe it, the sort of crap …

JS: I'd really thought, naively, that all that had stopped. I thought people didn't push people around in that way …

MA: Well, they do. People do try to get you to do ridiculous things – whether you do them or not. I refused right out to have anything to do with being Hunk of the Month in *Oh Boy* or *Heartbeat*. Like the press department said 'Please do this for *Heartbeat*' … I mean, Christ, I'm hardly a teen idol type or little girl type. It's so fucking stupid. And yet I still pick up what was it today – *Oh Boy* – and see something about us, comparing us with the Human League. The Human League were one of the first, and one of the newest is Soft Cell – which is such shit anyway because we're hardly new … oh, I hate those magazines, I hate 'em. And those colour photos they have are so bad because they always show your spots! They do!

JS: Are you expected to be a performing seal now?

MA: Yeah … they would get you to if they could … but the thing is, we're so awkward.

JS: Did you think it was going to be like that?

MA: I … quite honestly, I didn't know what it was going to be like.

JS: It just happened.

MA: Yeah. I mean, we didn't sort of sit down, here is our plan for the future, here is what we want to be like … because we don't think like that, we just do what we do. If it's successful and a massive hit, then it's successful and a massive hit. If it's not it's great because we're going to it anyway – whether you like it or not. I never knew what it was going to be like. Sometimes I find it very very … all very amusing and great fun, and sometimes I just find it very very depressing.

JS: Have people's attitudes towards you altered?

MA: Yeah, I've just lost so many friends.

JS: (*Surprised*) Lost them? Why?

MA: Because … people say success changes you, I sometimes think that it doesn't so much change you, it changes people's attitudes towards you, around you. Like people who I thought were good friends of mine, who would buy me a drink and who would like to come round to my house and play records … or who I could be generally trivial with, I can't be trivial with them any more. They come up and say, 'You must have loads of money now.' A lot of friends that you gain by success aren't friends because their interest has been directed towards you because of your success, which is not a genuine basis for friendship. You keep a core, you know … but you lose many, especially sort of living in Leeds. I was thinking earlier, I think I'll have a party. I was sitting here and I was writing a list of who should I ask to a party and I was really stuck.

JS: I'm sure you're worse now!

MA: Yes, I am worse. But I put it on. I'm still the same … temperamental little shit that I was before.

JS: You're under more pressure now?

MA: (*Musing*) I was worse at school.

MARK & FILTH: 2

HUW: My favourite thing in porno mags is dirty fingers and dirty nails.

Wonderful. You can always pick up a magazine and there'll always be a photo of someone with the most revolting nails. I'm sure they forget about them.

JS: Or else it's a fetish.

MA: I bet that a lot of gay blokes commit suicide over *Honcho* and *Him!* I do. Well, when they look through *Honcho* and *Him*, they're reading about muscles … huge … wonderful … glistening … torso. And I bet they just sit there going uuuh … I mean they're obviously designed for a wank, but I sometimes think there must be some poor little gay bloke lost in his little attic saying I can't stand it any more.

JS: Like being gay and stuck in some small, northern town …

MA: Actually Wakefield is really funny because all the sort of blokes … I hear so many stories about Wakefield and the people I meet from Wakefield, from what I gather … full of miners, and really butch types … they're all gay, they're all the types that would go with anything. They compare penises in the toilets and all that, and get interested in each other's bodies and things. Call them poof or queer and they'd beat your head in, but they'd probably go to bed with their best friend.

MARK AS HUNK: 2

HUW: It's like everything is going back to Victorianism, the press seem to be slamming everything that's slightly …

JS: What, the music press?

HUW: Not just the music press! I'm particularly thinking about the article in the *Daily Mail* where two very self-righteous people slammed the Top Ten.

MA: Yeah, the *Daily Mail* … they talked about a discussion programme on TV, they said there seems to be a lot of homosexuality creeping into our charts and for example the number one record is obviously about a gay relationship gone wrong.

JS: Which was your record?

MA: (*Muses*) 'Tainted Love'. I wonder if Gloria Jones realized it! Can you see that though?

JS: I thought it was very ambiguous, and if you express ambiguity, you're automatically hated by a lot of people. At least it was honest.

MA: Do you know that the thing I find really creepy is when I read things like Sexy Super-Hunk Marc Almond! Like in the letters to *Record Mirror*, there was a barrage of letters from girls saying how could you call super sexy gorgeous super-hunk Marc queer? It just makes me feel really cold. I feel … always feel hypocritical about what people write about me now, now that I've had a number one record … anybody with an ounce of sense in their head would *never* call me a hunk. I've looked in the mirror and I can't even say that I'm good-looking. I'm rather an odd-looking person if anything.

And that's why I find it very strange. They're talking about someone else: they're not talking about me.

MARC & FILTH: 3

MA: I really don't know how Freddie Mercury gets away with it … I never see him being slagged off as a poof.

JS: Why, have you been?

MA: God *what!!?* You haven't seen the *Record Mirror* scandal over the last few weeks? Cartoon to go with it … I thought that was a bit pathetic really … they've got this new little sort of snipe column in now, where they put everyone down …

JS: Are you gay?

MA: I'm … experimental. I don't think … I'm not … OK, so I'm not into politics or making stands, you know, right? I've had girlfriends – permanent girlfriends – and maybe other things as well. And that's what I'm saying because if I start saying I am I am, then I'm making stands … and also, living in Leeds, in a place like that you don't go and make stands like that if you want to go on living there.

JS: Why's that?

MA: Because like living in Leeds a lot of the time is very much more sort of survival than it is in London. And especially the area I live in is full of lads that go, er, you're fucking queer mate, fucking poof, sort of thing. If I wanted to be a figurehead and make stands then I'd go straight out and say it. But I want to carry on surviving in Leeds, cos I like living in Leeds. I enjoy going out to the clubs and pubs and I want to sort of keep my face intact. Do you see what I mean? It's not that I'm chickening out, and saying I refuse to … do you understand?

JS: Yes. It's just a shame … I do think that our society's attitude towards sex in general is disgusting, as opposed to its attitude towards violence, say, and could do with some re-education.

MA: Yeah, I've got good, sort of emotional friends of both sexes, and in my life whatever happens, happens. It just happens and who gives a damn, you know? With whom … if they've got tits or what. It doesn't matter. I've never been very good at being macho – the thing is, why should I be? I'm quite happy, I have a good life. I wear what I want, I do what I want, act what I want, do exactly what I like. It's just that I tend to get branded a lot because people say I act a bit effeminate, and I'm slight …

JS: But then you don't always do what you want now, do you?

MA: I try hard, drive hard to do what I want, and that's why some people at times … if I'm awkward or temperamental it's because of the frustration that I want to be able to do what I want and I can't.

JS: But the problem is, you're actually caught in a double bind, because the record company helps you to function to some extent and, say compared to a year ago, quite large areas of your life have been taken over by the record company organizing things – like flats and things in New York – that's the other side.

MA: It is a strange sort of Catch-22 situation: it's a situation that I do find particularly annoying, confusing and frustrating because I don't want to like them. I feel … I ought to have nothing to do with them, right? But I have to. I don't need to say to you – you obviously understand.

JS: You are working for a company in fact …

MA: I like what they do for me a lot of the time but I don't like liking it, if you see what I mean.

JS: And the thing is that if and when it stops, as at sometime it will, you'll have to do without it.

MA: Yeah, the thing is, as I've said, I'll always carry on doing what I want. If it stops tomorrow I couldn't give a shit. I'd still sing. I'd still carry on. Having hit records, or being a triple-underlined pop star is not the important thing. I always feel really funny and really hypocritical about being called that, because I look in the mirror and see myself and I look at what people say about me as if it's about somebody else. Which is why I feel very … funny about, strange and cold about things like people wanting to put a pin-up of me in a magazine … do you know what I mean? I find it amusing … but really really strange.

I went with no preconceptions, and was quite surprised at what I got. There is obviously quite a lot more to Marc Almond than meets the eye in glossy pin-ups. I liked him for his honesty and his sense of humour, and I like his records with Dave Ball: 'Bedsitter' has become the latest singles obsession.

His evasions, however, on the subject of his sexuality – which is quite understandable, as he has to live with it and its consequences – and the bitching and ambivalence about the record company – understandable too, and a favourite device of pop stars – may well by symptomatic of the situation he now finds himself in.

Fame, which is as sudden and unexpected as it is transient, must be very hard to cope with: work which was once fun becomes a job as the pressures of the company investment tell, and the other pressures of expectation and public life start crowding in.

For reasons that may come out in the interview, Soft Cell stand out a little in a bland conservative pop age: not a great deal, but enough to be conspicuous and thus marked. They'll need considerable luck and courage to cope with it all successfully, and I hope they have it.

The Face, January 1982

Teenage: The Teds

The overwhelming impression is of life, of movement, of desire – of people let out of a cage. The poses, the expressions are by now quite familiar, trite even: they have already entered a common language. But, consider the dates: mid-1953 to mid-1954. These boys are from a different age – not only temporal – but cultural. They don't think like we do. They have nothing to take for granted, explorers in virgin territory. Theirs is a blank cheque, a carte blanche: the first working-class boys to dress, quite literally, like aristocrats and, in so doing, to become aristocrats. However paltry that may seem today, after endless repetition and complete supersession, in 1953–54 this simple action had the force of a Molotov cocktail, the first salvo in the youth war.

For these boys are among the first Edwardians – a term abbreviated into the snappier 'Teds' from early 1954 – and, from the outset, they were Trouble.

This is evident from the context and framing of contemporary photos. Today, used as we are to language of posing and 'style', they present fascinating archive material – how can these people look so good, so, well, contemporary 30 years later? But then – most of these pictures come from *Picture Post* stories, which sought to at first explain and then understand the Edwardian menace fixed in the public eye as a major moral panic after a number of scare stories, best exemplified by the famous 'Clapham Common Murder' of summer 1953. Immediately, the news angle was set and everything else trailed after: Teds went round killing people.

Why Edwardians began is simple enough – the sudden enfranchisement of a section of the youth population to become good citizens, not by civic works, but by consuming to suit the needs of a fumbling post-war economy. Where they came from first is rather more problematic, shrouded in the mists of working-class criminal folklore, and by the murky twilight of that forgotten, oblique age – the late 1940s and early 1950s.

Unreliable, flawed sources like *The Blue Lamp* (Ealing, 1949) and the hysterical *Cosh Boy* (Romulus, 1952) point to a fair moral panic about youth crime filtering through to movie-makers – as the kids who had been hyped to kill the Jerries (like those terrible 11-year-olds in *1984*, or rather, '1948') at a formative age suddenly found that there was no socially acceptable outlet for all that aggro.

Add this to a different pattern of social organization, based not so much around age as we are today – for this was before that behemoth 'youth culture' – but around territory, and an entirely different social climate, of numbing boredom, of sacrifice for 'our country' several years after the war had ended, of disapproval directed against the youth that were too young to have been in the war – and you have the rich, rotting compost out of which the Edwardians bloomed.

Most sources, both oral and received, seem to agree that the Edwardians 'began' in the Elephant and Castle area of south London, an area that before its hideous, award-winning redevelopment in the mid-1950s had a long-rooted history of working-class crime, flash, marginality. People around the Elephant were a bit … lairy. The date is vague, but generally fixed at around the start of 1952 – certainly, by a year later, Edwardians were big news. Their conception may seem immaculate, but wasn't quite.

The style was the thing that marked them: working-class boys simply weren't encouraged to dress like rainbows, they were expected to know their place. Most people assume that their style was taken directly from a curious upper-class fashion of the late 1940s, the Edwardian look – a throwback to days when there was an empire, there was no socialism, and to the days when men were men.

Although aimed high, it was adopted by Guards officers and then went seriously askew: by 1949 it had been taken up by some more obvious homo-sexuals and that, in 1950, was that. Some early Edwardian must have seen the style, on a trip on the bus up west, and thought it fitted the bill: the style was really killed off when the lower orders started flaunting it.

But, in fact, the youth Edwardian style, as John Taylor has pointed out, is not so much a direct cop of its posher version but much more of a grafting of the external Edwardian details – the collar, lapel, waistcoat – on to the *shape* of the Edwardians' true spiritual ancestors, the Spivs: that inversion of the wartime VIP who had also inverted the drab, enforced puritanism of those post-war years. Yellow socks, chokers and nylons flowering among the bomb-sites: what a statement of desire!

Into all this, the first Edwardians arrived as Martians, or at least, as monsters worthy of the most paranoid possible, cold-war science fiction plot. Then, the magnificence of Colin Donellan was not a source of admiration but of fear, at best disapproval: 'It is no good saying he wants a good hiding. He has had plenty. Donellan is the type of young, post-war criminal which is the most dangerous, for the root causes of his immoral behaviour are most difficult to find.'

These harbingers didn't stand a chance: 'Edwardian Suits, Dance music and a Dagger …' screamed the *Daily Mirror* of 15 September 1953, where the full details of the Clapham Common Murder were plastered over the centre pages, to be feared, and, for this was the first round of the still whirling media/youth cult dance, to be emulated.

For most early Edwardians, crime was the only action, their only environ-ment: they bathed in it. The Clapham Common Murder (later immortalized in Tony Parker's *The Plough Boys*), where a John Beckley was dragged from a

Teddy Boys, 1954 (Keystone)

Alex Cruickshank at the Tottenham Royal, early 1954 (Hulton Deutsch)

bus after a running fight, kicked and stabbed to death, was absolutely typical of the early Edwardians' motiveless, manic, frustrated thrashings. In linking the attitude and action so strongly, the press made for the first time that simple, perennial equation: Youth=Funny Clothes and Big Trouble.

By 1954, the papers were finding new angles, still pegged to the news angle of Ted murder, people started trying to explain them. In May 1954, Llew Gardner stated, confidently, 'They just want an aim in life' – a nice war, presumably – while the *Picture Post* packed staff writer Hilde Marchant off to the Royal in Tottenham, to see the denizens at play. Just to make sure, a psychologist was in tow: 'They have also been called dandies and mother's darlings.'

Whatever slurs were cast at their masculinity, and indeed one writer, T. R. Fyvel, went on to accuse Edwardians of being not so latent homosexuals (lack of dad's presence during the war ran the simplistic Freudian line), the boys in these pictures look straight, defined, extraordinarily sharp, packing their territory like tigers.

Alec Cruickshank – 'It makes a change': such a simple explanation for such a display of desire – is posed up a pillar, immaculate to the angle of his Oxfords, to the cutting jib of his houndstooth knuckle-length, three button. Another resplendent Ted totem is picked apart by his mates in an astonishing fit of mass narcissism, so exactly that John Inman comes to mind – are you being served – but no! look at those expressions: you know that Fyvel was off beam, and that the Teds' real achievement was to make posing, if not macho, then at least non-specific, like the first mods.

Finally, three youths burst out of right frame, away from the dance floor to some unknown end, their mouths open in a rictus of everything that they've never tasted, the world theirs for that instant. In the end, *that* was all that mattered: a mass existentialism – I want the world and I want it now! – without nostalgia, without self-consciousness, without even any verbalizing. They just did it.

Such intensity couldn't last. Prototypically the Edwardian style became Ted, Ted became diluted and changed as it hit its mass audience in late 1955/early 1956, crossing over with the explosion of rock 'n' roll music. The Edwardians were heartening if only for the fact that, to them *music simply wasn't important* – unlike later Teen Ages, bonded by music and that commercial culture, they were bonded by class, clothes and attitude.

A pic from July 1956: these Teddys from Tooting, emerging after a crown court case, show the changes. Although elements of the pose, and the walk are the same – the hand buttoning a jacket in mod-like obsession – the styling isn't so fanatically exact, or as peacock, and the overall look is muted, greyer.

The Teds were still Trouble, of course, all those nice psychologists notwithstanding, but wound and ritualized into a sour pattern caught at its nadir in the Notting Hill Gate race riot as factionized by Colin MacInnes's *Absolute Beginners* (1959):

> … and what I liked least of all was that the oafo nearest me was carrying something wrapped in a science fiction magazine … 'We sore yer,' said an oafo. 'Darkie luvver,' said another. When I glimpsed the SF number unwrapping the chopper, I whipped my keys across his face …

The Edwardians were, quite simply, the first Teen Age – the first of all those youth groups who, through their economic power, were eventually misled into thinking they had political power as well. These pictures show the Edwardians to be simultaneously contemporary and archaic, a revealing disjunction: the relation between these demi-gods and the clumpy, debased Teds of today is tenuous, to say the least.

Now, they appear historically frozen, dumb, heroic explorers who cannot speak across time; war babies all, salute them as you would a Greek marble, in their peacock magnificence, flawed violence and fatal self-absorption. The true disjunction comes in the alarming freshness of these photos – circumscribed as they are by the current prejudices of the time – and the apparent modernity of their costume.

What do they say?

That this was the way that they chose to walk first down that road to the promised land of the Teen Age, and now that those promises have been betrayed and have been shown up to be the worthless trash they were, a mere convenience of consumer capitalism, they mock and accuse us by their return – at Teen Age's end.

From cannon fodder to cannon fodder runs the cycle, as 'those who do not remember the past are condemned to repeat it'.

The Face, June 1982

The Age of Plunder

A Beatles 12-inch flops on to my desk, sporting a rather fetching colour pic of those well-known faces in their velvet collar Burtons and their famous pink tab collars. The record contains their first – not very good – single 'Love Me Do' with an *alternative take*. Train-spotting sleeve notes and a facsimile of the original label add up to a product that is perfectly anachronistic (they didn't have Beatles' collectors *or* 12-inchers in 1962, but that's another story). It's perfectly aimed: backed up by a clever campaign on the London buses – youthful pics of the Four with the captions 'It was 20 Years Ago' and 'Did You Know that John Lennon was in the Beatles?' – the record charted and peaked at number five. I thought it was shit in 1964, but now?!

This alerts me, and I start noticing things. A few days later, I'm on a quick shoot: Manchester's Christmas lights are being switched on in the city centre. There's a bit of razzamatazz: a brass band, an electric organ, appearances by the stars of *Coronation Street*. What gets me is the large crowd, and how it's behaving: this is, after all, only a low-key event but there are thousands more out than have been expected and they're ravening.

The crowd is pinched, cold and in sections obviously very poor. As they surge and yell, I catch a note of real desperation and chilling frenzy beneath the surface jollity that could turn any which way. Things are nearly out of control. And, supreme irony, this crowd, which has been ground down by Tory policies reinforcing the divide between the two nations, starts singing and bawling between the carols; *Beatles songs*, those songs of hope from another age: 'She Loves You' and 'A Hard Day's Night'. 'Help' might have been more appropriate.

You wouldn't catch them singing ABC songs. Back to the wonderful world of pop, I turn to the *Daily Mail* of 16 November. A full-page feature trumpets

Mari Wilson as 'The Girl Behind the Return of the Beehive'. The piece adds, revealingly, that 'Mari, 25 … is dogged by the fact that her hairstyle has always been bigger than her recording success'. *Quite*. A few days before, she has appeared on *The Old Grey Whistle Test*: a quick interview reveals that she's done all the homework necessary on the beehive and the late 1950s/early 1960s, that she rilly wants to emulate Peggy and Judy and that she is going to perform one of her fave songs, 'Cry Me a River'.

She perches on a stool, surrounded by her violinists, the 'Prawn Cocktails' – so *Ealing* – who actually look like punks. It's not bad, but nothing like Julie London. But then Mari is one camp joke that has transcended as things tend to at present. She records for a very studied little label called Compact which has also done all the necessary homework: silly cod sleeve notes by 'Rex Luxore', silly inner sleeves with 1950s curtain patterns and a name taken from a cruddy early 1960s television serial that is hip enough to drop.

The thing that really floors me is that in the same *Daily Mail* of 16 November there is a tiny news item: *Compact*, the twice-weekly TV serial set in a women's magazine office, is to be brought back by the BBC in the spring of 1984. The original series was killed off 17 years ago. Clearly, we are dealing with something quite complex, that is beyond the bounds of parody.

We are inundated by images from the past, swamped by the nostalgia that is splattered all over Thatcherite Britain. Everywhere you turn, you trip over it: films, television series of varying quality, clothes, wars, ideologies, design, desires, pop records. A few more examples, to make your hair really curl: the Falklands War – so Empire, so 1940s war movie; *Brideshead Revisted* and *A Kind of Loving*, two Granada serials that looked at the 1920s and the 1950s respectively through rose-coloured glasses with the design departments having a field day with all this 'period' nonsense. *The* British film of 1982 that has the Yanks drooling is *Chariots of Fire*, a 1920s morality play. There's a rush of public-school and working-class boys into the army, an event unthinkable 10 years ago and a new confidence in the middle classes, just like the 1950s, with the rise of formerly moribund magazines like *The Tatler*, and the runaway success of *The Sloane Ranger Handbook*. It's all underpinned by a reinforcement of the old class and geographical divisions by the most right-wing government since the war. And I haven't even *mentioned* the 1960s.

Craving for novelty may well end in barbarism but this nostalgia transcends any healthy respect for the past: it is a disease all the more sinister because unrecognized and, finally, an explicit device for the reinforcement and success of the New Right.

Part of this is a response to increased leisure. Because we don't produce solid stuff any more – with the decline of the engineering industries – we are now all enrolled in the Culture Club. In the gap left by the failure of the old industries comes Culture as a Commodity, the biggest growth business of the lot: the proliferation of television, video (especially in the lower income groups), computers and information. But this flow of information is not unrestricted: it is characteristic of our time that much essential information is not getting out, but is instead glossed by a national obsession with the past that has reached epidemic proportions.

Pop music, of course, reflects power politics, and it is fascinating to see how it has toed the line. As elsewhere, 1982 has been the year of the unbridled nostalgia fetish: consumers are now trained – by endless interviews, fashion spreads, 'taste' guides like the *NME*'s 'Artist as Consumer' or our own arch 'Disinformation' – to spot the references and make this spotting *part of their enjoyment*. It is not enough to flop around to 'Just What I've Always Wanted', no, you have to know that Mari has done her homework and you should be able to put a date to the beehive. Thus pop's increasing self-consciousness becomes part of the product and fills out nicely all the space made available by sleeves, magazines and videos.

These days, it is not enough to sling out a record: it has to be part of a discrete world, the noise backed up by an infrastructure of promotion, videos and record sleeves that has become all-important and now is in danger of making the product top-heavy with reference. Basically, it's mutton dressed as lamb: do ABC *really* have to dress up (badly) as country squires to promote 'All of My Heart'? Of course not: but it sells the product like the wrapping on a chocolate box. But this is ABC's third or fourth image: when do they stop, and when does the audience have enough?

Record sleeves have been an integral part of this tendency towards mystification and an overloading of meaning: in this Tower of Babel the designer, too, has become all important. Designers even have two books to celebrate their role – the *Album Cover Albums* – and they win design awards and stuff like that. If – like me – you remember when records came in plain white sleeves, it's nice to see people trying, but it is getting a bit silly when the sleeve is more important than the record. Or maybe not: here is perhaps the ultimate recognition of the disposability of today's pop *music*, an acknowledgement of the victory of style over substance.

Here we refer, as always, to punk rock: because in those turbulent nine months the ground rules were laid. Punk always had a retro consciousness – deliberately ignored in the cultural Stalinism that was going on at the time – which was pervasive yet controlled. You got the Sex Pistols covering Who and Small Faces numbers and wearing the clothes from any youth style since the war cut-up with safety-pins; the Clash wearing winklepickers and sounding like the Kinks and Mott the Hoople on *better* speed; Vivienne and Malcolm buying up old 1960s Wemblex pin-collars to mutate into Anarchy shirts. Partly this was a use of deliberate reference points – an age before country-rock, session musicians and dry ice. It was also a reflection of the revivalist groundwork already put in by labels like Stiff and Chiswick, who were the first to reintroduce picture sleeves and customized labels, just like those French or Portuguese Rolling Stones EPs you'd find in *Rock On*.

Thus you will find items like the *All Aboard with the Roogalator* sleeve, at the time much more interesting than the record itself: a direct crib of Robert Freeman's famous picture for *With the Beatles*. Or, rather more wittily, the sleeve notes written by Paul Morley for *The Good Time Music of the Sex Pistols*, a 1977 bootleg, which are a word-for-word steal from *The Pretty Things* album of 1965: 'Exactly one year ago, as we write, the Sex Pistols were raw, unexposed and latent. They were like the atom, ready to ecstatically disclose to the world punk rock, a religion of fast moving people ...'

By this time, picture sleeves were, like 'Limited Edition' 12-inch singles or coloured vinyl, an established part of the record company come-on to the consumer and, thanks to designers like Jamie Reid for the Sex Pistols and Malcolm Garrett for Buzzcocks, an integral part of the way the product was put over. Sleeves like Jamie's 'Holidays in the Sun' and 'Satellite', and the Buzzcocks' 'Orgasm Addict' (designed by Garrett around a montage by Linder) complemented perfectly what was inside, as nostalgic and found elements were ripped up and played around with to produce something genuinely new.

The energy that had created punk and, as an unintentional by-product, revitalized the music business couldn't sustain: by the time the channels were fully opened, there wasn't really very much left to say. Punk's quite careful, instinctive constructions were unravelled stitch by stitch in a series of revivals, renewals and plain fads as every youth style since the war was paraded for emulation and consumption. The references that had been a means to an end became an end in themselves. Instead of trashing the past, pop music started to celebrate it – an act formerly unthinkable in such a tawdry, transient medium. The Age of Pillage had begun: so many sleeves to fill, so many images to construct – where better to go than pop's *own* rich past.

This was and is simple enough. Images from pop's unselfconscious past are invoked as some kind of ritual, or key to a time when pop was still fresh and all a gogo: money, sex and fame beyond measure. Key figures recur: thus you will get the Ray Lowry sleeve for the Clash's *London Calling* directly imitating that of Elvis Presley's first HMV LP, or the sleeve for 'Armagideon Time' reproducing the blithe young dancers that are to be found on any pre-1958 HMV single sleeve. These references are further compounded by genuine reissues, like HMV's own *It's Only Rock 'n' Roll: 1957–62*, which reproduces the dancers again, but in a different context: Collectors' Corner.

The Beatles are also ripe for plunder. The *With the Beatles* sleeve, perhaps the most famous and monolithic piece of cover art – a symbol from the exact moment when pop went *mass* for the first time – reappears everywhere. Little stylistic devices like the white band on top of the front sleeve, with the name of the group and a mono/stereo designation or silly sleeve notes surrounded by ads for 'Emitex' and notices that this is 'Microgroove' or '33⅓ Extended Play' have become so familiar as to be hardly worth remarking upon. What the Beatles signify also becomes a matter for comment: thus the Residents felt it necessary to graffiti-ize the *With the Beatles* sleeve for their own insect ends to make the *Third Reich'n'Roll* point: that pop music as epitomized by the Beatles has become a dread, totalitarian hand upon the minds of the youth. Perhaps they protest too much, but then a group like Haircut 100 will invoke the rear sleeve of *Rubber Soul* to reinforce their 'pure-pop' Monkee pretensions.

It is worth pointing out the difference in meaning between the original and the copy or homage. When *Rubber Soul* or *With the Beatles* came out, the design was innovative: not shocking perhaps, but thought-provoking. Its invocation by Haircut 100 or even the Residents shows how the Beatles have taken on, with time, a meaning very different from their original one and how falsely current pop views the past, redefining that past in its own contemporary image. Similarly, when the Elvis HMV sleeve appeared, it was simultaneously surprising and instinctive – not a matter for comment. Lowry's sleeve captures the *feeling* well – mainly because he is a genuine obsessive – but there's no getting away from the fact that the Clash are putting themselves in the 'Great Rock 'n' Roll Tradition' with all that *that* implies. It's ironic for a group that had said 'No Elvis, Beatles or Rolling Stones in 1977', but even that was giving the past a little too much credence.

Another example of the way this plunder works can be seen in the sleeve for the recent Bauhaus hit, 'Ziggy Stardust'. The group's pretensions in naming themselves after the architectural school – particularly when their work has no conceivable reference to it – can be dismissed as another example of pop's demented pillage of all twentieth-century art, but the mechanics of this particular 'revival' are quite interesting. The record was an unabashed tribute by the group, as they admitted, to glam rock in general and Bowie in particular and an astute choice as the Great Single that Bowie himself never released. The packing reflected this: the Bauhaus 'corporate' logo – another recent trend, this – was overlaid by the *Aladdin Sane* flash, typically inaccurate and out-of-sync, as 'Ziggy Stardust' came from the previous album. The package was then topped by lettering taken directly from Edward Bell's *Scary Monsters* sleeve, thus matching three different periods of Bowie into one 'authentic' package. The group made a very good job of it on *Top of the Pops* – all of David's mimetic gestures, and 'Ronno' lurches – but by then it was all beside the point. This *was* glam rock for 1982.

Pop's own past has not been sufficient: perhaps the most irritating manifestation of the Culture Club is the way that the whole of twentieth-century art and – more recently – any amount of ethnic material have been used with increasing desperation to tart up product that has increasingly less meaning. In this, Bauhaus are only small offenders.

Take the spearheads of last year's obsession with style, for instance: Spandau Ballet, before they got wise and changed direction, connived in sleeves by Graham Smith that peddled the worst kind of neo-neo-Classical pomposity in their frank debt to John Flaxman's lithographs. Or consider Chris Sullivan's poor Picasso – cubist period, please – pastiches on any Blue Rondo à la Turk sleeve. These were obvious enough and made the mistake of being much too 'fine art': anybody with an Athena poster on the wall could see where they came from; just like all the progressive groups used to do bad Dali in the early 1970s. Much more clever and systematic is the work of Peter Saville, perhaps the best-known sleeve designer in England today, and one whose work on the new Ultravox album gained, hardly surprisingly, more comment than the record itself.

Saville began work on designing Factory posters and sleeves, where his frank debt to Futurist posters and typographer Jan Tschichold fitted in perfectly with Factory's 'industrial', 'machine' image. Tschichold published the book that is regarded as the foundation of modern typography in 1928: *Die Neue Typographie* proposed a new, almost classical simplicity and a rejection of Victorian ornament – like the Futurist movement in Italy, it was a celebration of the age of the machine. Thus it comes as no surprise that Saville's brilliant sleeves for Factory Records – *The Factory Sample*, New Order's *Movement* and 'Everything's Gone Green' – reproduce Futurist and Tschichold designs fairly closely. They gave Factory one of the highest, if not *the* highest, graphic profile and made Saville's name.

If on occasions the sleeve became not an ornament but a prison, then it was because the product didn't come up to the Factory 'specification': a very good example of this occurs on Section 25's tentative, delicate *Always Now* album, which is all but swamped by a Saville sleeve that is an object exercise in over-design, and a clear indication that the designer has become more important than the group.

With time, this process has become clear as Saville becomes more important and more influential: his recent designs for Ultravox's 'Quartet' and 'Hymn' are perfect examples of cover art that matches the interior product in a way that is far from flattering. Like Ultravox, these sleeves are grandiose, cod neo-classical exercises perfectly executed for the erection of false pillars of worship. Like the ABC sleeve for 'All of My Heart', which has them parodying the classical grandeur of a Deutsche Grammophon sleeve, they represent some kind of nadir of style over content. Boys, my congratulations!

The past, then, is being plundered in pop as elsewhere in order to construct a totality that is seamless, that cannot be broken. It is a characteristic of our age that there is little sense of community, of any *real* sense of history, as the present is all that matters. Who needs yesterday's papers? In refashioning the past in our image, in tailoring the past to our own preconceptions, the past is recuperated: instead of being a door out of our time, it merely leads to another airless room.

The past is then turned into the most disposable of consumer commodities, and is thus dismissable: the lessons which it can teach us are thought trivial, are ignored among a pile of garbage. A proper study of the past can reveal, however, desires and spirits not all in accordance with Mrs Thatcher's mealy-mouthed ideology as it spreads like scum to fill every available surface, and it is up to us to address ourselves to them.

With thanks to Neville Brody.

The Face, January 1983

Malcolm McLaren: Svengali Steps Out

Into a pop climate of ethnic exotica comes another glossy product: Malcolm McLaren's new LP *Duck Rock*. It continues the themes begun by McLaren's last two releases: the huge hit 'Buffalo Gals', which, by introducing scratching to the UK, threw down a musical gauntlet which has not since been taken up, and 'Soweto', which, although backed up by the best promotional film of the year, failed to do as well. But McLaren isn't deterred: after years of other projects – from Teddy fashion to the New York Dolls, the Sex Pistols, Adam and the Ants, Bow Wow Wow and Boy George – the project is finally himself.

McLaren is best known as some demented Dickensian figure – the Fagin, dressed in leather, who foisted those foul-mouthed yobs the Sex Pistols on the nation while making them speak his lines. Or as the child-molesting puppet manipulator of Bow Wow Wow. But this trouble-making is only the logical, although to some surprising, outcome of his background in 1960s radical politics. What is interesting is not that he has retained those politics – a personal mixture of various anarchic, libertarian and situationist elements, those elements that fuelled the riots in Nanterre and Paris in 1967–8 and were popularized in books such as Richard Neville's *Playpower* – but that he has been prepared, like few of his generation, to adapt them to the times. For a generation that is schooled on ideas of 'authenticity', this is scandalous: and McLaren has never been afraid to home in on the first pop law – whatever society's greatest fear is at the time, mash it up, ram it down people's throats and turn it into trash and, in one of his best slogans, Cash from Chaos.

Perhaps his greatest problem since the Sex Pistols has been that of *diminuendo*: that was such a powerful coup that everything else must have seemed stale, for a while. The Sex Pistols, and the clothes for Sex and Seditionaries which his then-partner Vivienne Westwood designed, were a perfect, practical expression of his own contradictions: in Fred Vermorel's words, 'the vision of an artist, the heart of an anarchist, and the imagination of the spiv'. But later projects like the Ronnie Biggs Sex Pistols, Bow Wow Wow and mid-period Adam and the Ants seemed half-cocked: good ideas ruined by bad timing or bad application, with the spiv taking over at the expense of anything else. He's recently solved that problem by simply doing it himself. His latest incarnation as travelling pop star and Pied Piper is both surprising – very few pop managers ever cross that divide – and effective, as it seems to have given him a new lease of life.

Each fresh McLaren project comes with a separate theory, and this one is no exception: one central idea on *Duck Rock* is that by presenting music from around the world in a fresh context, you can stimulate interest in travel and tourism, and thus the exchange of information. It is this belief in content, however much you may disagree with the nuts and bolts of it, that differentiates McLaren from other pop hustlers; behind every apparently cynical stroke has been a firm belief in the idea of getting information across: content in culture. And thus, while many will criticize the new record as another example of cultural imperialism – just more exotica soma for *Top of the Pops* – I don't find this approach very interesting. What I do, though, is that McLaren, in a profoundly tawdry and amusing medium, isn't concerned with authenticity, but with research, polemic and effect.

This combination of huckster and visionary – McLaren is one of the few pop theorists (although he'd hate that term) or social motivators to celebrate the tension between art and commerce in pop – has already succeeded in tapping the nation's subconscious several times. And one key to the way McLaren works comes from the way he talks: he is concerned with the links between things, both temporal and physical – like the links between Dickens and the Sex Pistols – and those links are expressed in bursts of talk that appear like flights of fancy but, when examined, have a logic of their own. People have always worried about being conned by McLaren, but that's only because he's so blatant: it's never worried me, and I'm quite content to hear the latest stories he has to tell. 'The Zulus like the idea of the Sex Pistols not being able to play but being able to steal all the money off the record companies. They used to roll up in fits of laughter. Each Zulu guy, when I saw him the following day, would go, "Eh man, tell my friend the story of the Sex Pistol, man." That's all they wanted to know, they thought it was just hilarious. They love the joke of someone being conned. They loved the trickery of it all. And when I finally recorded that song, "Punk It Up", that was the only song they were ultimately interested in. It was their favourite, because it was something that was bringing in a whole new story to their lifestyle.'

McLaren's latest batch of stories on *Duck Rock* is, initially, a bewildering mixture of Zulu, rap and Cuban music among others. I wondered whether a common thread had emerged.

'I think it made sense in that everything could be termed anti-Christian. I thought there was a common parallel between the square-dance caller in Tennessee who is still living out his European culture and the rapper in the south Bronx and the Zulu chief who banged the drum, who called the announcement of the dance, and the Peruvian Indians in the hills of Lima. I found that in all cases it was all about things that go bump in the night. It's all about your magic. Dance suddenly becoming pantomiming of animals, and all the things that are very magical about this. I felt that related very much to, I suppose, the excitement felt when somebody saw Elvis doing his dance on stage in the 1950s. We live in a Christian society concerned with order: rock 'n' roll was always concerned with *disorder*. Punk rock promoted blatantly the word chaos. Cash from Chaos.'

He chuckles at a successful slogan, and throws a question back: 'Do you think that pop music is very pagan in its outlook?'

I mumble some agreement, and this sets him off on another story: 'Do you know, the origin of the square-dance is traced back to ancient Rome? It's just about the old round dance and someone finding a partner for the night, in a society where marriage wasn't the ordained thing. And I was interested when I talked to this old cattle man in Tennessee. He said to me: "You know what? I used to come into town and we'd go into a barn dance, sometimes called a beehive." And I thought right, it reminds me of the old song: *"I'm a king bee, baby, buzzing round your heart."* He said, "The local parson would always try to ban the barn dance or the square-dance because in a square-dance you get to dance with everybody, and he was afraid we'd get his daughter. And we used to hold these girls real tight and when the caller said, 'Swing 'em up', we would swing 'em up and they would squeal and we would pull 'em up and brush our bodies against theirs. We used to love the square-dance," he said, "and the fiddle player would play this real high note and the local parson and the church would call the fiddle the instrument of Satan." And I thought that well, there was an absolute common parallel with those irate fathers in those clichéd movies of rock 'n' roll in the 1950s when they called the electric guitar the same thing. But it was all about the same thing: they disliked the pagan attitude. They disliked the disorder. They disliked the idea that you could exchange partners. It was the classlessness of the square-dance that was the problem.

'Well, that was the beginning for me of this album. When I found that out, I thought for the first time I'd found something that I could actually say was European in concept. It wasn't borrowing from some black tradition. It was European, and had as much rock 'n' roll in it as anything else I was likely to hear later on in my travels.'

Pop's other master plunderer, David Bowie, has also turned to ethnocentricity for inspiration, albeit in a different spirit. The 'Let's Dance' and 'Soweto' videos caught brilliantly this latest pop fancy. I ask Malcolm about what many – perhaps more prosaic – souls see as a lack of constancy, a dilettanteism which is part of the way he makes pop music work for him. Is this ability to change his ideas fast necessary to stay alive in this medium?

'Depends on how strong your original idea is. I don't know whether you have to change your ideas very fast. It appears that's what David Bowie does. I think to me … I'm a punk rocker – it's just evolution really. I don't see the difference in what I do now as to what I did then, except I must say I expect a major difference in this project as to any other project was the fact that I've become very interested in the music. That's the real, real difference.'

In a different way, and not with as much power, McLaren's latest record subverts the current pop norm. It exhibits a wit – in the collating and editing – and joy – in the performance – which exists rarely in today's calculatedly fizzy pop. But he's up to his old tricks: taking the conditions of the moment – in using the producer (Trevor Horn) as superstar archetype, and in packaging this glossily exotic product in a typically fizzy sleeve – like he did with Bow Wow Wow or, more forcefully, with the Sex Pistols. If it's difficult to conceive of anything that he could do that would match that power, it's remarkable that he's still functioning, and still on the ball.

'Well, I think if you didn't accept those conditions the game would be up, you see. And you have a problem in convincing people. I suppose that's all. You're playing the game.'

Does he agree to bide by the rules?

'To a certain extent, yes. But the rules are very tiny. At the end of the day they aren't the things that govern your ideals or your ideas. They should only be used, I suppose, as a selling point. They give you the framework on which to operate. They also make it – oh, I suppose that if you live in a room with finely coloured walls, you wish everything around it to relate to it. If it doesn't relate to it, you're not too sure whether you should have had your walls finely coloured or not. If you're a hobo on the road, you don't need those highly coloured walls and you probably don't need the gloss. You just need the content.'

I protest that he could hardly be called a hobo, but this sets off another train of thought.

'No, but I dream of being one, and I think that spirit is probably the most noble and the most modern spirit going. There's not really much point, you see, in staying in one place. There's a great deal of point in moving from place to place. I got very inspired by an American, Harry K. McIntyre, who was better known as Haywire Mac. He was probably the first communist in the United States of America. He was the inventor of the American folk tradition; he wrote the song 'Big Candy Mountain', which was nothing to do with the Burl Ives version, but was far more vehement and far more to do with the idea of the hobo, you see. It's a bit like Fagin and the Artful Dodger, and he had to invent a song in which those young kids would be solicited or seduced by the Lemonade Spring and the Big Rock Candy Mountain.'

Dickens has been a recurring theme in McLaren's iconography, from the view out of Glitterbest's grimy windows to the poster he concocted for the Sex Pistols' Christmas Day appearance in 1977. Why was he so fascinating?

'Because I think myself that his most famous character, Fagin, and the Artful Dodger was such a potent classless and powerful political force in England at the time that it gave great freedom to a lot of people; the thought that to go out of your way to grab hold of what you need by hustling was something totally opposite to all the class structure. It was a very subversive idea and Fagin and the Artful Dodger became such colourful characters because people loved them. Dickens therefore summed up everybody's noble dreams in England.'

And didn't he cut into the nation's consciousness in a way which most modern novelists don't?

'Well, weren't the Sex Pistols great storytellers? They gave journalism such a new lease of life because for the first time they could write pages and pages about these storytellers. The music was irrelevant. They gave these terrific ideas every five minutes – "God Save the Queen", "Anarchy in the UK".'

He snaps his fingers. 'I think the biggest problem with the Sex Pistols is that people tried to make you believe they were working class. The fact of the matter was that they were fairly classless. The Sex Pistols became to my mind more important when they had the money, because money is the critique, you see. It was really then that I noticed a vast change with the Sex Pistols: when they were able to have money behind them, they were really important and the more money we could house behind us, the more we flaunted that idea.'

It's now six years, a generation and a cultural and political world away from punk rock. Does McLaren have a perspective?

'I think it had an enormous influence internationally. I think punk rock is more alive in Harlem, in some respects, than it is perhaps in Bracknell. I think you'll find in probably a year from now, punk rock will be something that has been seized upon by the black culture in Harlem and the south Bronx and you'll be hearing a lot of punk rock-type songs with punk rock-type lyrics with all the funk and the New York rhythms. I think there's a song actually coming out in two weeks' time called "Punk Rap Attack". They loved the word. They'd heard of the Sex Pistols. It was extraordinary: I was in the Bronx, and I saw a boy and a girl hand in hand, two black kids from the south Bronx, and they were walking down the street and they were both wearing "Never Mind the Bollocks" T-shirts. He laughs. 'Great! That's what's extraordinary. In 1982. That's what's amazing. Now they may not have even known of the Sex Pistols. They liked the look of it. They homed in on it. They saw something. They liked the words.'

In November 1977, a record shop owner was prosecuted for displaying that slogan on record sleeves in his window; now it's on T-shirts throughout the world. An interesting contradiction. The whole English idea of pop culture is one McLaren still has fondness for since, despite seeing London from abroad as a 'muddy hole', he thinks it can still run counter to the existing power structures.

'We never seem to know where our roots are, what our origins are, who the hell we are. We tend to borrow from hither and thither, we have a tradition of being the greatest pirates, the greatest plunderers, the best presenters of other people's ideas. If that's the case, then we can use that notion. Then they should engineer the situation of allowing kids to go out and explore other cultures internationally and bring them back to England, to give a whole new feeling towards tourism in England, and pursuing the idea of England being a cultural centre.

'There's a whole new class building up of unemployed that really genuinely desire to get out and experiment and create a new lifestyle for themselves. But in order to do that, they've got to travel. They have to pick up and exchange views. I think if England had that potential, that would create jobs in the world-wide industry of tourism. But that idea isn't allowed to generate itself simply because people aren't allowed to feel that international about anything.

'We are always told what is right and, I suppose, basically, the suppressed part of our class structure has always agreed that the King and Country and the ruling class know best. After all, they are gentlemen, and that tradition is something that people have prospered by in this country and suppressed the whole idea of a classless viewpoint in our culture. I think that people like Adam Ant and Boy George and all those other characters that have cropped up from time to time in pop culture are important and have been important because, basically, they are classless. And that's why the whole motion of style and dress and fashion is exciting in England: because it demonstrates a classlessness.

I mention that the idea and style of the gentleman has become very chic recently, after *Brideshead Revisited*, *Chariots of Fire* and other such phenomena.

'Well, it's coming back only because it seems a security valve, doesn't it? It's a way of … within a field of unemployment, everybody's been made to feel very depressed and irresponsible if they don't have a job. In order to acquire a certain status, in order to be able to open doors, you must look like part of the system or the status quo or what people understand to be respectable. You're within it rather than stepping out of it, and so suddenly the punk rocker becomes the hillbilly of society. He's not something to be acknowledged.

'But what is the most wonderful thing that I find when I come back is seeing how colourful all these characters have become. How they've gone from being very black and full of chains into adopting New Guinea-like make-up and the girls have a mix-up of Mexico and the Appalachian mountains and you can wear father's bashed-in boots and have the whole spirit of the hobo, the modern nomad. We are the most international-looking nation of all.

'That's why free transport is the biggest platform that the Labour Party could ever have to win this election. There's a huge dispossessed part of this population that's unemployed who need a sense of knowing that they are important. And they are important in my opinion because it's within that part of our society that is going to be created the most exciting new ways of generating cultures.'

Isn't this new record a much better realization of what he was trying to do with Bow Wow Wow and even Adam?

'Well, in those cases I was acting as a mercenary and I suppose that, after a while, a mercenary has to finish. Being a mercenary in the form of a manager was …'

Boring?

'Not even just boring, it centred me in a position which I didn't want any part of. I didn't want to be the go-between for the band that was desperate for success and I never knew what the success was. I preferred, in the end, to opt out. So I took it upon myself to be controlled by some record company to make this album because it was a way of resolving what my thoughts were in the past few years. My own inspiration, I couldn't secure that in Boy George or Bow Wow Wow. Maybe I never had the confidence at the time. Maybe I never saw myself as a presenter in terms of being the artist.'

In the end, it's only logical: instead of telling Johnny Rotten what to say or, more accurately, guiding him in what to say, say it yourself and cut out the middleman. If a Svengali is to be honest, he'll do it himself. But very few do.

So what's next, Malcolm?

'I've just discovered opera. And I'm working on something with Puccini. I'd like to use that and put it in the context of the street and maybe even introduce it to the south Bronx.'

I'll leave *that* to your imagination.

Later, I reflect that McLaren seemed much more relaxed than the previous times I'd met him – in 1979 and 1981 around the time of Bow Wow Wow's *C30 C60 C90 Go!* cassette – and less bitter. Perhaps the fact that he doesn't have to rely any more on other people – with all the charges of exploitation that seem to fly around him – has taken the pressure off. The other thing I noticed was his apparent respectability: certainly, his new venture hasn't been preceded or accompanied by the usual air of scandal. That doesn't lessen the impact of his ideas, just makes them less 'newsworthy'; and the time is not one for 'shock tactics' as a device for getting your ideas across (as the pathetic exploits of Michael Fagan and Specimen show) – more for guarded integration. In the end, McLaren is obsessed by England in the widest sense – in its politics, class structure, outward appearance, and in its own peculiar product, pop culture; his latest guise as world traveller is fulfilling yet another quintessentially English dream, but the perspective it has given him still makes him one of our most acute and exciting commentators.

Time Out, 27 May–2 June 1983

Androgyny: Confused Chromosomes and Camp Followers

Excessively polarized personality types thrive in a culture that demands the repression of certain natural tendencies while people are developing the so-called 'masculine' and so-called 'feminine' traits which society considers to be appropriate for each sex.

June Singer, 'Androgyny', 1977

Turning on *Top of the Pops* during the last few weeks has been like tuning into a tart's flophouse. Apart from limp tulips like Kajagoogoo (objectionable, let's get this straight, not because of their sexual orientation, but because of the way in which they operate) we've been thrown these bouquets: the Human League – which one Phil? which, Joanne? Oh, the one with the deep voice; Twisted Sister – simple, acned breastbeating enlivened by a leading singer who looks like a cross between Shirley Temple and half the Coleherne; the Eurythmics – Annie Lennox turning all severe in the back of a limo, so severe that she has to produce a copy of her birth certificate to prove she's a

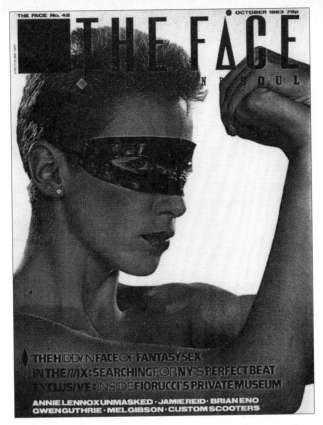

Face 42, October 1983

girl; and, of course, the transcendentally masculine George.

Gender bending is big news: quite apart from pop stars, whom you'd expect to go in for that sort of thing, you can have transvestite villains – David Martin in frock and wig; transvestite TV stars – George Logan and Patrick Fyffe as Hinge and Bracket; transvestite film stars – Dustin Hoffman as Tootsie; and, double bent, a transvestite Prime Minister – Mrs Thatcher as more of a man than you'll ever be. No wonder the Americans are worried. And perhaps that's why they are so interested in Robert Mapplethorpe's book of heroic poses of bodybuilder Lisa Lyon.

Naturally, much of this is pure vaudeville and disguise. And, just as the current wave of fizzy pop invites – and courts – comparison with the Age of Glam, then it is also seen as spineless, collusive and is summarily dismissed as without content – inducing the less-than-attractive phenomenon of bitching Boys Town writers and camped-up butch throwbacks like the JoBoxers. What all these people ignore is that there is something, as usual, going on behind pop's latest penchant for dressing sideways: what it is, is nothing less than a restatement of popular music's power, however oblique, to inform and comment upon the relationship between the dominant – what we are *told* to feel – and the subconscious – what many of us are actually feeling – at the same time as it shifts units.

Thus it was as heartening as it was amusing to see Marc Almond and Boy George mincing about all over the nation's TV screens – with as much courage as self-absorption – at a time when 'Our Boys' were warring with the 'Argies'. Small beer, yet their manifest camping was a relief from all the bellicosity that I, for one, refused to believe in. But they also hinted, in their different ways, at a pop perennial: the expression of divergent sexuality – whether homosexuality, transvestism (men dressed as women, and vice versa) or plain camp (an all-pervasive way of looking at the world, *not* a specific sexual orientation) – which always holds as its ultimate principle the theme of the androgyne that has recurred throughout history.

As a socially marginal area of expertise – with its roots in the entertainment business of variety, music hall and pantomime – pop has always fed off and expressed different attitudes about sex, and moral panics about the subject have marked the appearance of each new pop sensation. Consider all the fuss about Elvis in 1956, when the fact that he had a dick was so revolutionary that he was filmed only from the waist *upwards* on one TV show of the time. Or those pictures of Mick Jagger under a hair-dryer in 1964, or David Bowie's famous 'Hi! I'm bi!' announcement early in 1972. Whatever your worst fear of the time, pop will ram it down your throat and turn it into fool's gold.

The way that this has worked has highlighted pop's simultaneous (and uneasy) status as 'youth expression' and as kinky commerce – a function of its parodic, hyperventilated status *vis-à-vis* capitalism proper, which, of course, exploits all kinds of sex without facing the consequences. And pop's attitude to sex, as to other things, is contradictory and confused: part knowing, part deliberately innocent, part libertarian, part repressive.

Sex is as serious a business as it is big business: an integral, yet often occult part of pop's mythology and pulling power – so the playfulness of a sex song like Buzzcocks' 'Orgasm Addict', or the directness of Tim Buckley's 'Sweet Surrender' is rare. And it's just as serious to the world outside the playpen: ever since teenagers were defined as a race apart in the 1950s (just like homosexuals, or 'gays' were in the 1970s), ripe for exploitation as a consumer fifth column yet also a subject for envy, then sex of all kinds has been a central component in the anger against 'Those Bloody Kids' (© *Daily Mirror*, 1983) because, quite obviously, they're having it off non-stop. If simple heterosexual sex is still so irritating, and so much of a target of prurience, then divergent sex is so much worse: pop's solution to the problem is either disguise or blatant display.

If a certain low camp and institutional homosexuality was present in pop from the beginning, then its traces are to be found even in the primeval explosion (in more senses than one) of rock 'n' roll. Elvis is a good case in point: taunted at school and after because of his effeminate style of sideburns and flashy pimp suits, he used that same style to become the sexual molester of middle America and the world – the fact that he was a sexual mess (as well documented in Albert Goldman's biog) is incidental. In flaunting both sex and hitherto taboo styles, both racial and sexual, he opened the floodgates of all sorts of other mutations – the wildly camp Little Richard, or Jerry Lee Lewis with his child bride – who made rock 'n' roll simply too hot to handle. The next generation of clean cut stars swapped switchblades and weird shrieks for teddy bears and a return to a comfortable, pre-pubescent innocence.

This same innocence was found to be in British pop at the same time, where, not for the last time, the path to fame started at the casting couch. If the role of homosexual manager was vital at this time – bedding his clients while shovelling them out to be adored by teenage girls – then all was kept below stairs: sex was disguised by a veil of pretty boy passivity or plain neuterdom, and merely to have breathed deviation would have brought a flurry of law suits. If pop was becoming a sickly hothouse, then the mid-1960s blew it wide open: if, initially, the Beatles were the lovable mop-tops, not so much real human beings, but cute dolls with fluffy hairdos, then they and the countless groups that followed them were part of an androgynous flood which, although part of the same dubious showbiz milieu, handled sex matters in quite a different way

Both of the central mid-1960s groups, the Beatles and the Rolling Stones, were androgynous in a fairly direct way: they hinted, however obliquely, at the breakdown of traditional masculine codes by admitting the feminine side of their nature – all that long hair and, on occasion, deliberately camp behaviour – and their espousal of drugs later on contributed to the general chaos, what with the confused dandyism of the mods and much talk of the permissive society.

If the Beatles were innocent and contained – although John Lennon later took the principle to one logical conclusion by giving up his career to become a male housewife – then the Rolling Stones played the flaunting game, deliberately courting a bisexual image while singing songs of direct hetrosexual lust. Yet the outrage had a point: like Malcolm McLaren later, Andrew Loog Oldham divined that the route to social chaos and immortality was achieved through the use of sex as an irritant.

The activities and attitudes of all these pop stars reflected and fuelled what was going on in the outside world: in 1967, homosexuality was legalized, and the androgynous culture of the period found political expression in the Gay

Movement, which began in the late 1960s. Glam rock was partly a product of this greater freedom – Bowie came right out and turned bisexuality into a marketing device – yet the harsher times gave the treatment of divergent sex an apocalyptic, disinterested edge picked up off the Velvet Underground. Their material came fairly and squarely out of the Factory milieu: the playpen which Andy Warhol established for all worlds to meet – the straight and the gay, the rich and the poor – to meet and form the new aristocracy. Because of the Factory's specificity, much was lost in the translation to English mores when David Bowie and the punks got hold of it in the 1970s.

Glam was all about sex as disguise, as a mask: heterosexuality in drag because that was the fashion: yet through Roxy Music it also introduced the idea that is all-pervasive now, the idea of camp gone mad – an ironic, intensely referential way of looking at things where you'd refer to models and never quite admit to meaning what you were doing. Everything lay within inverted commas, as it has done in today's Age of Plunder.

> A new consciousness is rising out of the morass of a declining society that has bent too far towards rationialism, towards technology and towards the acquisition of power through unbridled competition – or whatever other means have been considered necessary by those in charge to achieve dominance and control over less sophisticted people. The new consciousness takes note that our society has become overbalanced in favour of the so-called masculine qualities of character.
>
> June Singer, 'Androgyny', 1977

If the New York Dolls were exemplary in moving from sexual to political drag, then the punks – through the same channel, Malcolm McLaren – retained an option on sexual divergence as part of the generalized shock package. Yet while they might have hung out at Louise's in Poland Street, any too obvious and public manifestations of sexual diversity sat ill with the cultural Red Guard: early photographs of the Sex Pistols by Peter Christopherson weren't used because they were a bit too near the knuckle, while the aggressive, rigid form of much early punk music precluded any camping except the political.

What the punks were concerned with – at first rightly, and later to their cost – was a moral, political view of a world that was rotting and on the edge of collapse: love, gender and matters of sex were simply not an issue – what with all that amphetamine, there was simply not time or inclination for it (although, paradoxically, it was through this lack of stress that a rash of gender-integrated groups occurred). In a typical gesture, Vivienne Westwood changed the name of the shop from *Sex* to *Seditionaries* in 1977, but even the aggressively fetishistic clothes of 1976 weren't worn in a sexy way: the point was confrontation and commitment – both highly masculine virtues.

Punk's furious belligerence was a direct response to the cultural and social vacuum it appeared to face under the Callaghan government. Now we have the strong, rigidly masculine government of Thatcher, it's perhaps not surprising to find feminine virtues being reasserted in pop music, not least the quasi-feminine image of the love object, whatever the gender. Glam is back: much of it is, like a decade ago, the surface style for a new age of frivolity – making videos while Great Britain polarizes – and musical soma: what better to accompany the synthetic allure of fizzy pop with surrealist, dream-like images of weird sex out of the ad factory.

This is Blitz culture gone admass: the Blitz obsession with the 'feminine', a fast-moving, sharply defined exotic surface image with a total *lack* of any commitment has defined the terms under which most modern pop groups operate – whether Duran Duran swapping countries with each video or Boy George wearing a dress and *not* being thought ludicrous. Blitz took up the gauntlet thrown down by the punks and turned it into an all-pervasive camp – whether as a way of looking at things or a way of wearing your clothes. The result has not been without humour, and complete confusion.

If the pretty-boy stereotype abounds – with the feminine regard of the love object, if not actually displaying feminine characteristics – then the correspondence has been the more assertive femininity of such as Grace Jones, Annie Lennox and Joan Armatrading. Yet such assertiveness is not without its own paradoxes: just as Marc Almond takes his 'feminine' images from the Factory milieu and such as Edith Piaf, then Annie Lennox is reminiscent of Marlene Dietrich, and Grace Jones a mere twinkle in the eye of a man, Jean Paul Goude. More amusing has been the tendency of obviously pallid white soul groups – like Heaven 17 and ABC – to beef up their sound with earthy, assertive black session singers – who usually steal the show.

The same device has been used by both Soft Cell and Culture Club – with Cindy Ecstasy and Helen Terry respectively – but the two transcend the limitations of fizzy pop in an interesting way – if, not the least, because Marc Almond and Boy George have the highest androgynous profile.

Even so, they still play the traditional pop game – and it's hard not to: their refusal to commit themselves on matters of specific sexual activity is partly born out of self-protection (as homosexuality is still the kiss of death) but is also down to an understanding of the way pop works – not by specifics or slogans, but by hints and inferences loose enough for the imagination to leap in and resonate. And it's interesting to note their current status: George is massively (and deservedly) successful with a mix of good songs and a carefully constructed image of pre-pubescent innocence – like the early Beatles, he could be a doll (and wait for the merchandising, Action Boys).

Soft Cell, on the other hand, have paid the price of flaunting their divergence: 'Numbers', an explicit song about promiscuous sex – its title taken from a novel by gay author John Rechy – died a death, and there's no doubt that many people find Marc Almond massively irritating, even if his camping relies, in true high-camp fashion, on existing models. The audience's tolerance of divergence only goes so far, and usually evaporates when things get a bit real: and fizzy pop is not, ultimately, about reality but cloaking it.

If pop's attitude to all kinds of sexuality has been confused and contradictory then its expression of sexual divergence has been, in part, the history of the androgyne principle – the breaking down of society's codes of what is 'masculine' and 'feminine' in favour of a less rigid, forced sexuality – making itself heard in one of the only places it can: exactly where it is thought not to matter, because, it's only pop.

In its current form, this expression may appear timid and slight, or plain silly, but the fact that it is occurring in such odd ways across the board hints at an attitude to human relationships that is in people's minds, however obscurely, and makes a mockery of our proposed return to Victorian values – a proposal, backed up by a moral backlash, which is anachronistic, dangerous and ultimately unwanted.

> Sexual alienation, one of capitalism's foundations, implies that the social body is polarized in masculinity whereas the feminine body is transformed into an object of lust, a piece of merchandise …
>
> Felix Guattari, 'Polysexuality', *Semiotexte* 10

Thanks to the following for their help in compiling this article: Christina Rose, Simon Frith, Kris Kirk and Marek Kohn.

The Face, June 1983

1983: Pop in a Plural Phase

In August of this year, a well-known jazz musician, Herbie Hancock, released a single which became his first hit for some years. Called 'Rockit', it was a jazz-funk instrumental which made liberal use of the 'scratching' technique, whereby tapes and discs are wound and rewound at high speed to produce a noise like a skipping or a scratched record. It's a technique that was pioneered by DJs like Grandmaster Flash in south Bronx street parties about five years ago, but has only surfaced in the pop mainstream during the last couple of years. But Herbie Hancock didn't hear scratching at its source: he got the idea for 'Rockit' after hearing a record made by a white Londoner, Malcolm McLaren, whose 'Buffalo Gals' was the first big hit (December 1982) to use the technique. In one completion of the usual process – white steals from black – black now steals from white, who steals from black.

This is one indication of a new twist in pop music's convoluted history. The old arguments about 'authenticity' and direct stylistic exploitation don't quite apply. Like dress codes – which sometimes appear (correctly) to trample all over previously held ideas about function and status – musical styles are now being cut-up and reassembled like a Burroughs pamphlet, a Dada montage, or like the paradigmatic Grandmaster Flash 12-inch, 'Wheels of Steel', which mixed together five existing records into a new whole. The effect is like that of sitting at home flicking the remote-control switch on your television; individual items can still hold your attention and, somehow, a pattern emerges.

Pop music is, at present, about plurality. There is no one dominant trend, no one dominant fashion, no one dominant source of energy. This works both for and against the music industry, which, at present, is firmly back in the driver's seat after punk's wresting away of the steering wheel. While it gives record companies more leeway to fabricate stars out of nowhere, it also makes it very difficult for the larger companies to plan, in the way that large companies will. No one can predict what the market will be like six months ahead, let alone plan artists' development.

EMI's Kajagoogoo are a perfect example: created at the end of 1982 as a group for the target 10–16 audience, they had three hits of decreasing chart importance, and then split apart with much acrimony six months later. The noise they made was axiomatic: a carefully produced, smoothly textured form of white funk – black music of five years ago made palatable for a white audience that has only just caught up. So rapid is the turnover that within that short period of time they seemed boring; lacking resources, they were carried away on the tide of novelty. Amid the chaos, the music business is playing as safe as it may: certainly the controlling companies demand a quick return on their investment and content themselves as much with a succession of novelties as with their established breadwinners. At the heart of every pop star, the insecurities are ever greater, and this gives the companies the upper hand.

Pop ideologues moan, quite correctly, that there is no single energizing force like punk rock abroad, and that this creates an atmosphere of cultural impotence, parallel (it is argued) to the labour movement's loss of confidence. This is only partly true. The simple fact is that pop music now sounds much better than it did six years ago; the problem is, it doesn't appear to mean as much. But in many ways, punk was a false sunset on the 1960s pop dream; although couched in the most nihilist imagery possible, its sophisticated critique (not of the world outside, as it would have been in the 1960s, but of its own world) was couched in often over-simple music, which had the opposite effect to what was intended. Instead of rendering the whole cycle of pop music consumption meaningless, it gave it new meaning, as people all over the country thought that they could pick up a guitar or a synth and say something worthwhile, or that by buying a record, they could be part of something greater. Now that this meaning has evaporated by the passage of time and the activities of the music industry, pop appears to some as bereft of any meaning beyond that of soap powder. It's just about marketing. It's all style and no content. It can't say anything now and, anyway, it never really could.

I sympathize, but don't ultimately agree. The problem for anybody trying to make sense out of the situation is that the market for pop records has diversified to the nth degree. Although the market for what are called pop records has extended to the late thirties age group, as the 1960s generation who were brought up on pop grow up, the target audience for the record companies has been the lower teens and under: a fickle but highly lucrative audience. This appears to be for various reasons: market researchers decided that this group had more money to spare than the traditional 'teenage' target audience of Mark Abrams's important 1959 study, *The Teenage Consumer*. That 16–24 age group was now perceived as being economically less potent as the unemployment figures came out. This concentration has had an effect upon pop music as a whole: the music industry is now geared to producing music that will satisfy that market, like others, in the way it feels happiest – angling for the lowest common denominator. National radio then follows its master with a wagging tail.

But that wasn't simple: part of the industry's problems is that the young pop generation is media-literate and surprisingly choosy: the house magazine, *Smash Hits*, which sells between 430,000 and 500,000 copies every two weeks, provides a variant on the usual fan fodder with terse text, excellent colour photo reproduction, and an irreverent approach. This sophistication is paralleled by a change in attitude in those 16–24-year-olds who stayed with pop music. Many disillusioned punks had turned to politics, or other less commercially pressured art forms, but those who remained began to nurture a culture high on self-consciousness and on outward style. Packaging and nostalgic references became all-important, and part of the consumption process. Magazines like *The Face* and, more democratically, *i-D* illustrated this perfectly, while the knowing synthesis of groups like ABC provided the soundtrack.

The unspoken assumption behind all this is that pop music has become culture, and that the audience has time to pursue these purely cultural (as opposed to overtly political) meanings. As is the habit of any dedicated opera buff, the minutiae of content and performance are exhaustively discussed in weekly papers like *New Musical Express*, while the outward appurtenances of style are paraded in *The Face*, whose layout and way of working recalls a more on-the-ball version of *Ritz* or *Andy Warhol's Interview*. And that importance of papers throughout the market shows just how far pop music has come from its tawdry beginnings at the fringes of showbiz: it now consumes acres of newsprint and commands a massive service industry.

The most successful attitude to this overload of words, images and sound has been an instinctive synthesis. Apart from the 'ethnic' markets (which themselves are more sophisticated than they are always given credit for), there is now nothing approaching the 'natural primitive' ideal which fuelled the beginnings of rock 'n' roll. In the current climate, the most successful artists are those who have incorporated plurality into the way they hear things, and have orchestrated, however unwillingly, its celebration.

The dominant pop sound of the moment is best exemplified by Culture Club, whose 'Karma Chameleon' is the best-selling single so far this year. It's the radio sound to which countless other groups aspire: so much so that the producer who could get this sound, whether Trevor Horn (Yes, ABC, Dollar, Malcolm McLaren), or Steve Levine (Culture Club) or Arthur Baker (New Order, Freeze), has become just as important as the artist. Above all, it's smooth, and not too disturbing; its model is, as with Kajagoogoo, a flattened version of black styles from the last five years with the wildness and harshness taken out. David Bowie had a stab at it with the *Let's Dance* album, using as collaborator Nile Rodgers of the successful disco group Chic, and countless idiots with floppy fringes and long manifestos followed in his wake. None, however, could touch the real thing: records like Shalamar's 'Disappearing Act' or C-Bank's 'Get Wet'.

Culture Club takes elements from rap, funk and ethnic music and mix them together into a melting pot, seasoned with Boy George's extraordinary image and knack of writing excellent songs that are able to affect his audience. Thus they satisfy both a current consumer demand for danceable, catchy songs and the ideological demand, much in favour at present, for 'soul' and 'passion' – catchwords to describe a romantic white view of the way that blacks go about making records.

This sound has had many imitators, and aspirants. But the other successful styles have been an integration of, on the one hand, ethnic influences – particularly African – and of the synthetic 'crunch' beat of New York 'scratch' and electronic 'hip-hop' music. The decreasing popularity of reggae music among both the mass audience and the avant-garde was matched by the increasing influence of African music. This summer, the Hammersmith Palais was packed out by a predominantly white audience for appearances by the leading 'high-life' group, King Sunny Ade and his African Beats. But the shifting rhythms of the music took time to percolate through to the mainstream. The ever-prescient Malcolm McLaren had tried with the extraordinary single 'Soweto', but it took a young, mostly white group, Jimmy the Hoover, and the Tamla Motown pro Lionel Richie to have hits, with 'Tantalize' and 'All Night Long' respectively. The video for the latter reiterated its broad market, showing punks, black bodypoppers, policemen, disco-goers out of 'Staying Alive', and African dancers all weaving in and out of the dance line: here was the synthesis and Richie made it work.

The history of current New York styles illustrates further the complex forces at work. 'Scratch' was a pure 'primitive', street style, but one which used the materials to hand in yet another kind of synthesis. You didn't have to make music, you just brought along your favourite records, scratched the best bits so that they repeat endlessly, and motor-mouthed over the top. This fragmented approach was then harnessed by Afrika Bambaataa's 'Planet Rock' in 1982, to the melodies and sound of a kind of music that you might have thought light years away from the experience of New York blacks, the robotic music of the German synthesizer group Kraftwerk. The resulting chemical reaction has been fertile, to say the least.

In 1982, Grandmaster Flash and the Furious Five had a top 10 hit with the pop protest song 'The Message': six minutes of pure, hypnotic claustrophobia with an extraordinarily impassioned vocal, which has now been turned into a road safety advert, such was its impact. The producer Arthur Baker, apart from Club successes with Afrika Bambaataa and others, has had top 10 hits with Freeze ('IOU' and 'Pop Goes My Love') and with New Order, a lugubrious Manchester group ('Confusion' and 'Blue Monday'). And in October, the perfect New York novelty record became a top 10 hit: 'Hey You' by the Rock Steady Crew. Over the opening computed electronic maelstrom, a young Puerto Rican voice declaims, with exemplary seriousness, the word 'Digital'.

The limits, however, of elevating basic dance-hall functions to a way of running your life are illustrated by the video for the Rock Steady Crew. After dance and singing sequences that capture the record's breathless one-off enthusiasm brilliantly, the camera pulls out to find the group flashing their limbs and teeth among several rather expensive missiles and warplanes belonging to Uncle Sam which just happen to be lying around. New noise; old attitude. To me, it emphasized the drive to conformity behind every 'soul boy' – which in English terms means groups like Spandau Ballet and Wham!, even Duran Duran. In the long run this conformity may prove their commercial downfall. The ideological backlash has already begun: groups like Wham!, the darlings of the press a year ago for their unforced 'soul', are now cast aside as panderers, while the market leaders, ABC, have opted for a tougher, rougher approach.

But then, Wham! had proved too one-dimensional, captured in too high a definition on all those colour posters. Artistic and/or commercial survival in pop now demands synthesis and adaptability, and the best records of the last 12 months have shown either a consummate awareness of this demand, like the records mentioned above, or a dogged determination to adapt this demand to a more serious or personal content. Many of the best white records of the year were those which refused the advertising gloss of many pop lyrics (like, say, Wham!'s 'Club Tropicana') and gave some hint of trying to get to grips with the realities of the world outside. Records like Elvis Costello's 'Pills and Soap' and Robert Wyatt's 'Shipbuilding' tackled 'political' subjects with a fair degree of commercial success, while the Specials' folk/dance tune 'Bright Lights' tore away some of the more obvious tinsel from today's pop dreams. And the Smiths 'Hand in Glove' and Soft Cell's LP *The Art of Falling Apart* indicated a rare vulnerability. In the context of the dominant radio texture, all sounded, well, rough.

But these, and others, were mavericks, perhaps harbingers of the future; there is a strong undertow of rougher, more personal styles. But this year, the dominant pop sound and style have been about a plurality that may stray towards the trivial but which at its best has provided a relief from mundanity in the same way as a peacock punk: an attitude that trusts its own senses and aims to create its own order, grandeur even, out of chaos.

New Society, 22–29 December 1983

The Smiths: Deliberately

These are exciting times for the Smiths. A top 20 record with only their second single, 'This Charming Man'; a non-stop stream of interviews that have confirmed their status as music press darlings of the year; frequent television appearances, including *Top of the Pops* and an *Old Grey Whistle Test* Special; and now a chance even to produce a personal icon, Sandie Shaw, singing one of their songs 'I Don't Owe You Anything'. All this from a group that live in Manchester, don't pretend to be soul boys, record for an independent label, Rough Trade, use conventional instrumentation and have been together for less than a year. Next week, they release their third single, 'What Difference Does It Make?', with an album on the way. What is their particular magic ingredient?

'Writing, as far as I'm concerned, is a human necessity – like having to brush your teeth. It's really that simple.'

Singer Morrissey and guitarist Johnny Marr sit quietly in a corner of the studio and talk with a firm confidence which might seem overbearing if it wasn't tempered by bitter experience and a hint of self-mockery.

'The long period of isolation I had meant a very long period of self-development. If you're going to produce something of value, you have to think about what you're doing – you can't dive into it – and I gained a lot from being isolated. If I'd had the usual uncomplicated adolescence I wouldn't be here now, writing. It's odd how terrible things can be part of a learning process.'

The Smiths can be seen as a supreme effort of will: emerging out of the grey, deadening anonymity of Manchester with a passionate vision of the way things should be. The immediate sign of this lies in their image: in contrast to the norm, they present themselves neither as androgynous soul kittens nor as carefully constructed teen idols, but in a typical move will regularly bombard their audience with gladioli or other exotic flowers. Further, their current single sleeve reproduces the famous still from Cocteau's *Orphée*, with Jean Marais gazing into his own reflection. It could all be very precious, but Morrissey insists that this iconography is carefully worked out.

'They're two very nice symbols to be attached to. We're interested in beauty because it can be very positive. It's good to be associated with non-violent things; we'd rather it was Cocteau instead of broken bottles, or any other kind of garishness.'

Marr: 'We're humane people. But if you are humane, people tend to think you're weak. We're aggressive about things we feel strongly about, but the reason we're friends is that we are humane. That sleeve and the flowers explain how we are as people: it's depressing when people call us hippies, because we're not. We're just trying to come up with a fresh approach.'

Morrissey: 'Iron Maiden or Saxon are never asked why they're here. If you show any signs of intellect, you're asked loads of questions, why this is like that and so on, and you're made to sound insane.'

The Smiths' approach, and the central thrust of many of their songs, have provided worse misunderstandings than simply being labelled hippies. In the summer, the B side of their first single, 'Handsome Devil', caught the tail end of a moral panic about paedophilia and was 'exposed' in the tabloids. This was absurd, but the Smiths' lyrics, written by Morrissey, express an ambiguous and complex attitude towards sex, admitting a vulnerability rare in these days of simple, idealized love codes. But while it may have got them into trouble, in the long run it is the backbone on which their success is based.

Morrissey explains: 'Many people who go into this business think that they have to have a very aggressive machismo or a very aggressive stage image. It's time for a different version: not everybody is like Ozzy Osbourne. The normal rock 'n' roll terminology sounds like a chant of agony.'

Marr: 'The lyrics about sex – "Handsome Devil", "Pretty Girls Make Graves" and "These Things Take Time" – deal with sex in three completely different ways. One might say it's great in an ideal situation, but another might be saying, Oh no! I can't handle this! Morrissey does have sexual politics and they inform what we do.'

And Morrissey sighs: 'I think we need more brains in popular music.'

Sunday Times, 8 January 1984

Frankie Says

The statistics are impressive: 1½ million copies sold of 'Relax' – now the best-selling single of the 1980s, and over a million copies sold of 'Two Tribes'. These, the first two singles by Frankie Goes to Hollywood, have both gone to number one – where 'Two Tribes' has rested for the past eight weeks. 'Relax', meanwhile, has been in the charts for over six months, travelling up and down the numbers along the way to its current top five position. These two records have earned Frankie Goes to Hollywood, and their producer, Trevor Horn, a place in history and a place, at present, as the current music industry sensation.

What interests me – apart from the excellence of the records – is how little all the superlatives actually mean. Despite their radical packaging (and, to a lesser extent, radical content), Frankie Goes to Hollywood epitomize the vacuum that lies at the heart of much of current British pop and, by extension, the music industry. Despite all their attempts to become engaged, Frankie remain curiously anonymous behind the screens of expensive studio production, slick video-making and instant ideology: all the blanking devices necessary to the current market-place.

Frankie's success is, in fact, a commercial imprimatur for these still fresh industry channels. They and the team at their record company, ZTT, have used these devices brilliantly and with a rare wit: the videos for 'Relax' and 'Two Tribes' have both been 'controversial' enough to warrant selective banning, while, the ban imposed by BBC Radio 1 on 'Relax' earned Frankie front pages and set the seal on their commercial success. 'Relax' works as both a classic smut song – for teens to snicker over and shout at authority figures – and as a populist invocation of the commercial gay subculture. 'Two Tribes' broadened Frankie's commentary to the wider arena of nuclear politics as the 'relevant' video made clear, with its bloody cockfight between a lookalike Reagan and a lookalike Chernenko.

If this was not enough, Frankie have secured – along with Katherine Hamnett – the year's fashion coup. If you're in any big city, you will have noticed people wearing outsize T-shirts with outsize lettering on them: mostly, they say 'Relax', echoing Frankie's advice about how to ejaculate. The style was first designed last year by Katherine Hamnett, to promote a variety of humanist slogans like CHOOSE LIFE and WORLD NUCLEAR BAN NOW. This was copied by Frankie for a limited-edition T-shirt to promote 'Relax' and was quickly recopied by the East End rag trade. The streets are now full of this, and other Frankie slogans – FRANKIE SAY RELAX DON'T DO IT!, FRANKIE SAY BOMB IS A FOUR LETTER WORD. Three-quarters of Frankie-wear is now pirated. There are even 'answer' T-shirts: WHO GIVES A . . . WHAT FRANKIE SAYS is the latest permutation.

Yet Frankie remain curiously characterless, in the style of the disco culture that underpins their market. Rarely do you meet committed Frankie *fans*. People just like the records, or think they're a laugh. Part of the problem is that the group have hardly played live. There is little sense of the projection or involvement with an audience that, say, the Sex Pistols, or even Duran Duran, had after only a small number of live performances. Frankie are a product of the recording studio or video sound stage: good for media manipulation but not for identity.

These days, if something moves, it gets wallpaper publicity. The effect is to force the development of young groups and yet to arrest them, like flies in amber, at a primitive stage of development. There is no sense of a gradual career development, as in the 1950s or even the 1960s: more a frantic desire to milk the golden calf dry as soon as it produces the goods. Frankie's blanket press coverage, the endless remixes, their success itself, has turned the group to cardboard cut-outs – high-lit shadows which, in passing, leave little of themselves.

The way pop is currently consumed has robbed Frankie's provocative media play of any potential power. When their FRANKIE SAY ARM THE UNEM-PLOYED T-shirt makes the concourse of Bond Street tube station, this is not an index of consumer radicalism. It only indicates how quickly even the most extreme ideas are made harmless unless they are rooted in reality. In the end, Frankie may be too clever for their own good.

New Society, 9 August 1984

Is There Life After *Smash Hits*?

Spring is here and a young person's fancy turns towards ... playing media musical chairs. The music press is currently going through one of its periodic clear-outs, with editors departing from three major titles – Mark Ellen from *Smash Hits*, Neil Spencer from the *New Musical Express* and Eric Fuller from *Sounds* – with a question mark hanging over Phil McNeill at *No. 1*.

As befits the corporate structure of these magazines, Ellen and Fuller are moving within publishing groups EMAP and Spotlight respectively, while Spencer is leaving IPC for pastures unknown, in all probability not too far away from the Labour Party.

The scurrying about in their wake has invested this tight world with an uncertainty that goes beyond a seasonal friskiness. People say: Is it the end of an era? (*Note*: In the music press, an era is a year.) Is the music press in crisis?

From the circulation figures, no. The Audit Bureau of Circulation (ABC) statistics for 1984 suggest that the dominant trend of the past few years – the drift towards glossy mid-teen chart-based publications like *Smash Hits* and *No. 1* – is peaking, leaving *Smash Hits* with half a million (its circulation having doubled over three years) and *No. 1* at just under a quarter million.

Conversely, the inexorable slide of the three late-teen tabloid weeklies – *Melody Maker*, the *NME* and *Sounds* – is slowing down, although all three are well down on their peaks within the last 10 years.

These figures must be a relief to editors and publishers alike, but they conceal a deeper unease: after all the excitements and struggles of the last decade – the current music press being almost entirely a post-punk phenomenon – what place is there now for the weeklies?

The main challenge to the established music press comes not from comparative newcomers like *Smash Hits* or *No. 1* – already well assimilated into the marketplace – but from the explosion or pop coverage that has occurred throughout the mass media during the last couple of years: from pop groups on prime-time light entertainment shows like *Wogan* through an endless sequence of Channel 4 music programmes, to Fleet Street's well-documented current obsession with pop gossip.

As the media expands into one of the UK's new industrial bases – just walk through Soho and figure out where the money is coming from – there is more space to fill without an endless supply of new material to fill it. (Of course, in the real world there is, but the increased demands of circulation and/or ratings in this competitive market would seem to preclude any such experimentation.) More media seems to mean worse media, as the same rations of jam are spread over an ever-increasing mountain of processed white bread.

Pop has proved perfect fodder for this process. Turned into video, it's a handy tool for TV producers to slot into sagging formats and attract the elusive 'youth market'. Turned into gossip, it provides the perfect spectacle for Fleet Street, which, with true suburban envy, consigns the former habitués of the gossip columns – the plutocracy and aristocracy – to the bargain basement of tacky heroin sagas.

Hand in glove with this, the industrial base of pop music has been changing, as record companies – dragged screaming into the video age – have been forced to encounter, and work with, the film and television industries. Pop music, and the music industry, are now no longer a separate entity, but a part of a huge industrial conglomeration which spans the whole media.

The age of pop being an unimportant sideline – where the children can play in their own dirt – is over. Executives, and by implication musicians, have to answer to the demands of an increasingly sophisticated and powerful corporate structure. With all this, it's not surprising that pop – or at least the music that is generally being made available – has changed considerably.

A decade ago – when *Top of the Pops* was the only network music programme – tabloid weeklies were a vital source of pop information; that function has been largely superseded.

They also had another function: the interpretation of a type of music, e.g. punk rock, whose grainy texture seemed to fit their black and white newsprint. Today's pop is heavy on style, thin on content; there simply isn't as much to write about, yet the weeklies largely persist in the formats that served them so well in the past.

Yet the weeklies do have another function which is rarely covered elsewhere in the media: criticism. Even though they're frequently undercut by their service relationship to the music industry, the weeklies still provide a place for young writers to surface and for a kind of writing that deals with pop more thoroughly than Mike Smith or John Blake.

Thus when Henry Porter complains in the *Sunday Times* that the only thing pop hacks can find to say about Eurythmics is that Annie Lennox is getting divorced, that needn't be the case; it's just the low quality of most pop media that makes it seem so.

In this climate the weeklies seem to have two broad choices: either to chase the pop end of the market or, alternatively, to broaden out as much as they dare from pure pop coverage and to develop a coherent critical position and an identity that can shield them from the worst demands of the weekly deadline.

I know which one I'd choose, but then, I don't have to answer to IPC or Morgan Grampian about my circulation figures.

The Face, June 1985

Humpty Dumpty
and the New Authenticity

True pop aestheticians will have noted, not without some alarm, a disturbing 'new' trend creeping in from the United States to find a welcome on these shores. It's been best summed up (by Simon Frith in a recent *Sunday Times* article) as 'The New Authenticity', and its harbingers are as follows:

- ❐ an influx since the new year of American rock (REO Speedwagon, Bryan Adams, etc.) in the British charts …
- ❐ the increased prestige granted to the latest season of the *Whistle Test* …
- ❐ the wish of previously pervy synth pop duos like Eurythmics and Tears for Fears to appear like 'real rock groups' in their videos …
- ❐ the increased popularity of venues likes the Mean Fiddler in Harlesden …
- ❐ a whole troop of American country bands who come over here to advertise their 'roots' …
- ❐ and finally, of course, the current mass media salivation over the 'last great rock 'n' roller' … Bruce Springsteen.

Guess what? It's the Pub Rock Revival.

Musicians like Los Lobos, and their UK equivalents, are always playing in bars, so why all the fuss now? Partly this is due to an Americanization of public taste caused by events in the media.

Despite the romantic pirate image, both Laser and Caroline have been highly influential in beaming to millions of listeners in southern England a mix of 'classic rock' taste and up-to-date American MTV-ization of the world is reflected in the growing number of pop groups who tailor not only to their videos but also their music to the MTV median – a 16–17-year-old American Midwesterner.

As the music industry becomes further and further integrated with the new media economy, it is not surprising that America, the place where this economy is centred, should start cracking the whip: the exportation of US pop parallels an industrial colonization.

It is this sensibility that has stepped into the vacuum created by the collapse of post-punk, or, more accurately, postmodern English pop. Several factors account for this: the frantic music industry rush last December oversaturated the market and effectively killed the goose that laid the golden egg. Several careers, most notably Culture Club, were ruined through force-feeding. In this the Band Aid project, although outwardly affirming the confidence of English pop, was symptomatic: its naked transparent use of the hooks and devised through which pop worked (you can see the joins in the song but it doesn't matter because the cause it good) removed their mystery and their potency. In the real world, Band Aid had some practical effect: culturally it now appears not as a celebration but as a swansong.

Into this breach then step the 'real' people. Like adolescents constructing an Airfix kit, they painstakingly build up an 'authentic' likeness – to scale, of course – of 'real' men making 'real' music.

This return to 'authenticity' marks a twin reaction: a reassertion of 'traditional' masculine virtues as against the perceived 'femininity' of the Teenybop Gender Benders, and a return to a peculiarly 1960s way of looking at pop.

In this version, music must exist by itself as real expression, as live performance, as a relationship between audience and performer that is untainted by marketing or any other suspicious pop trickery like theory or politics. Anything that is constructed is fake and not real.

This view of 'pure' music was at its most potent in the late 1960s, when it served the idea of a 'youth community': at Woodstock they listened in their thousands to the authentic visions of Arlo Guthrie and John Sebastian. The latest interpretation of Bruce Springsteen as the 'last rock 'n' roller' places him as the only person who can rekindle pop's sense of community, the only person who can reunite a fragmented pop audience. Yet to what higher ideal is he uniting this audience? In a classic piece of reductionism, Springsteen is exalted because he made 'less compromises than anyone else'. What a slogan for the 1980s!

This view of popular music entirely omits the developments of the last 10 years, which can't be simply written off as tawdry marketing exercises. As the carefully groomed 'new men' of *Whistle Test* (can't they find *one* female in the four presenters?) drool over Bruce, they fail to mention that he is just as heavily packaged as Frankie were, only better: you can't see the joins.

It omits any analysis of the current technological and industrial situation of popular music, and, even worse, it has to be said, if reflects a chauvinistic Americana which – post little events like Reagan/Cruise/Star Wars/Nicaragua/et al. – can no longer be isolated and uncritically lapped up.

The dominant American view of pop is of something separate from the rest of society. This complements perfectly the increasing economic and social power, not of Teens, but of the 35-year-olds who remember rock like it used to be and can't face up to present-day reality. You can't put Humpty Dumpty back together again.

The Face, July 1985

Culture as Commodity

Video art is so great. Music video is so perfect.

> … just got this new tape of all the latest videos and it makes you want
> to go WOW! There's the new Bowie – bet they spent a *lot* of money on
> that set. And there's '19' – don't like the tune much, but it's really
> effective when the rockets go whoooosh! in time with the music.
> There's a couple of boring black ones – well all they do is stand and
> sing, don't they? It's got Tears for Fears in a really cool old car, Power
> Station with a really gorgeous chick and Go West – they've got some
> great effects, like when this creepy hand comes down over the
> picture, and the group flies through the air. The best, though, is this
> one set in an old theatre, but instead of an audience there are
> hundreds of TVs all in the seats, and the camera goes whoooosh!
> right over them. And then it turns round and it's Bryan Adams! He's
> really cool on video: did you ever seen 'Run to You'?

Music video has, in fact, become the great unmentionable. Everyone wants
to make money out of it (fair enough: Old Compton Street would grind to a
halt otherwise) but nobody wants to face up to what it's actually *for*, or what
it's actually doing.

Just consider Bowie's latest extravaganza for 'Loving the Alien' – a peculiarly hollow romp through some stale Dr Chirico sets and some video surrealism – and think: was 'Ashes to Ashes' only five years ago? What has music
video come to?

As you goggle at the video jukebox, what soon becomes clear is quite how
standardized and patrolled music videos have become. The real importance of
music video is not creative but industrial. By accident rather than design, and
through a slow process of interlocking factors, the music video now stands at
the epicentre of a new inter-media industrial agglomerate (of which
Band/Live/Kool Aid is another signifier): this brings together the music, television and film industries into a situation which is still being defined, and
whose effects are still unclear.

This is the situation presaged so accurately by Debord in *Society of the
Spectacle*: 'Culture turned completely into commodity must also turn into the
star commodity of the spectacular society … In the second half of this century
culture will hold the key role in the development of the economy, a role
played by the automobile in the first half, and by railroads in the second half
of the previous century.'

This new media agglomerate clearly favours – and indeed, is patrolled by – existing multinational, multimedia cartels. As Andy Lipman pointed out in a recent *City Limits* article: 'Four huge multinationals dominate the world music markets – WEA and CBS in the United States, Polygram and Thorn-EMI in Europe. Each is yet again part of larger conglomerates with wider industrial, military and media interests. WEA is part of Warner Communications, with film and publishing interests, and with American Express makes up WASEC – owners of MTV, the American 24-hour pop music cable network.'

You could add, by way of further example, Thorn-EMI's interest in records, music video commissioning and video-cassettes (through Picture Music International), and the major European pop music cable channel, Music Box, among others.

It is, therefore, hardly surprising that the influence of music video on the music industry has been baleful. Control over production, packaging and outlets has been vested in the major multinationals as never before and their policy, quite naturally, is to consolidate, bet on certainties, and use their muscle to dictate the market. Sharp-eyed watchers of the pop press will have noticed a recent spate of 'Royal Rock' headlines just like in 1975 – 'Hell's Royals: Viscount Linley and Girlfriend Salute the Boss', or, 'Di Joins Live Aid' – but this time there will be no Sex Pistols. The music business has mutated from a tawdry, unrespectable industry on society's fringes into an integral, shiny part of the new media industrial complex – one that fits well the government's plans for a new, gleaming, privatized leisure economy. Hence Live Aid's international importance as the music industry's new face: fun, caring but above all adult and responsible – the Death of Pop.

Another result of music video's spreading influence has been further American control. Whatever anyone says, the eventual target for most music videos is MTV, who currently claim between $24\frac{1}{2}$ and 25 million viewers.

In comparison, Britain's only cable channel, Music Box, reaches 50,000 people (5 million throughout Europe). The MTV principle dominates music video making: the lack of any equivalent UK outlet on broadcast TV (added to an ever-decreasing profit margin on domestic videos) has resulted in many UK production companies making music videos for the American market and, by extension, UK record companies grooming groups like Go West for American success. America is investing in UK talent. While on the one hand this is welcome, on the other hand there are danger signals.

What sticks in the craw about Springsteen – as the ultimate exemplar of Americana – is the yawning, terrifying gap between his own intentions and his cooptation by the industry (and, in a recent speech, by Ronald Reagan). The music industry has dovetailed so strongly with power politics that what is attempted as a critique, or at least an affirmation of shared experience, comes across in the wastes of Wembley as an uncritical celebration. The most successful man in pop is simply not in control of his destiny; nor are we, in any way, in control of pop.

If pop is dying, it is clear that the eventual positive effect of music video will occur in the film and television industries. In its current context, music video has failed to develop – indeed, the general standard today is worse than it was in 1983, the high point – due to its very industrial importance, and due to the music industry's short-sightedness in failing to see it as anything more than a quick answer to an immediate problem: how to promote the latest single.

Yet there are encouraging signs that the form may develop into the – euphemistically titled – 'long form' video, a catch-all phrase for anything that is not a promo, and, more importantly, that the cross-fertilization between the three industries that has already occurred will bear strange and ripe fruit.

Music video's impact on television, for instance, is only just beginning. The slow crack-up of the existing BBC/ITV duopoly has left TV wide open both to music industry investment and to new forms of programming. The only channel so far to be sensitive to this is Channel 4, with the Phonogram-financed *Mirror Image* and Chrysalis's *Max Headroom*. These two shows illustrate neatly the downside and upside of music industry involvement in TV. Within this interface operate the new media brats.

Consider the Duvet Brothers' 'Blue Monday' or Gorilla Tapes/Luton 33's 'Secret Love', which scratch New Order, Malcolm McLaren, the Shangri-Las, Reagan and Thatcher into devastating curses on the power politicians' domination of the media. 'The individual talks back to television' is the Gorilla Tapes definition of scratch video.

Or consider Birmingham Film and Video Workshop's 'Giro', which demolishes both the Tory and the Labour lines on unemployment with style, tenacity and wit: 'The Right to Do Something We Enjoy and Get Paid for It' runs their slogan. The community talks back to television: you *can* take control is the message.

The Face, August 1985

The Sour Smell of Celebrity

Perhaps he should beware of celebrity, a word rather difficult for which to locate derivation. The prefix *cele* (or, more accurately, *coele*) means cavity. The suffix *brit-y* probably comes from *brat*.

Thomas Thompson, *Celebrity*

Whatever it may mean, celebrity is moving to centre-stage in our culture. No longer a means to an end, or even an end in itself, celebrity becomes an abstract principle which underpins opinion and aspiration. The mainstay of the new age of mass media production, celebrity winks at us from may quarters. Fleet Street proprietors issue calls for royalty and pop on the front pages, while chat shows (together with soaps) cheaply fit the requirements of a new, privatized era of broadcasting.

Success used to be what you aspired to. These days – like the terminology (from star to superstar to megastar to black hole?) – this has been superheated to become celebrity. The two states are quite different.

> Forty years ago, when I was growing up, the word 'celebrity' was almost never used in print, in conversation, in any sort of discourse, civilized or casual. Most of the people one reads about in the papers, or hears about on the air, or who were the subject of magazine profiles, were 'successful' or 'famous'. Sometimes they were both, but not necessarily. And if they were both, it was understood that there was a logical progression to their achievements.
>
> Richard Schickel, *Common Fame*

In contrast, celebrity is now seen as a thing in itself, rather than a process; it is the result not of interaction and achievement, but an entirely privatized ambition.

'*I'm gonna live for ever!*'

This incantation from *Fame* captures well the streamlining of the Me Decade that the spread of celebrity culture represents. Success now has to be highly visible so as to negate unemployment's anonymity – and thus it turns into celebrity. The lurking threat of unemployment is used as justification for grabbing at the ring of celebrity in the most competitive area of all: the music/fashion/media industries. With such a quick turnover of ideas, of people, and with the pack right behind, how could you look after anyone else but number one? And so – as in power politics – the preferred solution to our profound social and spiritual problems is not solidarity (or community in its widest sense) but selfhood, and a particular kind of selfhood at that: a take-away immortality.

The state of celebrity is the ultimate current expression of the self-styling at the heart of today's pop culture. Once a celebrity, you too become an abstract, existing principally in the battlefield of the electronic media, prone to psychic disorder and takeover.

> When you look in the mirror
> Do you smash it quick?
> Do you take the glass
> And slash your wrists?
> Did you do it for fame?
> Did you do it in a fit?
> Did you do it before
> You read about it?

Poly Stryene's chilling 'Identity' catches one ultimate downside of celebrity perfectly, that one is rarely mentioned in its currently exalted state. And if you think of Clint Eastwood in *Play Misty for Me*, you'll get a clearer picture of the psychic waste.

Celebrity's hyper-capitalistic selection of the (number) one out of many creates, in the main, a double loss of control, a double powerlessness. The star is trapped by the expectations and demands of his or her audience; the fans are caught, in a vicarious, physically distant yet enormously intense emotional relationship with the icon of their choice – as the spells and curses collected from fans in Fred and Judy Vermorel's *Starlust* make only too clear.

That is not to say that the relationship is always devoid of pleasure or power: many celebrities manage equanimity, while the use by fans of the spaces that celebrity opens up is well described by Dave Rimmer in *Like Punk Never Happened*:

> Gangs of girls were organizing themselves into formations to beat the bouncers. Long before Duran Duran appeared, they were standing in groups, chanting and waving and taunting the security guards and having a whale of a time just being there, together, and feeling the power of their collectivity.

Yet despite such phenomena, the keynote of celebrity is a loss of power and frustration which result in the occluded violence of a 'change in image' or a 'comeback'; or, on the other hand, the desertion of a star by 'fickle fans'.

In his novel, Thomas Thompson uses the three principal characters to posit three different types of celebrity. One is a closeted film star, another a highly successful reporter/novelist – traditional enough types. But the king-pin, and, to be sure, the character *maudit*, is a murderer, hustler and failed pimp who is transformed into a born again preacher so charismatic that he and his army of followers become figureheads of the rabid New Right. Such a fiction neatly ties up the current political and religious implications of celebrity culture – as well as being disturbingly close to real life.

> Everybody's looking 4 the ladder
> Everybody's wants salvation of the soul
> > Prince, 'the Ladder'

> We're more popular than Jesus now
> > John Lennon, 1966

It's not hard to see pop (and the entertainment industry) as an ersatz religion, its stages the secular Olympia. But those looking for the New Church hit celebrity culture's ultimate frustration: the granting of a quasi-divine status to highly fallible icons. Despite the rituals contained in modern mass production, our new Olympians are but mere mortals. If they hurt when they fall it's because they have forgotten their mortality. And for the fans, the spiritual investment they make founders only too often on the machinations of a crude industry. Sometimes when that happens their knives flash.

The current problem with celebrity lies not in its existence – which is, in part, the result of a century-long process – but in its primacy in our culture. Celebrity as aspiration, as lifestyle option, is everywhere now, yet despite the Vermorels' warnings – 'no excess, no absurdity, no subversion seems to effectively challenge it' – it must be remembered that it is not, as many would have it, a true expression of human nature.

Rather, celebrity is an ideological fool's gold, marking the successful colonization by the New Right of so many intimate areas of thought and motivations. As the parallel moral panic over AIDS attests, the New Right works through fear, envy, frustration and blind alleys of belief; many similar impulses lie behind our current celebrity culture. It is possible to recognize these feelings and to reject them; after all, not everybody wants – in Iggy Pop's immortal words – to hop like a frog.

The Face, October 1985

The Toytown Nihilists

People have been nibbling away at the subject all year, but it's time for a firm bite: the Age of Style is over. Longhand it as the Age of Style over Content, but the cultural situation under which we have laboured for the past five years is fading fast – and its products will soon be judged in a harsher light.

Consider these harbingers of style's decadence and demise:

❒ the reported death of the Yuppie in the US ...
❒ the inevitable slide of 'peoplewatching' – a key style phenomenon – through the very minutiae of snobbishness (*The Poser's Handbook, The New Georgian Handbook* and *People Spotting*) into a final nullity (*Glad to be Grey*) ...
❒ the rapid absorption of the 'style élite' by both Fleet and High Streets ...
❒ endless articles about café society and slumberland Soho ...
❒ 'style' overkill going into hyperspace with endless *Face/i-D/Smash Hits* derivatives; too many glossy magaines chasing a finite market with the same old stories in a desert of ghastly, full-colour good taste ...
❒ the fate of *Absolute Beginners*, already hoist with the petard of idiot 'Beatnik' fashion spreads six months too early ...
❒ the increasingly obvious heartlessness of elevating celebrity and success into a dominant cultural principle – Tory elevator boy Jeffrey Archer being the easiest example at hand.

And into the interregnum come a whole hole of morbid symptoms.

This was made clear to me recently at a sodden party (for one of style pop's finest hours) and another in a series of dull affairs that above all marked style's degradation. This was memorable not for the density of fashion victims, not for the cheerless brutalism of the location, not even for the amusing cocktails – but for the sight of two boys parading their cultural impotence by wearing swastika- emblazoned jackets. Sporting these loaded symbols in a worsening climate of institutionalized racism is idiotic and irresponsible enough. Underneath, however, lies an even deeper vacancy: the utter failure of ghastly bad taste – and by implication its intellectual counterpart, nihilism – as a way of looking at and dealing with today's world.

Saying 'No!' in whatever way you choose can have two functions: negation or nihilism. The difference is important. Consider Greil Marcus's definition in *Artforum*, November 1983:

Negation is not nihilism. Nihilism is the belief in nothing and the wish to become nothing. ['Have you got any drugs! I want to get annihilated,' said the Swastika youth.] Negation is the act that would make it self-evident to everyone that the world is not as it seems – but only when the act is so implicitly complete that it leaves open the possibility that the world may be nothing, that nihilism as well as creation may occupy the suddenly cleared terrain.

Nine years after the Sex Pistols's founding negation – 'You have to destroy in order to create!' shouted McLaren in a December 1976 interview – the idea of saying 'No!' has become an aesthetic and cultural end in itself.

Punk's negativity had an immediate liberating impact; it worked as negation. Much of its legacy is more malign, and works as nihilism: the circular rituals of the Goths, the tunnel vision of most post-industrial avant-garde groups and, worst of all, the outmoded idea of the swindler, the manipulator who takes industry, artist and punter for a ride. This is not cleverness but contempt.

The legacy of the style generation, whose own way of saying 'No!' in 1980 – asserting the supremacy of style – was equally liberating, is just as malign. Five years on, there now appears an emptiness at the heart of style over content, a philosophical black hole described by the symptoms above. There's too much media, and too much of its saying nothing.

Misapplied semiotics has a place here. The search for the symptom is all too easily translated into a peoplewatching consumer feature, or its language converted into the baffle-system used by the would-be *poets maudits* of ZTT. Such nihilist phenomena dovetail only too well with the voracious demands of the new media economy, and the symptom-laden saying of nothing falls in with the profound cynicism of the new media owners and the government which fosters them.

Style culture's biggest canard now appears to be the idea of unfettered mass media access, where it's enough to enter and to skate across the surface of the

Ray Lowry, NME 1986

media, without worrying about what it is you have to say and what effect it has on the people who consume it. Into this black hole rush the Toytown nihilists, whose 'novel' postures, in their more extreme forms, mask an old-fashioned and self-destructive pathology. Their ways of saying 'No!' are even worse than the saying of nothing.

'Your media is very greedy; you devour everyone and everything. It's the way you're made' – Brian Clough to Brian Moore, *Midweek Sports Special*, 23 October. One of the things emerging from the debris of style culture (viz. all this current discussion about fame and celebrity) is an analysis of media that treats this powerful area with the suspicion and selection it deserves, rather than as a mere prank or a means to self-promotion.

The real 'sickness' of our age is a disease of perception, which concentrates on endless consumption, personal vampirism and a debilitating lack of content. In a world where, as *Hollywood Wives* boasts, 'Nobody says no', negation is as important as ever – and within the media is where it should be applied.

The Face, December 1985

The Secret Public

10.45 p.m., Monday 9 September: Four young men attempt to cast spells in a cavernous Camden town venue. It's been a muggy, lowering day – the air inside is rank with wet heat, the packed crowd congealing with the smell of stale fat from the hamburger stall. The floor is wet with spilled lager and thwarted dreams; the atmosphere, in spite of the noise, is only dimly charged with flickering memories of negation and possibility. The instrumentalists on stage throw up the obligatory black batter batter as the singer throws himself foetal into the incantation 'JESUS SUCK! JESUS SUCK!' The audience gapes. Not satisfied, the would-be shamans depart abruptly after 10 minutes. The audience ponders this, then plays its part by forcibly dismantling the lighting rig and PA stacks. The static spectacle then dissolves into a surge of violence, though as if it were seen from behind a TV screen. When the police arrive on cue 20 minutes later, the 'raging' audience files out neatly, well pleased that their ritual has passed according to expectations.

9.30 p.m., the same evening: 'Looting began in the absence of any major police presence and continued in Lozells Road for two hours. Seven thousand rioters controlled the area and sealed it off with barricades of burning cars. Fifty shops in Lozells Road and surrounding streets were on fire … Electricity supplies in the riot area failed when water from firemen's boxes short-circuited cables.' (*Guardian*, 'How the Trouble Started and Flared', 11 September 1985).

Frustration is the keynote of these localized rituals. Any power they might have as cultural and social interventions is immediately deflected by press coverage that mocks with its implicit comparison to previous, larger-scale interventions that also failed. The Jesus and Mary Chain are swiftly compared to the Sex Pistols ('Sect Pistols', ran an early headline), whose power they approach … not at all. The hopelessness of their live performance is well illustrated by an 'apparently pleased' fan, who says, 'Why should I mind that they're too drunk to play? After all, I'm too drunk to listen' (*NME*, 14 September). Through their 'spokesperson', the group mutter that 'they want to be rich and anyone who doesn't want that is a lyin' fuckin' bastard'.

For Handsworth, coverage is streamlined into the picture that is emblazoned across every tabloid within 36 hours – a West Indian youth walking with a Molotov Cocktail in his hand is singled out as the hyperbolic image of the riot, a man who 'walks with a chilling swagger, a petrol bomb in his hand and hate in his heart' (*Daily Express*, 11 September). The reaction of the press centres on the pillage – a privatized anarchy – and informs us that the riots (this or indeed any other) are not the result of high unemployment, whose sufferers, in a simultaneous effusion, are termed as 'Moaning Minnies' by Margaret Thatcher, but were caused by 'major drug dealers', thus deftly matching up two moral panics and delivering them to the door of the law and order lobby. Indeed, it is their instruments – the police and not the judiciary – who are to chair the inquiry into the riots. Since one immediate repercussion is to have endangered Birmingham's status as a potential host to the 1992 Olympic Games, their position is at once made clear: 'There is absolutely no excuse for criminal damage to the region of £5 million.'

It is clear that the central modern rituals are those contained within the new condition of the mass media. If ritual is best seen as an organized way to harness power or to make things happen, performed along rigid lines but evading conventional 'rational' analysis, then its most powerful expression now rests in a place and form that is still barely recognized. Today's rituals are not carried out personally or from within a small community, but through a mass central repetition and reception. They are not performed by shamans or purifying spirits, but by venal agents – the right-wing cartel that runs the newspapers and is now setting its sights on broadcasting. The effect of these rituals is not to bring about a specific personal or group awareness, but to bring a seeping loss of power to the mass.

The modern licence claimed by novelists and short-story writers to use their imaginations as freely as they please prevents students of mythology from realizing that … storytellers did not invent their plots and characters but continually retold the same traditional tales, extemporizing only when their memory was at fault. Unless religious or social change forced a modification of the plot or a modernization of incident, the audience expected to hear the tales told in the accustomed way. Almost all were explanations of ritual or religious theory, overlaid with history …

Robert Graves, The White Goddess

The arrogance of this ritual is encouraged by the new mass media economy – the 'free press' of late twentieth-century capitalism. The hegemony of the media is the true spectacle – one might grimace at the sensational treatment afforded the Yorkshire Ripper case in particular, but not only is it portrayed in this way both on television and in the tabloid press, but it is carried through to every grisly murderer, rapist or child molester. This creates a class of 'ultra-criminal' to whose coverage future 'rippers' must aspire to. When James Huberty burst into McDonald's in San Diego in the summer of 1984, shooting indiscriminately, he was thinking as much of his headlines as the President does. Latterly, America has adopted the catchphrase 'serial killer' to describe its growth industry of motiveless murder, leaving one in no doubt as to its media's appetite for the psychopathic. However, in more straightforwardly 'political' situations such as the Handsworth riot, this destructive power is only briefly transformed when 'vehicles, including a police van and an ITN car, were overturned and set on fire'. Against the dominant ritual that diverts intervention into explicit shapes of 'private enterprise', law and order, celebrity and fear – whatever the daily fancy – the Handsworth rioters, head-line-hungry murderers, even the 'Chain-gang punks', are all assimilated into media folklore, ready for the cameras. Such challenges then become less of a threat and more a reflection of the dominant New Right culture.

AIDS is the modern Black Death. It is spreading like a plague, threatening to strike down millions.

Daily Mirror, 12 August

A Soviet newspaper said yesterday that the US media were whipping up hysteria over AIDS, 'harvested in the form of unbridled fear and unfounded suspicion' [*Sovietskaya Rossiya*]. It quoted an expert as saying that the disease did not threaten the Soviet Union.

Guardian, 14 October

When straws and half-truths are clutched at, an AIDS etiquette of self-preservation and survivalist instinct soon builds up, challenging any attempts at social cooperation. The fear is isolating, and as an inevitable topic of conversation with other gay men, it's a subject guaranteed to bring out the self – 'Oh, I hope I'm all right' – which further eats away at a sexuality that has taken time, patience and a small amount of courage to build. For two days after phoning a friend in New York and being on the receiving end of his AIDS rap, I have an anxiety attack that starts as blind panic: 'What about the time when … ?'; 'I wish I hadn't … '; 'What if I do get it?'; 'I don't want to die … !'

Yet after working out a personal position – to avoid 'casual' sex, to attempt to remain healthy – I start to get very angry. This time, the anger is not directed at myself, only to be translated into guilt and self-destruction, but at the way in which the disease and the ignorance that surrounds it are ritualized into a controlling agent, expressly designed to instil fear and guilt into its victims.

> … The developing gay way of life ran radically counter to the perceived sexual norms which the New Right was busy mobilizing behind in the 1970s. AIDS provided positive proof that the fault was in the essence of homosexuality, of which promiscuity and disease were the inevitable product. As one gay activist put it: 'No one blamed war veterans for Legionnaire's disease, no one attacked women over Toxic Shock Syndrome. But right-wing publicists are having a field day spreading panic and hatred against us over AIDS.'
> Jeffrey Weeks, 'Sexuality and Its Discontents', 1985

The first stage of the fight-back was to isolate, define and then quell the panic. The next stage was to look at my primary source of information about AIDS, the press, and to break that down. It soon became clear that most of the 'reports' were concerned either with spreading panic, reporting the spread of panic or isolating specific individual sufferers of the disease, or the panic. The lack of any concrete information was glaring, particularly about the medical condition and history of AIDS sufferers. 'Factual' utterances, such as they were, emanated principally from government medical agencies, which, despite their neutral language of authority, were obviously concerned to overstate the gravity of the situation *in order to get government funding*. These reports indicated not so much a future plague on the way, but a medical establishment as yet incapable of dealing with the disease. Instead of usefully discussing the part to be played by 'alternative' medicine, itself unpopular with the New Right because it invests control over health (*not* illness) in the individual, they suggested a science fiction scenario of powerlessness and vulnerability: a mock apocalypse. This is a feature of the media that exaggerates and distorts information where it pleases, so that when presenting 'grim truth' the 'true facts' become caught in a spiral of sensationalism that ever diminishes, while the disease infatuation itself grows more virulent. Once I had understood this I felt clearer and the spell was broken. I began to chop up newspapers, frantically constructing my own counter-ritual – against the real death wish.

> Trying to comprehend 'radical' or 'absolute' evil, we search for adequate metaphors. But the modern disease metaphors are all cheap shots. The people who have the real disease are hardly helped by hearing their disease's name constantly being dropped as the epitome of evil. Only in the most limiting sense is any historical event or problem like an illness. And the cancer metaphor is particularly crass. It is invariably an encouragement to simplify what is complex and an invitation to self-righteousness, if not to fanaticism.
>
> Susan Sontag, 'Illness as Metaphor', 1978

> It must be stressed, with a double-edged meaning, that AIDS should not be responded to with panic and paranoia, but with care. AIDS is just as political as the miners' strike or public spending cuts, and it is ironic that South Wales and their families understand this better than the customers of right-on London pubs.
>
> *Square Peg* 10, 1985

Being on the receiving end of media ritual is politicizing in the widest sense. The disease AIDS is now being used as a metaphor to signify the 'corruption' of sexual freedom and a lifestyle that steps outside rigid and highly constructed norms. The New Right are busy trying to roll back the advances of the 1960s, the key decade of New Right demonology, and AIDS provides the perfect metaphor. Its deeper meaning might well be a fear of that excess that is central to our culture.

I've analysed the AIDS ritual because it happens to directly affect my life, but the panic and lies on the subject are only part of a wider pattern of ritual performed against a whole host of 'deviant' subgroups of which West Indians, miners, socialists, witches and the unemployed are only a few examples. 'Normality' here is the true perversion. A construct used to justify a world view that is as evil as it is destructive.

> I did it on holiday like. Then all of a sudden I did it on page three and I can't see any difference. The only difference was that I was getting paid for it instead of walking around the beach, showing everybody (*cackle*). At least when you do it, when you're in the newspaper, you

only do it for the photographer and he's the only one looking at them (*cackle*). Except the next day, about 4 million people are looking at them, but at least I'm not in there with them.

Samantha Fox, LBC Radio phone-in, 8 September

The celebrity is a person turned into an abstract. Their status has little to do with their achievement (or lack of it) but everything to do with their ability to fit in with the demands of the new mass media ritual. This involves availability, accessibility, pliability and respectability – 'shocking' people are carefully dropped in to add spice or to populate ghettos. As they parade across the surface of the electronic media, celebrities become symbols (and, occasionally, victims), members of a televisual élite encouraging emulation and aspiration – the modern call to prayer.

The appearance of Samantha Fox on LBC displays well the celebrity function. Her achievement as such is to have an attractive enough body to feature on the *Sun*'s page three. Once it is established that she can string three words together – indeed, she oozes that cosmetic charm so beloved of advertisers – inter-media incest ensures that she slips into celebrity; regular appearances on LWT's *Six-o-Clock Show*, in the tabloids, supermarket openings, charity work … she's a busy girl. Her function is to reinforce traditional modes of femininity and gender, with a nod to modernity in the confused application of (God help us) 'healthy' attitudes to sex – *pace* the 'sexual liberation' of the 1960s and 1970s. Quite a devil's brew.

On air, the clichés pile up. Money doesn't talk, it swears. Models aren't dumb blondes, they gotta be streetwise (who doesn't?). Mass exposure is the goal to aim for, the more the merrier. A good pin-up shot is Art. Paul King is the only 'manly' pop star. She herself was plucked from obscurity by a talent contest: *anyone* can be Samantha Fox! As she chats away, who could disagree with her? Certainly not the male callers who ring her up, asking, 'Any film work coming your way?' and, 'What sort of men are you attracted to?' It is left to two brave (and angry) female callers to bring this question-and-answer titillation into question; against the assertion, 'you prostitute your body', Samantha weakly replies, 'It's 1985. The world's changed. I mean, let's face it!'

The driving seat in this phone-in ritual belongs to its immediate controller, the chat-show host. His background chit-chat, obvious when heard on tape but only barely detectable on air, provides a subliminal soundtrack of partiality, backed up by telephonists that screen calls in advance of transmission with the added fall-back of a seven-second delay system that effectively delivers time into the hands of this, the high priest, once the caller has been 'put through'. In this way, the so-called democracy of access media is revealed as a sophisticated, near faultless way of maintaining control.

Friday 18 September – Granada TV: ... the centre of this cheerless building is the bar, where the disillusioned drown their mental fatigue in subsidized alcohol. Each year, Granada takes in a few young graduates from the universities who are naïve enough to believe they will be able to express their vision of the world in these eggbox surroundings. Illusions are quickly dispelled, but by that time they are hooked on fat pay cheques; and soon they are media junkies trapped in a mediocrity that isn't even their own. Granada smells and looks like a school. On the walls of the corridors Francis Bacon replaces the map of the world. Everyone grumbles about his work and tears his colleagues to pieces. But woe betide any outsider who criticizes the place – then they turn like a savage pack to protect their unbearable existence.

Derek Jarman, 'Dancing Ledge', 1984

The same management techniques are applied as are used in a car-factory or in a factory producing television sets. The rigid hierarchies of industrialization ... lead to frustration and disillusionment in their victims.

Stuart Hood, 'On Television', 1980

Behind the glamour of working in the mass media, lies a familiar story of frustration and humiliation. Celebrities are not the only agents of mass media ritual – they are served, and frequently mocked, by the staff who make the programmes, who record the tape, who type the story, who interview the star. This, you would think, is where the system is at its most vulnerable. A chain is only as strong as its weakest link, and the people who serve these functions are only human after all, subject to the same desires, the same fears as the rest of us. Instead, they are not only the propagators but the most fervent subscribers to the ritual. To compensate for their relative powerlessness, they are consoled by a proximity to glamour, frequent perks, high salaries, and élitist status and its reflected glory. This weaves a considerable spell – to give up working in mainstream commercial television is exactly like giving up a powerful drug. All too often, this powerlessness of the media-makers is passed on in the production of media itself, revealed by 'self-censorship' and contempt for the audience – 'the public wants (i.e. will get) what the public gets (i.e. what we want to give them)' – which serves their masters well.

The situation of the media may well be the most critical political, social and spiritual question of the late twentieth century – reflecting its industrial importance to late (consumer/spectacular) capitalism. At present, the media perform a ritual that is embedded in the minds of a small but highly powerful élite, which, judging by recent governmental pronouncements on satellite and cable legislation, seems set to keep that power intact. This ritual is also set to the minds of the great majority – we view such 'information' and 'entertainment' through different eyes, but ultimately as just 'a viewer', there to be manipulated and translated into 'ratings'. The media's vested interests serve, in the main, to hinder any evolutionary step on the part of humanity at this difficult and dangerous time of transition. Yet this ritual is not invulnerable.

One answer is the willingness to attack this ritual on its own terms, to turn it around, but to use an additional set of criteria to the political and social. An added sense of the irrational, and of the possibilities that lie in refusal. The key interventions of the twentieth century – the age of the media spell – have all been concerned with the demonology of media, and have used negation as a powerful counter-blast. Refusal, whether social, logical or linguistic, can yet unleash a sense of possibility and of purpose to transform the world.

Touch Ritual, December 1985–January 1986.

Ooooh, Norman

A couple of days after two of the country's leading groups – the Smiths and New Order – had played a benefit in Liverpool in support of that city's Labour council, and a week or so after the final concert in the first series of Red Wedge events, the British music industry staged its own fightback against a creeping Red cancer. Its secret weapon? Why, Norman Tebbit.

The occasion of this drastic surgery was the staging of the annual BPI awards, held in London's Grosvenor Park Hotel. As the British Phonographic Institute is the established music industry's official body, the event was as self-congratulatory as you might expect, yet its framing displayed current music industry fantasies of how it sees itself, and how it would like to be seen by public and politicians alike.

The music industry would now like to think that it has moved on from those seedy beginnings on the fringes of established show business that are celebrated in films like *Expresso Bongo* (although, when BPI President Maurice Oberstein appeared in a schoolboy's cap, I paused to wonder) and from the shock to the system it sustained, briefly, during punk rock. It is now grown up, responsible, adult, able to take on charity work like Live Aid and, most of all, it is a serious dollar earner.

This spectacle, consequently, was peppered either with pop stars behaving like nice young businesspersons, thanking their mums and dads, management and record companies, or with endless statistics from the preening

Schoolmaster of Ceremonies, Noel Edmonds. Just as the most frequently quoted statistic about Live Aid is not the amount of money raised, but the number of people who watched it on television, so we were informed, several times, that this event was being watched by 100 million people all over the world; just as we were told that a half of the American number ones this year had been British, so were the awards concentrated on acts like Tears for Fears, or Phil Collins, who had sold most in America.

Seen from this angle, the music industry appears to dovetail only too well with government ideas and policies about enterprise culture and the future of leisure. It's all there: a highly competitive, patchily unionized industry; the celebration of success as a guiding principle; the subjugation to the almighty dollar; the staging of spectacles whose content decreases in direct proportion to the number of people who attend. The function of music here is exactly like that of a new type of car: it must accord to expectations, but must contain enough modernity of detail to satisfy the inbuilt consumer demand for change.

Within this context, Tebbit's appearance was hardly surprising and quite a propaganda coup. Heralded, as were all guests, by a snatch of music – his was 'Norman', a candy-floss 1962 hit by Carol Deene – Tebbit took the stage to present a 'special' BPI award to Elton John and Wham! for playing – and thus opening up the markets of – Russia and China respectively. He celebrated, in his usual charmless style, the international competitiveness of the music industry: 'I wish the car industry could be the same.' When he began to talk about 'the less fortunate', I wondered whether this was the public unveiling of the new 'caring' Conservatives; but no, he was talking about those poor people in communist countries who couldn't freely enjoy music.

As Tebbit lurked behind Wham!, I was reminded of Peter Watkins's flawed 1967 fantasy, *Privilege*, shown on Channel 4 the week before. In the film a pop star is used by the government of the day to shore up a cynical regime. Despite the talk of 'freedom', 'the people' and 'enjoyment' that was bandied about, it was clear that the music industry has firmly nailed its political colours to the Tory mast. It will be interesting to see how those who are attempting to encompass some sort of idealism from within the industry rise to this challenge.

New Statesman, 14 February 1986

Sigue Heil

The music industry has a new sensation this month: it's called Sigue Sigue Sputnik and it feels … sort of slippery. The group's first record – called 'Love Missile F1.11' – is already in the top three after only a couple of weeks on sale and may well be number one by the time you read this. Both record and group have been boosted by a carefully planned, heavily financed and adroitly executed campaign of hype that has had both the music/style press and the tabloids reeling. They've got the story: sex, violence and rock 'n' roll *theory*.

This may, of course, simply be an index of how desperate is the dailies' need for a pop fix. The Fleet Street 'pop wars' began in 1983 with the prominence of Culture Club and Duran Duran, two groups who took care to make themselves available to the dailies. The market has since become so competitive that pop has become an integral factor in the circulation wars; its coverage has recently reached a pitch of simultaneous aggression and triviality that cannot possibly last.

In this case, Sigue Sigue Sputnik have set themselves up for it. From the outset, they were meant to be post-Sex Pistols controversial. Assembled to specification by minor punk rock figure and holder of a first-class degree Tony James, they were unerringly aimed at a gap in the market: that for teenage trash. Reasoning that pop is now full of American music aimed at the older market, the 25–40-year-olds, James designed a potentially rich, all-male package of outrageous clothes, technological buzzwords for the 'computer generation' and endless slogans like 'Ultraviolence', 'Video Nasties', 'Rockets' and 'Excitement'. Sigue Sigue Sputnik, so we were informed in one of the myriad interviews the group gave before playing a note, were 'the fifth generation of rock 'n' roll'.

And, being a postmodern creation, Sigue Sigue Sputnik were as much about the process of hype as anything else. This is definitely post-Sex Pistols. In the film *The Great Rock 'n' Roll Swindle*, Malcolm McLaren reduced the Sex Pistols' short and scandalous story into a parable, 'Ten Lessons', on how to swindle the music industry. Subsequent entrepreneurs have taken this parable at face value and, in a series of imitations – Spandau Ballet, Frankie Goes to Hollywood – which became progressively less interesting, have turned a potentially devastating critique of the music industry into a device which simply reaffirms its power. Sigue Sigue Sputnik are fifth-generation swindlers: in signing to EMI for an unspecified (but large) sum, they became EMI's great rock 'n' roll swindle at the same time as they were celebrating their own cleverness. In doing so, they showed the idea of the Swindler to be totally bankrupt.

As the story of their swindle and its attendant hype became *the* story about Sigue Sigue Sputnik, they released a record which remains the most likeable thing about them. Produced by veteran Euro-disco producer Giorgio Moroder – whose 1977 hit for Donna Summer, 'I Feel Love', may lay claim to have invented the whole synthetic-disco movement – 'Love Missile F1.11' is an agreeable confection of throbbing pulse, found noises and sloganeering lyrics. In its flimsiness lies its charm. Indeed, it is reminiscent of nothing so much as a glam-rock hit like Sweet's 'Ballroom Blitz': Sigue Sigue Sputnik revealing their real roots.

The grand design hit problems when it encountered the real world. If you looked, the careful package was full of contradictions: the group's Russian name and typography clashing with its celebration of cheap American imagery, its attraction to 'domineering women' undercut by its 'boys with toys' fascination with gory death and nuclear explosions. And, of course, when Sigue Sigue Sputnik hit the general public with their 'designer violence', the public wasn't to know it was just a *concept*. After a concert where one of the two drummers allegedly threw a bottle and injured some fans, and the singer told appalling racist jokes, the next audience showed exemplary taste by attacking the lead singer with a glass. What did they expect?

New Statesman, 14 March 1986

The Real Thing?

During a recent visit to Manhattan I saw a peculiar sight. In a half empty club, somewhere in the mid-teens, a couple of young black men were breaking athletically to the usual hip-hop mix. Then the DJ put on Iron Butterfly's 'In-a-Gadda-Da-Vida', 17 minutes of white sludge rock from 1968, a record you would have thought utterly alien to them. The dancers cocked their ears, looked at each other, essayed a few tentative steps – and then, in total contrast to their previous style, began to snake in a manner wholly appropriate to the music's florid arabesques.

If it moves, try it; if it works, use it.

The synthetic properties of modern black dance music are rarely celebrated by white commentators. To many, there is something suspicious in its celebration of the 'artificial'. The Campaign for Real Soul prefers to isolate black music that is full of 'soul', 'passion', 'commitment', 'roots', music that is enjoyed by small but definable minorities, black or white. Much modern black music – especially hip-hop, with its fundamental reliance on electronics – is found to be devoid of these qualities: assembly-line fodder, tainted by impure elements.

You could be forgiven for thinking that this is only the latest twist to the classic 'authenticity' argument so enmeshed in conventional left theories about popular music, and you would be right. It's no accident that many white groups attempting to mix pop music and radical politics – like the Redskins or the Style Council – refer back to 1960s or early 1970s models of black music (much as, in the late 1970s, people referred to reggae), forgetting that, for their time, Motown and Stax were as synthetic – assembly-line even – as Tommy Boy or 4th and Broadway.

Hip-hop (using the term here in its widest sense, to cover a variety of black dance forms from the heavy metal rap of Def Jam, to the electro of Tommy Boy, to the new Latin styles coming from Miami) did not originate in Louisiana juke joints: it is a style – and an approach to culture – that has its origin in chance and paradox. There are few stranger stories in popular music than the way that black New Yorkers invented a new musical style from an obscure, German, synthesized disco record, but this is exactly what happened when Afrika Bambaataa and the Soul Sonic Force lifted 'Planet Rock' virtually note for note from Kraftwerk's 'Trans-Europe Express' in 1982.

A simultaneous development, popularized that same year by Grandmaster Flash's 'Wheels of Steel', could have come straight from the pages of William Burroughs. 'Wheels of Steel' cut up five different pop/disco records and introduced to a wider audience (as did Malcolm McLaren's 'Buffalo Girls') the idea of scratching, whereby the noise of a record scratched or reversed on a turntable became a piece of music in itself, usually a percussive repetition. This was a direct extension of live DJ expertise of the late 1970s/early 1980s, where obsolete funk records with percussive breaks were scratched into new rhythm tracks (for further explanation of this process, see David Toop's excellent *Rap Attack*, from Pluto).

Four years on, these twin developments have turned even the standard hip-hop record into a vertiginous amalgam. A 1986 12-inch like Janice's 'Bye-Bye' will, on the 'dub' side, feature a synthesized vocal line sounding not unlike the Chipmunks (old B-Boy favourites); Syndrums; a percussion break transcribed direct from 'Trans-Europe Express'; another vocal line that is a cross between a tribal chant and a doo-wop chorus; and, finally, a bridge that is a direct quote from *The Addams Family* TV theme.

This is a sophisticated mixture, immensely pleasurable and brilliantly constructed: as the term 'dub' denotes, much hip-hop is beginning to describe the same inner space as did reggae in the mid/late 1970s. And, if you look, the resistance so beloved by white critics is still there in hip-hop: the 'found' elements in the style come from a genuine 'folk' practice, and the very public use of the music – that annoying clatter of the street beatbox is *deliberate* – and its coded lyrics invert Anglo-Saxon expectations of both privacy and language.

The theory of authenticity is as irrelevant to modern black dance music as it is to pop in general: to insist on it is to deny that blacks can have an attitude to popular culture as complex and as cerebral as whites, and to ignore the fact that many people, both black and white, want a popular music that not only reflects but celebrates and amplifies the confusions and complexities of contemporary urban living.

New Statesman, 23 May 1986

Billy Idol: Fantasy Made Flesh

The person sitting opposite me is the subject of more sick rumours than I can forget. There's just one thing I have to check.

Billy, these stories …

'People have really strange ideas about me.'

Billy, you encourage them.

'I don't know.' Pause. 'Maybe. I don't *start* any of these rumours though, I don't make them up. Honestly, would you start a rumour saying you've got AIDS?'

Billy, that's not it. Two days ago, I heard you were dead!

'Aaah.' Another pause. 'That's a great one. Billy Idol is dead … Well, long live Billy Idol! Haw Haw Haw!'

Over the next two days, the Most Shocking Man in Pop speaks, in a variety of voices about many things. Only once am I brought up short: when Billy Idol, the scourge of America, Rock Rebel Supreme, the taker of everything that NYC, the baddest city in the universe, has to offer, quietly tells me that he spent five crucial years of his adolescence in … Worthing, Sussex.

Of all the places; there can be no seaside resort more faded, more geriatric. I am struck by the profound gulf. William Broad, you have come a long way …

If there are many tales to come from punk rock, then this is one of the more unexpected: how the singer of the punk rock group that everybody wrote off as a joke trumped each of his contemporaries with American celebrity; how the irony of his name, assumed as the perfect punk take-off of a vapid stardom, became reality; how a shy, rootless adolescent from deep suburbia became a peroxide bottle, a clenched, repeatedly brandished fist and *that sneer.*

Billy, what does it feel like to be flesh for fantasy?

'What's exciting about being me is that I don't have to pander to anybody.

I can just be me. It's really wild! I never thought I'd be in this position. Here in New York, I thought I'd be like everybody else. But now I can be more *me* than ever before.'

The 1986 Billy Idol is very different from the Billy Idol that I met nine years ago, when I performed my first ever pop interview with his group. Then, he seemed like a blank slate, waiting to be scrawled on: painfully self-conscious, trying too hard to live up to the ironies and implications of his name, and letting his cohort Tony James expound the theory behind Generation X.

Now, he is very much the star: confident, highly *present*, with all the aura of celebrity. He talks a lot, very quickly, about being himself, just as you would expect from someone who has to constantly re-create their identity. And then, just when you're tuning out, there appears, like sunlight through cloud, a shaft of pure Bromley: sharp, witty and as honest as it is possible under the circumstances. It's all very confusing. Not the least, I suspect, to Billy himself.

'I've been lucky. Really fucking lucky! I know a lot of this happens because of the way, I think, that I take risks; but I'm such a lucky bastard! It's hard to live with the idea that people are focusing on your sodding life a bit more than they used to. It's just weird. I've never really had it before so I'm sort of dealing with it. But I'm lucky, because in a minute music will transport me out of this horrible world where people can get at you, back to where you can really sing what you like.'

Since Billy Idol's breakthrough in late 1983 with 'Rebel Yell' and a subsequent nine-month tour, the horrible world has been getting back at him. Success brought more problems than it did solutions. 'A lot of bad things happened after "Rebel Yell". I really wanted the next record to be something important, so you have to live through that time, and a lot of wild things happened. It was awful to find out what everyday life is like again.'

Briefly, what happened was that Idol lost two important and long-standing relationships: his companion of four years, dancer/choreographer Perri Lister, whom he met soon after she had left Hot Gossip, and his manager of five years, the man who hoisted him to American success, Bill Aucoin. This seems to have precipitated a real crisis. 'There's lots of things commanding a change. That's what's been happening all the time we've been making this album. It's more, yeah, you gotta take control over your life. It's like, what do I want to represent? What am I? How do I want to live really?'

His laughter at the end of this is a release of nervous tension. 'Er, isn't this getting a bit too serious?'

It isn't what I expected from the man who posed naked, except for a leather jockstrap, and played the outrageous punk for a distinctly unamused female *Rolling Stone* reporter. This is not the Billy Idol of legend. What has happened?

The way Billy tells it, his childhood and adolescence made him perfect for becoming a rock star. Born to middle-class parents, who moved to America in 1959 for four years when his salesman father went there looking for opportunity – 'It made it easy to come back' – Billy returned to England and kept moving through suburbia: from Bromley to Worthing and back again. You grab on to pop because it's the only thing that seems real in that peculiar environment. Then, when you're old enough, it's time for subcultures.

In his early teens, William Broad grew his hair long – 'It annoyed my father; it was great' – and dodged skinheads in nearby Brighton. In 1971 he moved back to Bromley – 'No wonder I went wild when I got back to London, after Worthing' – and soon discovered Bowie. 'I cut my hair off in 1974.' And then, Roxy Music, Dr Feelgood, Kilburn and the High Roads; travelling up to town with another local, Steven Bailey. And then, just around the corner – as everybody now knows – there was punk.

In there somewhere, however, is an episode which Billy is less enthusiastic to recall: university. In 1975, William went to Sussex University to read English and Philosophy. The picture he paints is grim.

'It was too like school, too peculiar: nice people, but not real. They all fell apart in horror when I said "Fuck!" in the pottery class. Because I had short hair they thought I must be in the army.'

William cut out, back to London. With Steven Bailey and several others he started to hang around this new group, Sex Pistols. They got called the 'Bromley Contingent', became part of the new punk élite. Time to put theory into practice: Bailey becomes Severin, after Lou's 'Venus in Furs', Broad becomes Idol: it's a joke, right, but it's smart, and it's better than Vicious, or Generate. OK, now for the group: a bit of reshuffling, as nobody quite knows what works, and then LSD and Chelsea become Generation X. Finally the hair: dyed peroxide, it turns Idol from a name into an *identity*.

Trouble is, Generation X don't quite happen the way they're supposed to. People expect too much: they've been overhyped, and neither a more calculating Tony James nor their managers strike the right balance between politics – which they're *supposed* to take on board – and making records like the Sweet – which is what they *want* to do. On one song, 'Day by Day', they hit it, then, after a couple of top 40 hits, and a brush with the top 10, 'King Rocker', things start to fall apart … What a plight: their career stalls as they go into litigation with manager Stuart Joseph, the guitarist goes OTT and makes them sound like a heavy metal group (*not* trendy in 1979) and *nobody* takes them seriously. It's awful.

'It was just as if things were out to destroy me! It was really heavy, there was a lot of betrayal when Generation X ended. It was all very weird. Now I understand a lot more about things, I realize life is just like that. But then it was a fight for survival: it was easy for Generation X to disappear; we'd been made into pop puppets. And if we disappeared, that would be it.'

But it's not quite over. You make the best record of your career to date, 'Dancing with Myself' – with a new manager, Bill Aucoin, a new producer, Keith Forsey, and with Tony James and a pick-up band. It's too little, too late. On the eve of a tour to promote the final album, *Kiss Me Deadly*, you do what you did in another situation you didn't like – you cut out. Out of Generation X, out of England, and into your American present. And into some bitterness.

Tony, I just need to check a couple of points with you about Billy.
'What'd he say about me?'
Well, he didn't slag you off ...
'If I don't give you good copy, you won't have much to go on, will you?'
Tony, that's my problem.
'OK. Bill Aucoin was my idea; I met him and liked him. It was a very difficult time: Generation X were falling apart and needed marketing properly. Aucoin had the expertise through his Kiss days. Billy was becoming increasingly difficult and unreliable. His leaving was a scummy trick: he didn't have the guts to tell me. Although he writes terrific dance songs, what he's doing is basically Generation X with nothing new. What I'm doing with Sigue Sigue Sputnik is 10 years on.'
Tony ...
'Look I don't really want to be in a Billy Idol interview: who needs it? I don't really want to talk about him. I gotta go.'
As precious hand-baggage, 'Dancing with Myself' wasn't such a bad thing for Billy Idol to be taking on the plane to New York. Although an approach to culture that worked, punk had never quite got the moves right. 'Dancing with Myself' occurred at the moment when the mass post-punk audience was turning to the clubs; it defines, brilliantly, that ever-present, solipsistic *now*. Armed with a retainer from Aucoin and a bit of his own money, Idol plunged headfirst into NYC clublife and washed out the memory of England, with its self-consciousness, in a journey to the end of the night. His dissolute reputation – self-inflicted via quotes like 'All my life I've been wrecked' – stems from this period. People went so far as to name perky drug cocktails after him.

Idol also took with him the producer and the new manager from Generation X. Keith Forsey from Ilford, Essex, had worked with Giorgio Moroder on Munich Machine and Donna Summer records; his subsequent productions for Simple Minds, the Psychedelic Furs and Idol himself have established him as a master of state-of-the-art rock/disco – records that, often despite the raw material, are full of light and space. Bill Aucoin had already presided over one of the marketing coups of the 1970s: a group so larded with make-up that no one knew who they were out of costume. By the late 1970s, however, Kiss were fading – fatally they appeared out of make-up – and

Aucoin decided to go for something a bit more up to date, *new wave* even. He bankrolled Billy, let him run around, and put him in touch with a 'demon' American guitarist, Steve Stevens, and plotted ...

Together with Perri Lister, this team of Aucoin/Forsey/Stevens shaped Billy Idol, American superstar. First came club credibility – an extension of Billy's lurid lifestyle – with a cover of Tommy James and the Shondells' great moronic dance classic, 'Mony Mony'. In 1983, Aucoin realized that the way to translate steady sales and big city popularity into something rather more substantial was video, and in particular video suited for MTV, then a music industry sensation. He hired David Mallet – famed for his ground-breaking 'Ashes to Ashes' clip for David Bowie – and spent over £35,000, a big budget then, on 'White Wedding'. The record was a hit – number 10 on the dance charts – and Billy became an MTV staple.

Now that we're coming out of the video era, its effects seemed a little more prosaic. What 'White Wedding' and subsequent controversial clips, like Tobe Hooper's perverse, apocalyptic fantasy for 'Dancing with Myself', did was to cement Billy Idol as a celebrity – with an outrageous, fantastic image – out of proportion to his record sales. Idol has had only one top five single, 'Eyes without a Face', but everyone in NYC knows who he is. Billy Idol became a cartoon, a white prince from the 'Dirty Mind' era, an outrageous macho sex-pot with the generic punk sneer shared by Johnny Thunders and Sid Vicious. Through 1984 and 1985, it was *wild*: Billy storming out of Radio One's *Round Table*; Billy dropping his trousers in a NYC restaurant; endless press stories of sex and drugs and rock 'n' roll.

Billy Idol still seems decidedly capable of celebrating all those things, but, if celebrity is a contract, he's looking very hard at the small print.

'I just say things and people make it out to be much larger than it really is. It's amazing because nothing's going on. They've got a very conservative public over here and maybe I'm just a bit too open sometimes. And if they're going to say these terrible things, you might as well do them. It's silly, but that's the age we live in. Everything is demeaning and a bit trivial, even to the extent of what they've made records sound like – they're just pieces of consumer garbage.'

It could be that Billy realizes there's no point in playing the punk with me, but, reading between the lines, Billy the Sexpot Scourge of the US became a personality too hard to handle. As Richard Schickel writes in *Common Fame*, 'lives in the public eye implicitly encourage appropriation and aggression'. Billy's an easygoing, likeable guy. He's got a bit of an ego – 'I think mine's revolting, actually,' he says – and he likes a bit of a cavort, but it all got a bit *weird*.

And he's not *that* macho, despite the allegorical sexism of his videos (his explanation of the imagery is really too daft to consider) and his raunchy postures. Billy still retains that androgyny and plasticity essential for any male idol. But it got too personal, what with people expecting him to put up his fists all the time. 'People do mess about with you and do really silly things; you have to tell them to fuck off sometimes,' he sighs. 'All in all, it's … time for a change; time for a new me.' Billy gets serious …

Billy's in this tunnel right now. It's called: finishing the next album. *Whiplash Smile* is, by now, a year later; it's been over two-and-a-half years since any substantially new music. He's been right out of the public eye and the rumours, festering in the sickness of the celebrity cult, blossom into fantastic forms. The American market is more tolerant of long periods off – and the belated English success of 'White Wedding' and 'Rebel Yell' in 1985 was a nice boost – but it must be a bit worrying.

Billy just had a lot of stuff to go through. He doesn't want to talk much about Perri – except to mention that he's seeing her again – but he's more loquacious about Aucoin. He regards the split as a consequence of Aucoin's attempt to make him go Hollywood with a part in a projected film of Nik Cohn's *King Death*, playing a Billy Idol parody of a rock star. 'It was the wrong thing to do; we should have been doing music,' he now says. Aucoin won't say anything: Billy is now managed by an Englishman who worked for Aucoin, Brendan Bourke. He comes from Islington and looks tired, but he seems to be able to control Billy.

There's a lot riding on *Whiplash Smile*. But Billy doesn't want to talk about it any more; he believes in action, not reflection.

'It'd be great if you heard it,' he says. So I get taken to Right Track Studio to see Billy, Keith Forsey and Steve Stevens at work.

'We've been through several pain barriers on this one,' says Keith Forsey as he spools through seven songs. The musicians and the producer seem to treat my appearance as a chance to relax and enjoy the fruits of their labour. Billy leaps about and throws a few shapes, while Steve Stevens, dressed in full psychedelic punk regalia, cocks an ear and checks that everything is just right. The three are obviously a very close team. The songs are 'World's Forgotten Boy', 'Soul Standing by', 'All Summer Single', 'Fatal Charm', 'I Don't Need a Gun' and 'Man for All Seasons'. They are glossy examples of rock/disco, not surprisingly, a true NYC synthesis. A slow ballad, 'One Night One Chance', succeeds in extracting the edge hinted at by Billy's 'Raw Power' references.

'New York is seldom niggling,' writes Jan Morris in *The Great Port*. It is, however, a long way from Worthing. Billy Idol is no longer an Englishman – despite the pertinent observations he can still come out with when pressed – but a native of New York. He has acquired some of the brashness, hyperactivity, self-obsession and generosity of that environment.

He talks a lot in a language I don't really understand, about rock 'n' roll, which he seems to regard like a religion. The part of me that lives in the outside world is appalled at the restrictions of his life. Yet Billy Idol's rebelliousness and wish to control his own life are a lot nearer to the centre than he thinks. The twist in the tale is that he *does* have his own vision.

'See, that's the thing,' he says, after a long rap about 'me' as it affects the record. 'That's the chance you've got. They're calling this the age of "no reason". When things have no reason and you see that all around you, then there's all the more reason to *have* a reason. What a fantastic chance: it's like I'm still throwing myself into the void.'

The Face, July 1986

Boy George: Chasing the Dragon

It started with a trickle; it swelled into a flood. On Tuesday 10 June, the *Daily Mirror* 'broke' the Boy George and Drugs story: sadly, it could only manage cocaine. Never mind. Three weeks later, the *Sun* delivered the story for which thousands of pounds had been on offer – Boy George Takes Heroin: Official. Fleet Street then went nuts.

Twenty tabloid covers later, the whole affair leaves a bitter taste in the mouth: not only because of the tone of the material – few things are more revolting than Fleet Street in a fit of morality – but also because of the way the story has been gradually shaped to reaffirm 'traditional' moral values.

The Boy George scandal is the perfect convergence of two current Fleet Street obsessions: Pop and the Drugs Menace. Over the past four years, Fleet Street has gone progressively more pop, reaching a state of hysteria this year. Pop stories now routinely make the front pages, as the pop star system has taken over from traditional showbiz as the source of gossip and juicy scandal.

This trend has partly paralleled music's increasing integration with the mainstream media industries – but it also marks increased competition between the tabloids. 'Get them young' is their motto; in the *Sun* and the *Mirror*, pop now has a daily page. The 'quality' press is hardly immune: despite the more serious nature of its coverage, it still serves the music industry with reviews and brief, PR-style interviews. Instead of the young, its coverage is aimed at the 25–35-year-old bracket currently being courted by record companies.

Boy George, as Culture Club's public face, was the first modern pop star to take advantage of this trend. Even more so than contemporaries like Duran Duran or Wham!; he took care to make himself available, not to the traditional music weeklies, but to Fleet Street. As Culture Club's career prospered through 1982 and 1983 – with seven top five hits to date – Boy George became ubiquitous, first announcing himself as an outrageous dresser of ambiguous sexuality – 'Gender Bender' in Fleet-speak – and then attempting to dispel any sexual threat by declaring that sex was boring and that he'd rather have a cup of tea. By the end of 1983, he was almost as spotless and sexless as Cliff Richard.

Yet, with hindsight, this was a dangerous game. The mass access has its price. Like Ian Botham, Boy George must have wondered whether he had signed a Faustian contract. As Culture Club began to slip, ever so slightly, in 1984, the stories turned sour. The angle was not Culture Club's success but Culture Club's impending failure. A highly contentious *Sun* story by John Blake about George allegedly attacking a Duran Duran fan started the counter-attack; from then on, Fleet Street and George were at war.

As Culture Club's career hit new lows – caused by overwork and overkill – Boy George became an accident waiting to happen. The initial *Mirror* story only confirmed two widespread music industry rumours: that George and some of his entourage were in a bad way and that some of the tabloids were offering serious money to nail them.

Initially, the stories delighted in exploding the twin Boy George myth of sexlessness and distaste for drugs. The *Mirror* followed its cocaine exposé with the very old news of George's affair with Culture Club drummer Jon Moss. The *Sun* then injected a new note: not only was 'the family' invoked but the dynamic of death injected – hinting that this story had to have some final resolution. At this point, it ceased to be a showbiz story and passed to the news desk.

Being less used than their showbiz counterparts to the wild and wonderful ways of pop, the news teams imposed their own world-view on Boy George. Partly, this took the form of heightened resentment and prurience about the lifestyle of George and his entourage: the *Star* ran a most peculiar item about Marilyn being arrested – 'Stripped Naked!' ran the headline – in which he was fetishized like the woman after whom he takes his name. TV news items referred to Marilyn as 'moody' and 'impatient' – feminine attributes in news language. This was homophobia run rampant.

As importantly, the Boy George story became part of the wider Drugs Menace, as smaller items about pot-smoking soldiers and 'Crack: the New Drug Menace' were appended. This has been fertile ground for Fleet Street recently, as the recent headlines and comment about Olivia Channon and the hanging of Kevin Barlow, heroin smuggler, make clear. And then, the true resolution of the story became apparent; after a *Sunday People* story which purported to find the location of Boy George's pushers, the *Mirror* followed up with: 'What are the police doing about it?' And so the arrests, and George's

George just before the first flush of fame, Summer 1982 (Derek Ridgers)

quick removal out of harm's way.

In the end, and this may be the real tragedy, Boy George and Marilyn hardly matter as people, despite the superficial rhetoric of concern. In the words of a winter 1983 Culture Club hit, they have been turned into 'Victims'. The Boy George scandal has been shaped by the press to highlight the penalties of social deviance.

The story has everything: fading pop star, heavy drugs, not just homosexuality but hints of weirder sex and finally, as Friday's edition of the *Star* made clear, the ultimate taboo, AIDS. This is a devil's brew and, further, the fact that Fleet Street has brought the story to a 'resolution' in real life accords exactly with the 'solution politics' of Mrs Thatcher, where symptoms (as opposed to causes) like football hooliganism and drug abuse are singled out and 'dealt with'.

And, as the *Star* made explicit on Wednesday – 'For once I agree with Norman Tebbit; the BBC should ban anyone in pop music connected with drugs' – Fleet Street is throwing its weight behind the Tory fantasy of a return to 'traditional moralities'.

Observer, 13 July 1986

Sid and Nancy: Needle Match

As the title suggests, Alex Cox's film *Sid and Nancy* concentrates on the principal characters to the point of monomania, the two fusing into one horrible hydra of emotional and chemical addiction. In its final stages, where the claustrophobia of self-obsession and interdependence is translated, spatially, into an airless hotel room, the film contains some real emotional truth. The rest tells great whopping lies.

Part of the problem is that Cox can't decide whether he's making an (anti) love story or a genre pop film. Because both Sid Vicious and Nancy Spungen were involved with the Sex Pistols, Cox feels it incumbent upon himself to deliver a snap history of punk rock – spiced with a bit of subcultural theory – as seen through the eyes of Sid. The problem is, Sid's wearing sunglasses after dark, and Cox's own vision is similarly obscured by the rose-tinted glasses of nostalgia. The result is a fundamental mistake which reveals the poverty of Cox's fantasy.

It is, of course, extremely difficult to re-create the still-living past, and, in this case, a past so well covered up as punk rock. Punk's tenth anniversary is turning into one of the media events of the year, and, as is usual on these occasions, almost everything is covered except an accurate attempt at describing what went on. The framing of *Sid and Nancy* as a revival item, coupled with the inevitable publicity that a film like this heralds, only contributes to the misconceptions and misrepresentations.

The initial media codification of a sophisticated, art-school style into 'foul-mouthed' yobbery is swallowed wholesale. The foul mouth is made manifest: the first time we see 'Johnny Rotten', he is dribbling an unholy mixture of baked beans and champagne. The camera closes in. Such literalness – depressing in a director whose 'radicalism' is part of the package – is carried through to the script.

Here all the characters are presented as ciphers, revolving around Sid as his satellites: their sole purpose is to prop up the two main protagonists. Apart from being a waste of the dramatic possibilities that arise from the most superficial reading of the characters of John Lydon and Malcolm McLaren (to name but two), it is inaccurate: until the last American tour, Sid was himself but a cipher in the Sex Pistols.

Such nit-picking wouldn't be important if the film was a consistent, abstracted fantasy. Unfortunately, Cox gets just enough period details, incidents and locations right to give it more than a whiff of documentary. He claims otherwise, but these claims aren't backed up by the direction, which, apart from a few good belly laughs, is quite pedestrian in this part. Having decided to re-create several well-documented incidents, he lays himself open to a different kind of scrutiny, that of authenticity; he fails so resoundingly that it calls even the film's success into question.

Cox's vision of punk rock – a tiny element in reality – is of a boy's rock rebel fantasy. It's quite American in this respect – particularly in its crudity – which may give a clue to the film's commercial orientation. If that is the case, then even the emotional charge of Sid's disintegration is, like the Government's (anti-) heroin adverts, double-edged: you just wonder whether Sid Vicious isn't being presented as the latest example of the romantic myth – 'Live Fast, Die Young and Have a Good-looking Corpse' – that, in the person of James Dean, has been at the heart of youth culture.

Punk was an attempt to force open the window of an airless culture: any pursuant claustrophobia was as much temporal as anything else, there simply not being enough time. Time has now stretched: punk is consigned to the eternity of the museum, where people squabble over the exhibits. And *Sid and Nancy* is so hermetically sealed from the outside world that it makes you wonder whether opening the window is even possible.

New Statesman, 25 July 1986

Waugh Crimes

In a recent issue of the *Tatler*, a rising young *Private Eye* person was profiled. Beside his picture – peculiar in the way that only a 25-year-old trying to look 55 can be – ran the copy: '[He] is one of many who models his character on that of Evelyn Waugh.' On a recent *What the Papers Say*, outgoing *Private Eye* person Richard Ingrams castigated the *Observer* for its pop coverage. The writing itself wasn't perfect – a little too purple, perhaps – but what was more interesting was the way in which Ingrams's attack relied upon an implicit sneer: what business has a 'serious' paper got in writing about this stuff at all?

It is, of course, easy to generalize, but such arch fogeyism is very influential in British culture, and has been so for the last few years. It isn't organized, or a 'movement', because it doesn't feel it has to be; after all, it has the weight of history and wealth on its side. Just as the chocolate-box adaptation of *Brideshead Revisited* fully announced Reactionary Chic to the nation, so has fogey style derived from one person: Evelyn Waugh. His legacy and its interpretation describe the cultural gains of the New Right.

In 1981, Dick Hebdige wrote:

Like all militant reactionaries, Waugh delineates a previously submerged set of values, preferences and assumptions by attempting to exculpate them at the moment when those values and the interests they embody are in crisis, on the point of disintegration. In fact, they growing belligerence of Waugh's prose from 1945, the year in which *Brideshead Revisited* was published, onwards dramatically signals the extent to which the cultural consensus had shifted against him during these years.

Towards a Cartography of Taste 1935–1962

Waugh's elevation into legend – as the house god of literary London – has come at the same time as, and may have fuelled, a concerted ideological attack on the social gains of the whole post-war period. *Brideshead* is a key text in this, not the least because of its televising. It's a mixed book, in which extremely acute psychological *aperçus* are set against a self-admitted indulgence in nostalgia for the pre-war period. As Hebdige points out, the book was out of its time; what has happened since is that, in the received version, the *aperçus* are flattened out at the same time as the nostalgia, and snobbery is accentuated. This makes it a perfect text for these times.

It is interesting how 'history' has favoured one great novelist of the 1920s and not the other. Since the early 1970s (the late essay *Doors of Perception* being a psychedelic biggie) Aldous Huxley's reputation has fallen as fast as Evelyn Waugh's has risen. Starting from a similar black comedy, rooted in social observation – usually observed with an exact, dry wit – Huxley's books soon began to show signs of that which is forbidden: ideas, and politics. Some of them dared to address the subject of what the future might be like. For this cardinal sin, Huxley is now thought 'pretentious' – being shorthand for 'I don't want to think about what he's saying' – and has been cast out of the literary pantheon.

Meanwhile, the cover-up line for prose that says the most appalling things is, 'Yes, but it's so well written!' In his own style, Waugh moved from a self-conscious youthism, through brilliantly observed, clipped social farce, to a more mannered, consciously elderly and cantankerous style. It's a complicated mixture, boiled down today into a received style that revels in the minute dissection of upper-class social mores (leaving out the black comedy), or in flatly stating the unspeakable – opinions which, if offered in public, would result in a fight. It is here that you now find the peculiar phenomenon of 30-year-olds attempting to write like people twice their age: it is here that you find the insult and invective that passes for argument in the halls of the New Right.

The Waugh legacy is not itself responsible for this climate, but is a key to understanding the palsy that affects the British literary scene and by implication, ours being a literary culture, our culture in general. Nostalgia for the past is pandemic (and, I grant you, affects youth culture just as much); ideas and arguments are less important than style, spite and the correct caste. Post-war capitalism developed, for a time, illusions of a more pluralistic British culture: the "classless" society may well have been a chimera but it granted access to a wide variety of voices other than those of an exclusive élite. Because of the hierarchical nature of our society, most of these voices were concentrated – marginalized even – in pop culture, which still remains one of the few areas in Britain in which the present and future are even considered.

Today, upper-class junkie stories – with a plot straight out of Buchan – vie with pop star junkie stories. Queen and Country are reaffirmed, yet again, in the much-hyped wedding of a lesser prince to a Sloane; ITN fawn over a retired major. These are petty indices of a collective delusion concerning Britain's present internal state and its position in the world. It is extremely important that British culture develops a way of addressing the present and the future rather than the past, that recognizes our pluralistic, multiracial society and our position, finger-in-the-dyke of trends in world politics. It does depend on your point of view, but history may not be on our side after all.

The Face, September 1986

Chrissie Hynde: Hymn to Her

'People often say to me interviews, 'You aren't very prolific, are you?' The answer is, no, I'm not! I could step up my output, but who would want it? I don't think people *need* product from me any more than every three years.'

Chrissie Hynde's terseness – as much directed towards herself as towards everyone else – relaxes into laughter. The Pretenders are releasing their first record for three years and Hynde, as the group's spokeswoman, is back on the promotional chain gang. She doesn't like it.

'I'm very lucky to do what I do, but there's a *lot* of associated bullshit. I don't mind being a voice, but I never wanted to be a face. I don't want to promote my product, I don't want to advertise my product. If you don't wear high heels and lipstick in this business, pal, you'd better get on to a model agency quick. If it was up to me, I'd get Yasmin Le Bon to go on the covers of all my albums and I'd just call it a day, just stay in the studio all the time.'

Yet here she is, albeit mildly irritated after a photo session. Hynde is full of contradictions: the Kent State student, profoundly influenced by 1960s radical politics, making mainstream records for a conservative 1980s market; the former hell-raising punk now very concerned with the upbringing of her two daughters; the gossip column regular – often as the 'partner' of Jim Kerr, or, formerly, Ray Davies – who shuns celebrity; the 35-year-old woman who likes nothing better than to rock out with a tough male rock/soul group.

Such, of course, is the complexity of many people's lives; it's a tribute to Hynde that she carries off these contradictions with grace and a certain *attitude*. This attitude – tough yet vulnerable, acerbic yet romantic – informs her songs and gives them depth; the Pretenders are one of the few groups to make music that is both pop, in that it is accessible and attractive, and adult, in that it admits complexities rather than delivers certainties.

Hynde herself is a survivor. Her personal history is as turbulent as it is well documented: the rise to fame in 1980 after years of scuffling; the deaths of Jim Honeyman-Scott and Peter Farndon in 1982/3; the dissolution of her troubled relationship with Ray Davies in 1984. These experiences may have given her strength, but at a price; she speaks of *Get Close*'s long gestation with a protective brevity.

'First off, half the band was dead. Then I was trying to get a group together to replace the old one; you don't just pull musicians off the street and hope to get the balance right. Then I had a baby: that takes a year out of your life. You have to maintain a life outside the music industry: I did a year's worth of touring with *Learning to Crawl*, and after that you need another year just to clear your head, before you even think about writing new material. You've got to let the well fill up again.'

The Pretenders now feature the black rhythm section of bassist T. M. Stevens and drummer Blair Cunningham, a change reflected in the funk inflections of much of the material. According to Hynde, this just represents her R&B roots and is what she's been trying to do with the Pretenders all along; *Get Close* still retains a good deal of rock flash, strong songs and that expressive voice. 'Recording the vocal is like – I'm in tears out there,' she says.

Many of the songs, like 'My Baby', 'Hymn to Her' and 'Tradition of Love', reflect a fresh focus: Hynde sounds more disciplined and, dare one say it, contented. In common with many of her generation, the 1970s rebel is finding some stability and acceptance, in her own life at least.

'You've got to face the music. I'm 35; I'm doing what I want to do. Rock 'n' roll is no longer a youth cultural phenomenon, although it's there for the youth culture and I wish they'd pull their finger out. I find it healthy to be around children: the problems that you have don't relate to their lives at all. At the weekend I was playing with my kids: I was so wound up at having to do a lot of work around this album that I suddenly looked at them and got very sad that I couldn't get into the sandbox with them. It keeps your perspective.'

Get Close does contain a couple of stinging songs, however. 'Chill Factor' continues Hynde's sharp look at the power imbalance between men and women that she first explored on 'Thin Line Between Love and Hate'.

'Any woman who's had a kid – and I include myself here – will tell you that the man who's not at home has no idea of what's going on there. It's a *very old story*. The guy comes home and his life sounds very glamorous; but nobody knows that you need, essentially, a degree in physics to get two kids ready when it's raining outside, to get them to the shops when they're screaming at the top of their lungs, to get them out of the check-out counter when there's every sweet known to man at their eye level. It just occurred to me one day that the mother is, quite literally, too often left holding the baby; spare a thought for the single woman who's not going to do anything else but look after her child for 10 years.'

Hynde's spikiness hasn't entirely disappeared. She gets very voluble about America – her country – which she regards as 'out of control', and for a few moments the old venom returns: 'The media – TV, newspapers – are incessant and hysterical. It seems like the whole country is brainwashed and people's consciousness has gone to sleep. The book I'm reading at the moment is a James Baldwin book called *No Name in the Street*; it's from 1971 but it's like another world. 'How Much Did You Get for Your Soul?' on the album is pretty much pointing the finger at the black community. It's like you've got musicians who have careers as artists doing commercials. I'm talking about Grandma Turner, Whitney Houston, Michael Jackson and Lionel Richie. Where are the leaders? Where's Martin Luther King? Where's Malcolm X? Can you imagine Michael Jackson saying what LeRoi Jones said: "Wake up, niggers, or you're all be dead"? Somehow you can't say that after you've had a nose job.'

Observer, October 1986

The Orton Diaries: All the Rage

In December 1966, the month he picked up an *Evening Standard* award for Best Play of the Year, Joe Orton resumed a diary he had given up in 1951. This fresh diary was to reflect his new found fame and confidence: titled 'Diary of a Somebody', it was written for publication in as brutally frank a style as was possible.

But every diary writer, particularly so mannered a stylist as Orton, can't help but edit their experience and the hole at the centre of these diaries is Orton's lover and collaborator, Kenneth Halliwell, who appears as a brooding and developing absence. In this case, the editing proved fatal: after he had splattered Orton's brains all over the collaged walls of their bedsit, Halliwell wrote a suicide note: 'If you read his diary all will be explained.'

This Grand Guignol finale gives a terrific narrative kick to what is already a superb, sustained piece of writing. The teasing, rather camp student of the 1951 diaries is replaced by a writer at the peak of his powers: by 1966 Orton had both the discipline and the confidence to 'rage correctly', to turn his multiple social and sexual resentments into a style – and a vision – that is 'quite natural whilst in actual fact being incredibly artificial'. It's the same 'combination of elegance and crudity' that he ascribes to Genet.

Thus, after describing a busy afternoon's sex with two other men, Orton ends: 'I thought they were both very nice fellows.' The conjunction of explicitness and archaic, almost stilted language is irresistible. Orton brings out the humour and the politics of the sex that, whether in an abandoned Leicester building site, a luxury Moroccan apartment or a public toilet in the Holloway Road, permeates these diaries. Sex to Orton wasn't mere sensation but a deadly weapon, the focus of his scalding rage. 'Sex is the only way to infuriate them,' he writes at one point. 'Much more fucking and they'll be screaming hysterics in no time.'

Much of this diary is concerned with theatrical matters: the success of *Loot*, running in the West End throughout the period, or the revisions of his final play, *What the Butler Saw*. After years of rejection, Orton finds himself a celebrity in the mainstream of 'Swingeing' London, 1967; he notes its details and absurdities with the same eye and ear that he has for overheard dialogue. Descriptions of the irreality of everyday life – 'A very fat, pompous-looking woman reeled out of a pub shouting, "Melancholia? Ad nauseam"' – are interspersed with descriptions of buying Donovan's 'Sunshine Superman' or meeting Brian Epstein about the *Up Against It* Beatles film script. So is a period captured.

The diaries' centrepiece is a prolonged account of Orton and Halliwell's trips to Morocco in the summer of 1967. Orton turns his caustic attention to the swish homosexual milieu that is attempting, vainly, to co-opt him. One of the few sympathetic portraits is of a gentleman called George Greeves, who is, if anything, more scabrous than Orton himself: 'His stories are endless. He keeps up a constant stream of foul-mouthed commentary – life and death, nobody is saved. His chief target are the rich and pompous.'

Here is life flayed to the bone, in an equation of which Dionysiac delirium provides the other half. In Tangier, Orton sent himself ecstatic with hash cakes and a sexual timetable of formidable complexity: there he could find the distance and the release from the 'tight-arsed civilization' of England that he hated.

Orton wasn't interested in conventional politics. The legalization of homosexuality rates a terse, passing mention. His diaries – framed on publication by an excellent John Lahr introduction – contain enough wit, observation and social history to put many writers to shame. But their concentration on sex pinpoints an area of activity that our society finds fundamentally troubling: this ability to disturb gives his critique of English society a power that has reverberated, not only through the theatre, but through the activities of any cultural provocateur who, like the Sex Pistols, wants to see 'the old whore society really lift up her skirts'.

New Statesman, 7 November 1986

Megastores: An Exclusive Church

Within the past two years, about £9½ million has been spent on four new record shops in the centre of London. After the reopening of the Virgin Megastore about a year ago, they've come thick and fast.

In April, Smithers and Leigh opened on Oxford Street by Marble Arch; in July, Tower Records unveiled their £3½ million refurbishment of the old Swan and Edgar building on Piccadilly Circus; and last month, HMV held a rather overlit reception to herald their as yet unfinished £3 million Oxford Street store. The concentration of stock is massive: these four hold about 1½ million records and tapes in just over a square mile.

Megastores – note the hyperbole – are the latest idea in record retailing. They are different to the customary record shop – represented nationwide, say, by a chain like Our Price – in location, size and design. The positioning of these four stores on prime consumer thoroughfares is deliberate: they are meant to trawl passers-by already dazzled by the lights and frenzied atmosphere of London's Golden Mile. Once attracted, the consumer enters a new type of department store, dedicated to a highly particular range of products, and carefully designed as a stimulating, all-in environment.

Megastores are the result of a more aggressive sales posture on the part of the music industry, atavistically worried about falling sales figures. Like the CD – deliveries of which went up by near 400 per cent during 1985 – they enshrine the latest marketing fashion. This, as the 1986 *BPI Yearbook* advises, is to bypass the traditional 16–24 'teenage' market in favour of the 25–34 and 35-plus age groups, which, it has researched, have more spending money and are, demographically, on the increase.

The economics are plain: it's much more attractive to sell 20 copies of this autumn's hot 'adult' item, the Bruce Springsteen boxed set – at £25 each, £500 – than the equivalent amount in the form traditionally associated with the younger market, the seven-inch single: at £1.59 each, 315 copies.

This older age group is more sophisticated, 'lifestyle' consumers who have grown up with the idea that pop music is the focus for a culture. This is the megastores' big score: once you fight your way past the Springsteen boxes blocking the entrance, you'll find a bewildering, inclusive array of products inside.

All four offer a profusion of CDs, pre-recorded and blank audio and video cassettes, 'rock' books and magazines, hi-fi accessories, T-shirts, and, of course, records in endless categories: classical, rock/pop, soul, jazz, reggae, easy listening, folk, country, rock 'n' roll, international. All remain open until well after office hours. All, bar HMV, offer a café on site or nearby, where, with varying degrees of delight, you can work off that appetite stimulated by all that hard spending.

There are individual characteristics. Tower sells American magazines and has a very good soundtrack/showtune section. Smithers and Leigh has hi-fi equipment and a vegetarian café. HMV has obtrusive lighting, a full international section, a big sale and Samantha Fox jigsaws. Virgin has airline tickets, a pop/rock section strong on back catalogue, and design by Terence Conran.

This last shows the megastore ideal at its height: Virgin's dove-grey interior and hushed lighting give a suitably reverent tinge to what is, after all, a serious business. They cast an otherworldly glow on to the celebrants who, lulled by the hypnotic expectancy of the latest disco 12-inch, flick through the racks of records fetishistically encased in plastic. Here, the very masses of displayed goods is irresistible.

Megastores are temples to a new age of consumerism. They are genuinely popular, and they work well – it's hard to leave their credit card delirium without buying something – yet their mass appeal is an illusion. Just like the use of music in TV ads, they symbolize the corporatization of pop; their centralization of space and their market orientation make them not a populist but an exclusive church.

Observer, 23 November 1986

Madonna and Youth

Most successful pop records work on a strict principle of repetition. Like any basic ritual, the potency of a hit single grows from the endless reiteration achieved by its being played on radio, TV, in the disco or in the home and – as befits the nature of our society – from its very mass production.

Repetition also occurs in the structure of a successful song, usually through a metronomic rhythm (particularly when provided by a drum machine) and the familiar verse/chorus sequence. This is the song's outward layer. But in the best pop songs there lies a buried hook – a phrase, an instrumental line or even a nuance – which, in acting against its overt structure, acts as the song's deeper, real centre.

Madonna's 1985 hit 'Into the Groove' is a good case in point. Taken from the soundtrack of the successful film *Desperately Seeking Susan*, it's an excellent example of the breezy, assertive style that has become Madonna's trademark. The song's structure is fairly simple; its basic tension is drawn from the contrast between the simple, heavily accented beat, some slightly fussy instrumental fill-ins, and Madonna's projection of two voices: low and husky, high and excited.

'Into the Groove' is a basic come-on: a dance song with lyrics – 'We could be lovers if the rhythm's right' – that describe the experience to which it is meant to provide the perfect accompaniment. As you get deeper into the song, the verses pick out the difference between private fantasy and public fulfilment that is the pay-off: 'I'm tired of dancing here all by myself/Tonight I want to dance with someone else.' The tension is wound even tighter; you begin to await the resolution set up by all this aural foreplay.

The key part of the chorus deals with the moment when Madonna crosses the line between fantasy and reality and puts desire into practice. 'Touch my body and move in time/Now I know you're mine,' she commands. On the second time this occurs, she repeats the second phrase, 'Now I know you're mine'; there is a sudden break in the music's relentlessness, some sudden depth and space. She repeats the phrase again. On the fourth time, she trades, without warning, her husky whisper for a high, keening wail that splits the song wide open: its hook is a brilliant evocation of desire.

At first, Sonic Youth's cover version – called 'Into the Groovey' – comes off as a poor art-rock pastiche: the so-called 'avant-garde' using references and irony in a mocking yet parasitical relationship to the mainstream. It employs most of the standard tricks: fuzz guitar, doomy punk bass and harsh, distorted vocals. The tempo is slowed right down: what once swung is now a subterranean thrash, delayed even further by some heavily echoed, obtrusive percussion.

And then a similar alchemy takes over. About two-thirds of the way through, Sonic Youth strip away each instrument until only the bass is left, resonating: again, there is some sudden space in the murk. The song then goes off on to a different tack: the other instruments come in quickly as the tension builds. It is here that Sonic Youth pull off their masterstroke: at exactly the same point in their version of the song, they drop in Madonna's voice singing the key refrain 'Now I know you're mine'. The first, husky-voiced intrusion sets you up for the same keening wail that splits 'Into the Groovey' as surely as it does the original.

'Into the Groovey' is that rare thing: a cover version of a song that has already been recorded in a definitive version, which nevertheless gives it a new meaning. In this respect, Sonic Youth have pulled off an exemplary piece of musical pop criticism: they've isolated the key moment in the Madonna original and transposed it into a new context where its apparently uncomplicated lust is seen in a more sombre, some would say suitably troubled light. As good fellow New Yorkers, Sonic Youth know that Madonna's innocence is highly constructed, yet they are not immune from an innocence of their own: 'Into the Groovey' is a fans' tribute from people who still believe in an avant-garde.

New Statesman, 2 January 1987

Dead Soul Boys

The other Week I tuned into *The Tube* – more from habit than desire – and promptly lapsed into reverie, only to be roused by the sound of people chanting, shouting and generally showing signs of life unusual to that programme. They were heralding the last act: a group of four ugly, male, post-punk survivors affecting the long hair, coy references (at several stages it was possible to see, among the cosmetics, the 'DRUG' scrawled upon the singer's guitar) and apocalyptic rhetoric of the late 1960s 'underground bands'.

In themselves, the Mission (who currently have a top 10 hit with, suitably, 'The Wasteland') were distinctly uninteresting, but what they had, almost unwittingly, set in motion among the audience was. As their fans – both male and female – swayed on each other's shoulders, threw confetti-sized pieces of paper and generally wigged out, being a Mission fan looked like *fun*. To me, this represented a real change: the fact that the young audience, some of whom weren't even born in 1969, were prepared to lose control with such minimal encouragement was an index of a deep lack in pop and its current interpretations.

About seven years ago, I did an interview with the backline – good musicians all – of Malcolm McLaren's then pet project, Bow Wow Wow. In an exchange of mutual nervousness, the only thing that stood out was when the drummer talked, animatedly, about being a Soul Boy. In the midst of post-punk angst, this seemed liked a fresh idea. Pop was, at that time, hamstrung by introversion and depression (Ian Curtis, a key figure, had just committed suicide): the bright colours and simple, non-intellectual activity of the Soul Boy – with his insistent cry, 'Fuck Art. Let's Dance!' – seemed like a real alternative.

Since then, the Soul Boy ideal has bossed English pop. On a musical level, it has resulted in the white assumption of black dance forms that, in the hands of acts like Duran Duran, Culture Club and the Eurythmics, has been responsible for the world-wide spread of English pop (for the third time) since the early 1980s – a development most welcome to a domestic music industry bruised by punk. It has also resulted in the increased access that English listeners (and consumers) have had to a flood of excellent black American dance music and, by further implication, the broadening of taste that has made African and other world music popular. These are positive benefits.

It is important, however, to distinguish the music itself from the meanings and interpretations that are erected upon it: for this is where the problems start. The Soul Boy myth has also bossed Pop Culture, ever since the 'style' magazines that began in the early 1980s based their appeal on a sharp, streetwise, laddish, metropolitan view of the world, where the cut of your trouser, or entry to the correct nightclub, opened the door to a new world of desire. At the same time, pop history was rewritten to include accounts of Northern Soul and the mid-1970s east London Soul Culture that, apparently, was the 'true roots' of punk. The message was clear: cool, control, but above all pride – in the emblematic words of Curtis Mayfield, 'Move On Up'.

But Style Culture – like the definition of postmodernism that it embodied – had a fatal flaw: in its obsession with the minutiae of style, in its very aspiration, it failed to develop any political or theoretical distance from wider economic and social trends. Style Culture quickly became the cutting edge of a Thatcherite consumerism and social mobility: its traces can now be seen in pop videos, all over Fleet Street, and in what may be its pinnacle – the Nick Kamen Levi's advert. Perceiving this lack, the left has attempted to hijack Soul Boy 'community' and ally it to older ideas about musical 'authenticity' (whereby soul, like reggae was 10 years ago, is valued for its 'real', 'roots'-like qualities). But this account founders on the reality of black dance music production, on the base desires that are the heart of pop, and on the very conformity and conservatism of the Soul Boy (and Style Culture) myth itself.

Like Style Culture, the Soul Boy myth is dead, obsolete. In the space torn between styles, what interests me is not 'a return to rock' – in America, it never went away – nor the unholy mixture of hip-hop grossness with the received sexism of heavy rock, but the chance that there is for the voices that have been excluded by Style Culture's metropolitan bias to make themselves heard. There is a new, uncharted area of youth culture: it might include the rock fans or Goths that populate any provincial city or small town – it's no accident that the Mission come from Leeds, a Goth stronghold; it will also include another group which, because it is not organized in consumption, is not deemed worthy of media attention. These days, this means that it might as well not exist. Yet any observer will have noted a large amount of people – disparagingly called 'hippies' – who operate like the *Wandervogel* of the 1920s, moving from town to town, from festival to 'Peace Convoy', outside the bounds of 'normal' society. Any future account of Pop Culture must include their bohemian wanderings, which are a true index of consumerism in crisis.

New Statesman, 6 February 1987

Boy George: Reborn

'I had the worst Christmas of my life this year; I haven't had a good time. I just sat upstairs shaking. I still get it now; it's not like it's gone. I was told it would take three years, which is *eugh*.'

Just under two months later, George O'Dowd is back in the media harness: with a solo record released tomorrow – a faithful version of Ken Boothe's 1974 chart-topper 'Everything I Own' – and a new storyline to promote. As a soap obsessive and a media veteran, he knows it's time for a new plotline in his very own soap opera: Boy George, who was once fallen, is now reborn. It's a good script, and George delivers it with his old brio and manipulative charm. Every so often, however, there is a slippage which forcibly reminds you that this soap is for real, and there is a human cost to becoming myth.

George's withdrawal from drugs – by this time, he says, the heroin substitute methadone – coincided with Fleet Street harassment in its terminal phase: 'Christmas Day outside this house, my mother went to the door and said to this photographer: "Why are you here? It's Christmas Day – go home and have your dinner." He said: "*He* might die." As you can imagine, my mother freaked out: she thumped him in the face, just whammed him.'

George's house still bears the traces of being under siege. On the way in, there are two video cameras, barbed wire, high walls and a cluster of fans hovering on the cold doorstep. Inside, the large Victorian house is flamboyantly but comfortable decorated: behind the gold discs, framed George tabloid covers and props for the next day's video shoot, *EastEnders* is on the box and dinner is being cooked.

There is an air of suspension, of people taking one day at a time. George's parents, Gerry and Dinah – with whom he is living for the first time since he left home at 15 – have the polite but cautious weariness of people slowly awakening from a bad dream.

George O'Dowd looks pleased, no happier than when outlining his career relaunch and the battles he's having with his record company, Virgin. For somebody who is only now emerging from a prolonged bout of self-destruction and a particularly vicious trial by media, he is surprisingly healthy, if drawn; underneath his shaved blond hair, his eyes are prominent and clear, and his talk is animated and intelligent – a flow that, once begun, is difficult to control.

In the heightened reality of his household, George is the focus of all attention. Even in imperfect health, he is still a forceful, impressive character, very much the high-octane pop star. In a back room lies the evidence of his power to stir up fundamental emotions: the walls are lined with hundreds upon hundreds of Boy George dolls, handmade by fans from all over the world.

'When I became really successful, I felt, Why am I doing this? What is it all about? And there were drug problems within the band which were going on long before my problems started. I couldn't understand why a certain person was in a bad mood: I wasn't aware of drugs because at that time I was like the flying nun, you know, the puritanical sister.'

Culture Club's story has become mythical through the retelling. How the rebellious George, who as a teenager would hang around the houses of pop stars like David Bowie or Eno, formed the group that has sold well over 50 million records world-wide, that spear-headed the second English invasion of the American charts and that, incidentally, helped Richard Branson fund Virgin Atlantic. Culture Club were *the* success story of 1983, and George the biggest teenage pop idol of the decade, with an almost too clean image: a kind of living doll. In 1984, things started to go wrong.

'I found out that the dream became a nightmare. I became Bored George if you like: my life had become like a schedule. The worst problem wasn't success: it was very personal. I was having an affair with Jon Moss (Culture Club co-founder). In the beginning it was the most marvellous thing and suddenly things changed. It's the Abba syndrome, isn't it? How do you love someone who bangs drums out of time deliberately to annoy you on stage?'

Success, George says, made him 'envious and arrogant': this quickly rebounded on him. From August 1984, Fleet Street, which George had so brilliantly exploited in his rise to fame, turned on him with a non-stop series of knocking stories that brought out all the resentment about this 'gender bender' pop star.

Early in 1985, he suddenly took off to New York and immersed himself in its fast-lane lifestyle; soon after, he started taking heroin. 'People say to me all the time, How does someone intelligent like you get involved with drugs? It's like why do people get drunk and fall over? There's no answer to that.'

By early 1986, George's decline was so evident that it was an open secret in the music industry that there was a Fleet Street 'contract' out on him. In July

1985, chequebook journalism secured the first 'Boy George and Drugs' front page: the trickle became a flood in the screaming headlines that followed. This was a shocking morality story seemingly without end; it combined the Fleet Street obsessions of pop, drugs and deviant sex. And right in the middle of all this, George couldn't stop his compulsive flirtation with the press.

'I hate the tabloids, but I love them, because it is a big joke and it's a game. If you're really clever you can really manipulate that to your advantage. I remember, even at the worst time – it's not funny now but I was very high – I turned round to them and said I could run a marathon now, and I knew they would quote me on that. What am I going to do: cower in my house? I mean, they'd slashed the tyres on my car so I couldn't leave.'

Two events brought George up in his headlong flight to destruction: the deaths from drugs of close friends Michael Rudetski in August and Mark Golding in December 1986. This persuaded George to consider the consequences of his actions; he is still visible upset.

'When Mark went, that really was my cue to start pulling myself together. It was like, hang on, this isn't funny any more; this isn't a game. In a way it was all part of cause and effect: it was my doing, you know. I'm not saying I had a hand in it, but if you play with fire you burn. And it's a really sad thing that this has to be the cue to stop being the fool, but that's the way it is. We have to learn the hard way.'

After the much-publicized but abortive cure with Dr Meg Patterson – 'It was a real nightmare, because I was wired up to a machine and there was no aftercare. After that I didn't go back on heroin but substitutes, because my nerves couldn't take it' – George finally came off drugs last Christmas. He is now, he states, on a course of 'non-addictive tranquillizers. My treatment was basically cold turkey with a little bit of the edge taken off.' He is now living in a controlled environment – where certain people are 'contraband' – supported by his parents and the members of his family he is still close to.

But there are still pressures, both internal and external. George is living out the consequences of last year's behaviour in a series of court cases: when I spoke to him, he was fresh from a court appearance giving evidence against his alleged suppliers. There is still brother Kevin's 'conspiracy to supply' case to come, and a massive $44 million writ filed in New York by Michael Rudetski's parents. Despite his success, George still needs to work, being 'careless with money': his career in Japan is in jeopardy, while in America it is effectively over – he is banned from several cities and his visa has been revoked.

In Britain, his courageous re-entry into pop music and the media has been successful so far. His appearance at the BPI awards was a showstopper, and the advance reaction to his new single – in its reggae lilt a happy echo of his first success, 'Do You Really Want to Hurt Me?' – is encouraging. Still only 25, Boy

George is daily growing in health and confidence: despite his past disasters, he has retained his power as a performer and his pull on public sympathy.

Yet, as he returns to his media addiction and his restlessness drives him out to the nightclubs, how will he avoid a repetition of his compulsive behaviour? His answer is that these are early days yet. 'I'm still ill. It sounds like something out of *Coronation Street*, but I still get bad anxiety attacks. I still shake. I get up now and if I have a good day, I'm like "Wow, thank God for this, isn't it wonderful?" But there are days that I get up and think that I'm never going to be normal, that I'm never going to have a normal day again.'

Observer, 22 February 1987

Boys will be Boys

After an indecently long lay-off, pop has some new Bad Boys. The Beastie Boys' first visit to the UK since the extraordinary success of their debut LP *Licensed to Ill* (number one in the USA for six weeks) has been heralded by some vintage tabloid activity. With headlines like 'Brawling Beastie', 'Beastie Boys Go Bonkers' and, most titillatingly, 'Pop Idols Sneer at Dying Kids', we were on ground so familiar – and so reassuring – that it could have come out of one of those Comic Strip Enid Blyton parodies.

This is obviously as indicative of current tabloid diversions as it is of any real story: the Montreux Festival, where these outrages were apparently perpetrated, is a stage-managed, consensual binge where the press and TV collect to watch (and hype) a clutch of the current music industry products miming to their wares. In this context, the Beastie Boys were only doing what they're paid to do: live up to their deliberately childish name.

So consensual is this process that it's hardly worth remarking on, except that the Beastie Boys aren't just a hype but the most visible example of a process – both musical and cultural – that has thrown up some interesting contradictions. Their success marks the final overturn – some 10 years after it happened in the UK – of the post-hippie hegemony that has had a stranglehold on American pop; it has been accomplished by three upper-middle-class New Yorkers, all born between 1965 and 1967, who have integrated 1970s hard rock styles (whether punk or heavy metal) with the braggadocio of current black rap pushed to the point of total obnoxiousness. Their producer, Riki Rubin, is the most sought-after producer of the year, while their label, Def Jam, is the hottest on both sides of the Atlantic.

The opening track on *Licensed to Ill*, 'Rhymin & Stealin', samples the drumbeat from Led Zeppelin's 'When the Levee Breaks', integrates it into the song's structure and adds suitably piratical phrases (some lifted from the Sex Pistols) delivered in a hoarse whine. Elsewhere, 'She's Crafty' directly steals the guitar riff from Led Zeppelin's 'The Ocean'. This is postmodernist plunder in excelsis: both Rubin and the Beastie Boys have taken up the gauntlet thrown down by 'Wheels of Steel'. Using the new sampling technology, they've pushed this collaging to its logical, formal limit. Its influence has been immediate, not only on Def Jam acts like Public Enemy or Tashan, but on hip-hop and white pop.

This formal innovation is allied to a theoretical polemic. It's no accident that the most sensational Fleet Street fantasy homed in on the Beastie Boys' alleged mistreatment of sick children. The whole point of the Def Jam operation – particularly Rick Rubin's side of it – has been to create a new teen music, and that means driving a wedge into the current hegemony of pop as an adult, caring, sharing medium, epitomized by Live Aid, Farm Aid, the Sun City record et al. This means taking on not only the 'forbidden', gross style of heavy metal but also its attitudes. Given the almost psychotic self-assertion in modern hip-hop, this has resulted in the Beastie Boys and Public Enemy mixing music that is formally exciting with lyrics that are violent, misogynist and homophobic in the extreme. This is bad taste in spades, and doesn't end there: with new group Slayer, Def Jam has graduated to heavy metal, with explicit lyrics about murder in general and Auschwitz in particular.

The Beastie Boys (and most Def Jam products, bar the gentler, more affirmative Tashan) are at once liberating but oppressive: their studied lack of taste makes for excellent pop records but lousy attitudes. There is, of course, a wind-up here: if you get upset at the bad taste, or by people saying the unacceptable (as did Fleet Street in its stupid, moralistic way), then you fall right into the trap. But so what? This sort of double take is new in US pop but not here, where it's become degraded currency in the 10 years since punk. Now it just seems intellectually sloppy and socially ignorant – as lazy as the Beastie Boys pretending to be stupider than they actually are. Irony always evaporates on contact with the mass market and the Beastie Boys' white hip-hop has pandered to an all-American, all-male *Animal House*-type grossness at the same time as it's dragged American pop into the present.

It would be easy to dismiss the Beastie Boys (and Def Jam) were it not for the quality of their music, whose power gives one element of their polemic some weight. To attempt to steamroller pop into a purely 'socially useful', adult and caring role is to ignore its roots in base emotions and to impose a purpose on it that doesn't readily coexist with what pop is actually used for. The very refusal of base pleasures has been one of the historical failures of socially concerned pop. Yet Def Jam is no better, indeed far worse, because of an opportunity missed: its stylistic transformations are allied, not to a wish similarly to transform the world, but to a studied reinforcement of some of its more objectionable aspects. In this respect, despite its humour, Def Jam has swallowed the values of the political and cultural climate which it pretends to counter.

New Statesman, 22 May 1987

Roger Waters: *Radio K.A.O.S.*

'There's no interest here or in Europe and I'm too bloody long in the tooth to fight that kind of apathy,' says Roger Waters, about the reaction to his latest record, *Radio K.A.O.S.* and the stage show of which the album is a précis. At 43, he's terminally unfashionable. A poster on the wall of his Putney office, advertising his new single, quotes adjectives from a recent music press review: 'asexual', 'paranoic,' and 'unnecessary'. 'Arseholes,' he exclaims mildly, before getting on to the serious business of promoting his new LP.

'The dialectic involved in the possible access of technology and the attempts being made by people who are seeking to control it is central to the theme of *Radio K.A.O.S.* I see that as one of the most important things happening now. Information is power; access to information is power. The story points to the possibilities of modern technology and the advances that may be made in the next few years of releasing a group unconsciousness.'

Radio K.A.O.S. is no radical break: it's a concept album with a dense storyline about a radio DJ and a handicapped young Welshman who finds his power through a cordless phone. The music is not particularly to my taste but is livelier than usual and contains a moving finale, 'The Tide is Turning', where Waters's slightly lugubrious tones are lifted by the Pontardoulais choir. Perhaps as interesting as the music is the fact that Waters keeps thinking and is still motivated to make music. 'I like it,' he says simply.

Waters has been making records successfully for 20 years, ever since the success of 'Arnold Layne'. Pink Floyd have sold over 60 million LPs and, despite musical fashions, they're still selling: *Dark Side of the Moon* is still in the US Top 200 683 weeks after its release. They dominated pop music in the mid-1970s with a series of LPs that were more than just collections of songs but were carefully composed and sequenced, self-contained environments.

This was the approach to music first made popular by *Sgt Pepper* and it was there in Pink Floyd from the start. As Waters says: 'In 1967 we did a concert called "Games for May" in the Albert Hall. The whole idea was to perform a piece that was more than the band on stage singing a song, one which encompassed the audience.' It still sells because, although very British, it captures pop culture at its greatest period of world outreach: as the continued success of Dire Straits, the Beatles and Fleetwood Mac shows, a large bulk of consumers world-wide remain blissfully unaffected by any formal change in pop since the early 1970s – they want music that is rather neat, well played and technologically up to date.

This cultural hegemony made Pink Floyd prime targets of punk hostility: in early 1976 the Sex Pistols regularly appeared wearing a Pink Floyd T-shirt, crudely ripped, with 'I Hate' scrawled in biro. Pink Floyd seemed to emphasize a static, remote and spectacular approach to pop that was then obsolete; as it happens this view wasn't far from the truth. Waters now states that 1975's highpoint, *Wish You Were Here*, was the actual end: 'It should have been called *Wish We Were Here*; it was my way of describing that there was no longer a group. The Pink Floyd label was so successful and convenient and comfortable that it actually took far longer than it should have done for us to accept that it was over. Which was characterized by the fact that more and more I did everything. The more it was like that, the more uncomfortable it became, right up until *The Final Cut*.'

Waters is polite but cautious. He retains the authority and some of the autocracy of his rock star past – and present: Pink Floyd still live on in an acrimonious legal battle about the rights to use the brand name. He also has as the air of a man who has confronted his demons. *Radio K.A.O.S.*, if nothing else, has an optimism that is refreshing after the apocalyptic pessimism of the last few Pink Floyd records. He agrees: '*The Wall* was simply an exercise in therapeutic self-analysis. Maybe it was more useful to me than anyone else; that was certainly true of *The Final Cut*. It was a process I needed to go through. I got to the end when I finally accepted the loss of my father. I now think that, despite our short-term problems, we're living in exciting times.'

Roger Waters is an intelligent man in a peculiar situation. He's become an adult in an industry founded on adolescence; he's dealing, however awkwardly, with spiritual and philosophical concerns in a culture dominated by materialism; he's a wealthy man with a socialist background who, despite various public mistakes – Pink Floyd lost £1 million in a tax scheme masterminded by Andrew Warburg, jailed in June – is highly aware of his 'dilemma' and balances pragmatism with idealism. He still retains his politics – being particularly scathing about 'monetarism' and the Falklands affair – while not applying those politics formally or to his relations with the music industry. He's musically of his time, while being intellectually forward-looking.

Power contradictions, and I suspect that he deals with them by being deliberately ordinary and low-key. While this may result in unexciting, craftsman-like music and rather plain, unpoetic lyrics, the spectre of the other side is still apparent: Syd Barrett, the original Pink Floyd writer and singer, who, after a sequence of vertiginously brilliant songs, developed schizophrenia in the late 1960s. He is now a cult hero: a tribute album, *Beyond the Wildwood*, was released only last month. Waters is still emotionally involved: 'I dreamed about him only last night. It was in the open air and he was still gone, but I sat down and talked to him and it felt good. He was still saying things I wasn't in a position to understand, but I was supporting him and he was accepting it. We were both happy.'

Observer, 12 July 1987

Jimi Hendrix: Where You Going with That Gun in Your Hand?

One positive effect of pop's circular motion has been the many time capsules that have been opened up by record companies freshly aware of their back catalogue and the attractions of a non-teenage record-buying public. To regard this as purely nostalgic is to miss the point, and to collude with the commercial impulse which regards the past as dead time, revivable only in the zombie world of anniversary-pegged revivals.

The last month has seen two new issues of timeless pop material. The first – whose attraction should be self-evident – is RCA's *Elvis Presley: The Complete Sun Sessions*: all 16 released Memphis recordings from 1954–5, plus a couple of sides of outtakes and try-outs. It's fascinating to listen to the three musicians stumble through seven takes of 'I'm Left, You're Right, She's Gone', the first attempts at the sexy, unbelievably charged sound that created modern pop music. The second is a Polydor double-album compilation of three Jimi Hendrix concerts at San Francisco's Winterland in October 1968; taken together with a video-cassette and a record of D.A. Pennebaker's *Monterey Pop* footage, they provide a documentary of an artist who is still capable of surprises.

It's been hard to get to grips with Hendrix for quite a while. The necrophilia that ensued after he died in October 1970 was so nauseating – and so pervasive that it turned his life into some harbinger for the end of hippie. Turned into base metal, his guitar virtuosity became the staple for many imitators during the 1970s – one, Frank Marino, even claimed to be Hendrix's reincarnation – and unwittingly helped to create (along with his sexism) that most unattractive of styles, heavy rock. A sequence of shoddy issues and repackages during the 1970s reduced his reputation still further. Yet in contrast to the fixity of the Hendrix icon, what comes over most clearly from these 1967–8 live performances is how much everything was in flux and how much was up for grabs.

What is immediately apparent is Hendrix's playfulness and commitment to improvisation. The Monterey concert is one long, delighted, amphetamined squeal of electricity as Hendrix fused blues, soul and Dylanesque late mod pop with the new technology of public performance. Some of the material isn't up to much (at the beginning, Hendrix was a distinctly awkward lyricist) but its performance is so accelerated and gleeful as to be transcendent. Hendrix arrived at Monterey as a distinct hype – a black American guitarist who had played for years in jobbing soul bands transformed by English packaging and rendered exotic through the application of the latest Carnabetian finery – and left as a star. The performance ends with a moment of pure calculation, as Hendrix pours lighter fluid on his guitar and sets it ablaze, which on record translates into minutes of blissful feedback.

In comparison, the Winterland concerts are more considered: a song like 'Hey Joe' isn't taken at breakneck speed but stretched inside out with light-ning-quick changes of pace and emphasis. The flights of virtuosity and inspi-ration that are there at Monterey are now deployed more thoughtfully: at the heart of the record are the performances of a simple blues, 'Red House', and an improvisation called 'Tax Free', where Hendrix moves from the heavy, block chords of his pop/R&B style to a freer approach where feedback and dis-tortion are harnessed to describe a whole variety of late-twentieth-century noises, from city sirens to technologized warfare. At the same time as Hendrix's mastery becomes more apparent, the limitations of his group of English session musicians become more obvious: within six months, he had disbanded his white Experience and set off on a quest for self.

In the light of the extraordinary invention and self-re-creation of these performances, the subsequent events aren't so surprising. It was Hendrix's misfortune to appear to embody an age without boundaries. Today's perform-ers are protected and distanced, but Hendrix's life and music were indissolu-ble: at the same time as he was smashing boundaries, he was caught – as a black man packaged for a white audience at a time of increased black mili-tancy – in an appalling dilemma about race, while a distinctly exploitative management didn't help him to outgrow his Black Stud image. The conse-quence was psychic destruction.

The example of performers like Hendrix has helped to define the bound-aries that they so flagrantly ignored. Today, pop music is characterized by a calculation and a materialism that reflect its changed industrial situation, its place in the 'media squaring' produced by the new, multimedia conglomerate. Yet just as the late 1960s are beginning to come into focus, a decade after punk's critique, then the vaunting possibilities of Hendrix's work have become a touchstone for those attempting some kind of emotional truth, whether in Chrissie Hynde's cover of his song about the dissolution of false images, 'Roomful of Mirrors', or in Prince's explicit homage in his vertiginous black/white, male/female, rock/funk fusions. As he rips through his current hit, 'U Got the Look', the most protected of all contemporary pop stars pays tribute to Jimi Hendrix's continuing legacy.

New Statesman, 14 August 1987

John Lydon: The Bitter Truth

'It's wonderful that the most rebellious thing you can be doing at the moment is wearing a suit and tie and working in a bank. I mean, that's a real good piece of marketing, isn't it?'

At 31, John Lydon has lost none of his trademark sarcasm. Hardly anything or anybody escapes the lash of his tongue, not least the Britain he's returned to after several years of Santa Monica. 'It's a run-down dump, a one-party country. The opposition's all fragmented and pathetic. It's ridiculous.' But he's returned here. 'Of course. I *like* living here. It's pleasant. The police aren't after me at the moment and I seem to be in everybody's good books for the time being.' He pauses. 'Oh *dear*, does that mean I've become Cliff Richard?'

John Lydon is as much Albert Steptoe as he ever was Antichrist, his misanthropy partly genuine, partly ritual, partly parodic. 'I'm in a bad mood today,' he announces jauntily on arrival. 'I've got terrible hayfever: it makes my eyes itch and my face blotch.' Not that you'd notice a rash or two: his natural, indoor pallor is accentuated by a thatch of blondish hair that resembles a demented hearthrug. He sits down, rips open a beer, yawns and moves his first pawn: 'I'm just not in the mood for these *in-depth* revelations.'

His last media appearances here a year ago were marked by an extraordinary level of vituperation: some of the epithets directed his way included 'fat', 'smug', 'very ugly' and 'an oaf'. 'Go on', he invites. 'You can call me what you like. I've been called so many things now it's more power to my elbow.' When I don't rise to the bait, he adds mildly, 'Oh damn. I enjoy arguing with journalists.' And then, in Johnny Rotten voice, 'It's my favourite hobby.'

John Lydon has come to some arrangement with his past and his present, and with his private and public personae. 'History was my favourite subject at school,' he says, and it's his privilege (or millstone) to have made history.

If one person's physical or psychic state can appear to sum up an era, then in 1976 and 1977 it fell to Johnny Rotten's furious, scourging anger to sum up punk. The degree of media attention he received at that time would have been enough to sink most people. His work since 1978 has been a thorough-going attempt both to broaden out from this past and not to be imprisoned by it.

'I've definitely gone against the grain all my life. I always will: there isn't a neat little hole for me to fit into.' During the past nine years Lydon and Public Image Limited have released seven albums that, despite having a central core, have moved through a wide variety of styles: from the vast, cavernous sound that defined post-punk music on *Metal Box* to the mixture of experimental German and Near Eastern music on *Flowers of Romance*.

At times, the wish to confound expectations has seemed like a mania, most obviously on the disastrous 1983 tour that accomplished *Live in Tokyo* – 'That was a house-clearing exercise,' Lydon now says – but, after 1986's well-received *Album*, he seems to have found some stability in his life.

'The work is definitely the most important thing now,' he says. 'Celebrity is *not* the motivation. I love making records. All the musicians that I'm work-ing with now are strong characters and it feels permanent. I don't think I've ever felt permanence before.' Using the same group that toured with *Album* – John McGeogh and Lu Edmonds on guitars, Allan Dias on bass and Bruce Smith on drums – PIL have made a strong, melodic rock record, 'Happy?', which moulds Lydon's patented muezzin wall into a coherent whole. The sweep of songs like 'Seattle', 'Save Me' or 'Fat Chance Hotel' cloaks a distinct, and pleasurable, sense of unease.

As Lydon explains, 'It was fun making this record, a lot of fun. But the lyri-cal content is definitely lethal. It's full of twists and turns. You get sucked in by the melodies, then the bitter truth will slap you in the face.'

The lyrical content of 'Happy?' – which reflects an increasingly hazardous world – shows that Lydon is more concerned with life outside his four enclosed walls than he has been for a while. The anger is still there, but tem-pered by humour, irony and, behind all the backchat, some definite beliefs.

'The bitter truth is that life's bad because you make it that way. It can be quite difficult in England, but people moan about the difficulties and don't do anything about them. If something doesn't suit you, go change it. The last book I read, for instance, was by Anne Rice, called *Interview with a Vampire*. It pretends to be an interview with a real vampire, his lifestyle, how bored he is with everything. It just fits all walks of life and death: people shouldn't be bored because they don't have very much time left. You're bored only because you don't bother to do anything about it, because you don't use your brain.'

Observer, 13 September 1987

Sampling the Sounds: Copyright Complexities

Within the last couple of weeks, two events have pushed sampling into the spotlight. Pete Waterman – of production team Stock, Aitken and Waterman – has brought action against MARRS, whose recent chart-topping single 'Pump Up the Volume' contains, he alleges, a direct sound lift off his recent 'Roadblock' hit; MARRS have replied with a counter-suit alleging that Waterman's Sybil 'Red Ink Mix' takes sound from 'Pump Up the Volume'. In another legal flurry, English rappers the Justified Ancients of Mu Mu have been forced to destroy all remaining copies and master tapes of their *1987* album after Abba objected to their use of 'Dancing Queen'.

Records which include the sound of other records aren't new. As long ago as 1956, Buchanan and Goodman had a big US hit with 'The Flying Saucer', which strung together current rock 'n' roll hits with a commentary. Sued by 17 labels, the duo claimed that 'Saucer' was a burlesque – effectively a new work – and won. But records like this were novelties, or polemics: direct commentaries on existing hits. What is new about the current situation is the impact of new technology on musical styles and fashion.

Sampling takes advantage of the new digital and computer technology that has come on-line over the last five years. As Bill Drummond of the Justified Ancients of Mu Mu explains, 'Sampling is to digitally record a sound, any sound; by using a computer these sounds can be trimmed, EQ'd and pro-grammed before being committed to tape, which is the basis of making all "modern" records. The starting price for a sampler is as low as £400 – a Greengate – but rapidly escalates up into dizzy six figure sums.' Martin Young of MARRS takes pains to distinguish sampling from scratching: 'On "Pump Up the Volume" we did a lot of sampling on the bass, the piano and the drums, but the majority of the stuff on the top was scratched into a sound montage, which is a very different skill and takes a lot of technique.'

If, as Young suggests, sampling is the less skilful technological application of scratching techniques, it has certainly created far more problems in copy-right law. As John Brooks, legal and corporate affairs director at CBS UK says, 'The 1956 Copyright Act says that if you copy a *significant part* of any piece of music then that is a breach of copyright. This rule has an almost arbitrary interpretation in the courts because of its flexibility and the nature of music composition itself, as there are only so many notes to go around. But, as far as *sound* is concerned, technically even copying a sliver is an infringement although the courts will apply the *de minimus* rule and will consider other factors like recognition, skill, flair, loss of earnings, and so on.' In the Waterman v. MARRS case, for instance, Waterman argues that, although the excerpt from 'Roadblock' is a few seconds at most, it is highly recognizable (like James Brown's yelps) and central to the appeal of 'Pump Up the Volume'.

Sampling sets up ethical problems as well. Is there any moral difference in Eric B and Rakim sampling Bobby Byrd's 'I Know You Got Soul' and not crediting him, or the Justified Ancients of Mu Mu doing the same with Abba? It is arguable that because the Abba track is so familiar as to be effectively public domain, the new version is a comment on a historical artefact, and that will not deprive Abba of any revenue; as far as Bobby Byrd is concerned, the reverse might be true because of the obscurity of the track. Yet Bobby Byrd has received publicity and a career boost – his original version has now been re-released. And James Brown, who has been sampled in endless new tracks and who has complained forcibly (and successfully), has also benefited indirectly with a new compilation album now in the top 30.

In the music industry, ethics and aesthetics become blurred very quickly. As Charlie Gilbert of Oval Records says, 'Pop has always been an imitative, jackdaw medium.' He doesn't think there's much real difference between what's going on now and the wholesale plundering of R&B or Motown bass riffs that went on in the 1960s. The difficulty is how to judge whether sampling is used *creatively*, which, although potentially important in law, is very much a matter of personal interpretation. There are definite limits to sampling: it can never totally replace the sound of 'real' instruments and, as Martin Young says, it tends to be overused as 'a novelty or a gimmick'. At its best, sampling can produce a montage of sound that is an aural reproduction of the way in which we now are forced to process information at a faster rate, but at its worst it is used as pure effect without structure in a barrage of babble.

In the end, like most other things, these problems come down to law. Bill Drummond, for instance, thinks his artistic case crumbles in the face of his indisputable copyright breach. The immediate future is uncertain: John Brooks doesn't envisage any new legislation but thinks the common law can deal with it in a test case such as Pete Waterman seems to be angling for. But this may well reduce the issue to the even simpler question of who will control resources, in an industry already realigning its power base around copyright rather than production. The situation is addressed by the American group Culturcide, who, on 'Tacky Souvenir of Pre-Revolutionary America', record a vicious polemic over pirated tracks by leading rights owners like Michael Jackson and Paul McCartney. 'Plagiarism is necessary. Progress implies it,' they state, occcupying the opposite pole to an industry attempting, as ever, to catch the lightning and control the impact of technological change. *With thanks to Simon Frith*

Observer, 18 October 1897

All This and Less: Night-time TV

It's Saturday night and it's one a.m. You turn on ITV and you see a snappy title sequence that says *Night Network*, followed by a three-minute barrage of videos, captions, programme menus, etc., ending with the line 'All This and More'. Dizzy, you come to earth with a bump when you see the two presenters – cloned chirpy trendies – flanked by a pair of mildly puzzled pop stars. It's time for *Video View* – a sort of pop video *Juke Box Jury*. After a couple of INXS and AC/DC videos, and three or four rapid ad breaks, vertigo begins to set in. Where does the advertising stop? Where does the programming start?

Like TV-AM, *Night Network* spices up what is essentially a cheap, studio-based programme with cheap, bought-in programming. Both use *Batman*; *Night Network* adds cut-up gobbets of *Captain Scarlet*, a Gerry Anderson pup-pet drama from the early 1970s which falls the wrong side of camp. A similar Gerry Anderson puppet character is now part of an extended Tennent's lager ad campaign, just as an adaptation of the *Captain Scarlet* logo forms part of the *Night Network* titles. Another famous mid-1960s series, *The Prisoner*, appears cut up into an LBC ad shown during *Night Network*'s one to four a.m. run.

A further staple is the pop video, used throughout *Night Network* and other 'youth shows', like C4/Border's new *APB* and Thames's execrable magazine, *01 for London*. The pop video's frantic intercutting – usually used to disguise a lack of ideas – has had its formal influence: *Night Network* has taken much of its format from *Network 7*, itself developed out of *The Chart Show* and *Max Headroom*. The pop video is not programming but an advert: its influence can be seen in the ads which frame these shows, some of which – if they are music retail ads – actually show the very videos that have just been featured.

Watching most of the current youth TV shows is like watching a closed loop, as empty as the adverts they so resemble. There are several reasons for this. The idea of youth television is intimately related to television's expan-sion since 1982: first through C4's ground-breaking 'Youth Programming' department, responsible for *The Tube* and *Whatever You Want*, among many others. For a while, 'youth' meant experimental with interesting results. With the advent of TV-am and BBC *Breakfast Time* early in 1983, a new type of for-mat had to be developed to fill the massively increased visiontime of three-hour, rolling living shows. After the spectacular failure of 'the mission to explain', a suitable format was eventually found in the only other sector to engage with this sprawl: Saturday morning children's TV – where the plugs are indistinguishable from the programming.

The coming on-line of the pop video was important to this new television economy – useful to attract a youth market as, for the first time, TV stations began to think of market research – and *cheap*. Until 1985, TV companies had only to make minimal payments for videos, and even now they are cost-effective, as their comparative lavishness frequently disguises the tackiness of a cold TV studio. Already used on kids' shows, they spread throughout TV-am, cable channels like 'Music Box' and, eventually, through C4's youth sector, as the admittedly ailing *Tube* format was replaced by *The Chart Show*. This spread also reflects the increasing connection between the music and TV industries in the new TV economy.

The central problem of TV's expansion has been that revenue from advertising or licence fees hasn't expanded at the same rate. This means the same amount of jam spread over more processed white bread and a consequent lapse in editorial standards. As elsewhere in the media, documentaries and investigations aren't cost effective, aren't *efficient*.

This has resulted in the ubiquity of the consumer-guide frame throughout Fleet Street and TV. In television terms, it has also meant more market research (almost unknown in the early 1980s) and more specific targeting. For instance, out of C4's excellent *Solid Soul*, successfully aimed at hard-core soul fans, came the idea of *The Last Resort*, the programme whose success lies in catching and defining the market that advertisers now want.

Here is the reason for all that late 1960s/early 1970s kitsch now infesting TV. As Jonathan Ross has so winningly demonstrated, youth TV is now not aimed at the classic teenage 15–24 age group, but at 25–40-year-olds – the *real* consumers. *The Last Resort* is a well-made entertainment but its influence – like that of *Network 7* – has been baleful, as producers frantically look over each other's shoulders. A camp, ironic sense of TV and pop history have combined to create a self-referential, postmodern form where superficial references and nostalgia-triggers disguise its near total lack of content. This is not TV for youth – as true teenagerdom is no longer an automatic entry into consumerism – but the infantilization of a generation now coming into power.

This sort of TV has more to do with a failure to get to grips with the changes in the overall television economy than with any audience's specific desires. As the current events at TV-am demonstrate, deregulation goes hand in hand with a Tory free-market philosophy, where television loses its public service function and editorial content in the search for profit. Yet even so, there's no reason why the result should be as moronic as *Night Network*: C4's recent *After Dark* is a good example of intelligent, open-ended programming. Any future attempt at youth TV should recognize the diversity of the 15–24s, or indeed any age group – and that soon there will be the first post-consumer generation. Indeed, this extreme emphasis on consumption is the historical mark of the post-war baby-boomers and it is one cause of that generation's current failure to resist free-market pressures.

New Statesman, 29 January 1988

Leave My History Alone: Pop in Ads

The camera fastens on to a young man watching television. Short-haired, he is dressed generic trendy; the interior has that heightened 1950s/1980s look common to many ads and to films like *Blue Velvet*.

Against a Stockhausen backdrop of mediawash, the young man gets up to perform his significant gestures. We see him, lovingly shot with all the correct close-ups, making a cuppa – one tea-bag, milk on top – before returning to his 'avant-garde' environment. So, what's this selling? Milk? Stockhausen? Tea-bags? All is explained when the caption comes up for an ad agency: how to make your TV ads different from everyone else's, the Ogilvy and Mather way.

Here is the motion of commodity culture – as ads advertise ads in a seamless, apparently unbreakable circle. Music has an important part to play in this perfervid enclave: as money becomes not a means but an end in itself, in a closed, corporate system, music must follow suit. Thus the lush, seductive tones of modern pop or dance music – best heard in one of the many Megastores – is only slightly different from the trance-inducing Muzak traditionally employed in supermarkets, while the nostalgic trigger of 'classic pop' is a valuable hook for television advertisers attempting to snag your narcotized attention.

Television ads have long featured snatches of pop tunes but what was once a stream has become a flood. The Levi's ads with their careful 'period' detail and *original* songs define the smart end of the market, but they are unusual; much more common are ads which take a hit from the folk-memory bank and re-record it with new words to suit the product concerned. Samples from several days' random viewing include new versions of 'Yakety Yak' (Green Shield Stamps/Mobil), 'Get a Job' (Brook Street), 'Apache' (Tango), 'Mr Soft' (Trebor), 'Mr Moonlight' (Terry's), 'Whatever You Want' (Bella), 'Thank U Very Much' (Cadbury's Roses), 'Twist and Shout' as per the Beatles (Ross Foods), 'The Power of Love' (Durex), 'Mellow Yellow' (Yello Mello) and, most irritatingly, 'Da Doo Ron Ron' (Heinz Ketchup), 'Israelites' (Kraft Vitalite) and 'Tired of Waiting' (Nivea Shampoo).

This is not accidental. The music industry is at present undergoing a major, unseen shift from being a producer of *things* – i.e. records, CDs, cassettes – to being an exploiter of rights that exist on paper: principally performance and publishing copyrights. Although there are as yet no statistics, it seems likely that revenue from this 'secondary' income is now equal to royalties – i.e. revenue minus products and distribution costs – from the sale of goods. The result of his reification, which is also due to the music industry's successful integration into the multinational media economy, can be seen in the number of film sound-tracks, video-cassettes and pop TV programmes currently available.

And, of course, TV adverts featuring music. Alerted to the revenue possibilities – anything from £15,000 for a local ad to £250,000 for a multinational biggie, plus the spin-off potential for releasing – record company divisions of giants like Polygram now use their Music Publishing sections to tout their copyright material to agencies, following the work of specialist song-finders like Song Seekers. Under copyright law, the copyright holder can do what he/she wants with the song, whether or not the writer agrees. Despite well-publicized incidents like the Beatles' US case about the use of 'Revolution' in a Nike ad, in practice most writers are happy to receive fresh payment and publicity for a long-dormant asset.

From a business point of view, this works well and, since that is currently the sole justification for anything, very few people have questioned what violence is being done to pop history. It came home to me while watching 'Tired of Waiting for You' – part of *my* past – but, as advertising begins to plunder all pop time, it will rewrite most people's past. I deeply resent seeing songs like 'Israelites' rewritten to sell a paste, or 'Da Roo Ron Ron' reinforcing class stereotypes – the black train guard, the 'Ya' braying Yuppies with their Volvo estate – because those songs were originally part of an enfranchising social process. Their very success gave a voice and a face to the dispossessed: whether, in these cases, West Indians, white working-class punks or pure mavericks. Today, that success is turned against them as they are re-employed to prop up an oligarchic, outmoded but highly powerful value system.

The spread of 'classic' pop ads reinforces a current opinion – held on both left and right – that that's all pop ever was: marketing, consumption, a product. It's certainly true that pop's assimilation into the colonizing thrust of the multinational media industry marked the end of consumption as a democratic project, and that most pop music today is either explicitly or implicitly selling something. Yet to insist that this is pop's sole function is to surrender to the circle: to enter into a common dialogue of hopelessness. It is a feature of the politics of commodity capitalism and its culture that, despite its extreme aridity, it brooks no other view, no other possibilities. Yet the tighter the circle the more fragile it becomes. When the bubble of postmodernism is pricked, the voices on the margins will be there, waiting to seize their moment.

New Statesman, 25 March 1988

This is England: History Rewritten

History is being rewritten fast these days. It is a feature of our interconnecting media economy that loose ends cannot be left alone, that those moments – peculiar to pop – when some sort of emotional or perceptual breakthrough is made must be cauterized lest they spread their virus of freedom – or even truth. And this rewriting process is all the more obnoxious not because it is a conspiracy but because it is a climate.

Take the current Pepe jeans ad: to the usual meaningful boy-meets-girl-yawn stream of images is set Johnny Marr's shimmering, instantly recognizable instrumental track to the Smiths' celebrated 'How Soon is Now'. As it runs, you start waiting for Morrissey to come in, but of course, he doesn't. The storyline of 'How Soon is Now' – in quest for love, man goes to (gay?) club, fails to connect, goes home alone – has a lyric devastating and disruptive in its awkward, hurt simplicity. 'Shyness that is criminally vulgar' has no place in the idealized, heterosexual, white dreamworld of jeans ads.

There are other people to deal with besides the Smiths. Punk is still a slight, nagging problem – you don't see it in ads yet; it casts a dark, albeit dim shadow over the rejoicings of most postmodern popsters and in the memory of some music industry executives who can still recall an era when pop stars didn't just lie back and think of England, or didn't, indeed, sell many records. 1986's 10-year anniversary was a bit of a damp squib – and, in the nostalgia stakes was easily eclipsed by *Sgt Pepper*. So it's time to exploit the back catalogue. Since this is well nigh impossible to do with the Sex Pistols, as their affairs are still in such a mess, it'll have to be the next best thing.

The last month has seen a 'major re-evaluation' of the Clash. CBS UK has released a double album called *The Story of the Clash* – 28 tracks from 1977 to 1982 – and a four-track CD single led by their version of 'I Fought the Law'. The album made the top 10; the single the 20. Singer Joe Strummer, returned to England to promote his soundtrack album for Alex Cox's *Walker*, found himself faced by a barrage of questions about the group he'd disbanded in 1985. Initially reluctant, he saw the way the wind was blowing and quite understandably softened. For the same critics that had mercilessly panned the last few Clash albums were now granting him readmittance.

But this return to the pantheon has a price. The first, obvious point has to do with the way in which the repackaging and choice of material are slanted to reflect the present-day rather than the period whence it originated. And the single – 'I Fought the Law (and the Law Won)' – was a non-event. Hurtling freedom songs like 'Complete Control' or 'White Riot' might have been more appropriate but they hardly have the right message for the times; 'I

Fought the Law''s defeatism certainly does.

The album itself similarly ducks the issue, its four sides opening with two devoted to the post-1979 material, from the period when the Clash had become all that they could be by that time – a superior, American rock 'n' roll band, given an unusual empathy through Joe Strummer's furious underdog vocals. The nub of the matter is buried on side three, where there is a still stunning segue of '(White Man) In Hammersmith Palais', 'London's Burning', 'Janie Jones', 'Tommy Gun', 'Complete Control', etc., yet the original impact and meaning of these 1977–8 songs are deflected by their context. The side contains an extract from the March 1977 interview that changes them from being art students with Jackson Pollock spattered clothes into macho, working-class street fighters: a contradictory state from which the step to becoming rock rebels (as the sleeve shows them) was a short one.

It is as authentic rock rebels, the very quality for which they were once reviled, that the Clash have been celebrated by the dreadful rock critical consensus of all those Q boys. This ignores the real story. *The Story of the Clash* swallows the mythology wholesale and regurgitates it; here, there are no mistakes, no songs that don't quite fit. There is no account here of those vertiginous few months when nobody knew whether punk was going to work, when concerts might end in fights between the group and audience; no account of those early Clash songs junked because they didn't fit the 'political' image. The past is presented as a *fait accompli* rather than as a struggle.

The gaping hole in this record is the absence of any material from the last Clash album, 1985's *Cut the Crap*. In 1982, the Clash had a US top 10 hit with 'Rock the Casbah' – an event which precipitated a crisis within the group. Co-writer Mike Jones left (to form BAD) and Strummer began writing with manager Bernie Rhodes. The result was a loose, funny, bitter record that combined the by-then discredited punk guitar sound with hip-hop drumbeats, noises, found conversations. At least half the record was brilliant: one song in particular encapsulated both the failure of punk's utopian ambitions and their continued relevance. 'This is England' posited the gap between the optimism of 'London's Burning' or 'Complete Control' – written in the dying days of the post-war consensus – and the bleak oppression that followed.

Like the Sex Pistols' terrifying 'Belsen was a Gas', the pain and reality of 'This is England' renders it unfit for recuperation at present. There is no place for either song as punk is reduced from a moment that would transform the world into simple 'good old rock 'n' roll'.

New Statesman, 15 April 1988

What's So New About the New Man?

The camera pans inside a warehouse, funky but expensive: in these days of crammed two-bedroom flats, space is at a premium and an open-plan, ex-industrial environment is highly desirable – hence its heavily featured appearance in Euro MTV and *Streets of Fire*. Time for some significant moments. We see a scantily dressed young man – very good-looking but *not gay*.

His gestures are carefully, deliberately clumsy: he is hunting for his breakfast. There's no woman around to get it for him. As he grimaces at the cat, the only other occupant of this space, he reaches for … his reliable flexible friend. He slips on some clothes (American Classics) and saunters to the cashpoint through a post-industrial, post-vandalism street scene. The final long shot shows him at the window of his wharf with his cat, gloriously alone.

Our Wapping wonder, reading the newspaper that is produced just down the road, is a glaring example of the way in which advertising and marketers are attempting to get to grips with the young male market. He's glaring because he's not quite right: you end up by remembering the tableau, not the product, and by suspecting that the ad is not so much targeting the requisite market as displaying an internal adland ideal.

Yet he's nearly there: the image of the softer, almost 'feminized' male who does the shopping and dares to consume has become an adland staple – often called the 'New Man'. This creature does have some counterpart in reality: current advertising uses sophisticated social research to fuel its flights into fantasy. But it's important to remember that, despite its formal innovation, advertising is not concerned with creating new social trends but with reflecting existing conditions at a safe distance and in a palatable manner. It is by nature conservative, although not as conservative as its clients.

It is axiomatic that advertising exists in order to sell things. These days, the sort of products usually being sold are not necessary but convenient or highly specific, tied in not with need but with *lifestyle*. A late 1960s term from the time when increased consumer spending really began to bite, 'lifestyle' has become a UK advertising reality since the early 1980s, tied in with the new market research techniques like group discussions.

The key idea, often called 'market segmentation', is to do with specific targeting. 'It's recognizing that many people are sufficiently well off and trained as consumers,' says Peter York of market planners SRU. 'Many products now have what is called added value. It says *I* will buy this because it has something for *me*.' This idea – that you are what you buy – is of paramount importance: current in America since the early 1970s, it has spread to the UK only in the past few years.

The greater sophistication of market research and advertising techniques has gone hand in hand with the search for new markets. In this, the young/young middle-aged male market has been one of the last great unknowns. The history of teenage marketing originated in the immediate post-war period with successful campaigns and products aimed at young women. The young male market has been notoriously difficult.

Indeed, the whole business of selling things to men has been fraught with difficulty. There is not only the problem of getting men to spend money on themselves – not thought until recently to be a masculine activity – and training them as consumers, but the problems of targeting and representation. 'Advertisers have been looking to grab that sector for a long time,' says Lucy Purdy of RBL (Research Bureau Ltd). 'They know that the fashionable young man is out there, but they don't know how to address him. You can't address all men; there has been no one image specific to all men.'

There's no problem with advertising cars, car-phones, petrol and that sort of executive stuff. From the famous 'National' ads of the late 1950s on, all you have needed to display are the right social cues of the time; status, power, speed, beautiful women or whatever. Similarly, advertising cigarettes and drink has been straightforward, complicated only by the guidelines about cigarettes and the current lager wars. These are things that men *do*. It's when you get on to sex roles, narcissism and personal growth – the staples of the new male consumption – that the problems start.

It's all right when a man says, 'Look at what I have accumulated.' Far more difficult when a man says, 'Look at me!' The mere act of display is problematic and potentially alienating to the male viewer. As Andy Medhurst writes in *Ten-8*: 'To put a body on display is usually to render it passive, to make it available for the bearer of the gaze, but that runs the risk of conflicting with our ideologies of 'masculinity', of 'feminizing' the body in question.'

Within the terms of UK advertising in the 1950s, representation was hardly an issue. In contrast to the detail already to be found in America, British ads were very primitive, simply displaying the product or relying on an assumed knowledge of traditional class codes. It was very much a gentleman's wardrobe: ads for shirts would display an image of Regent Street behind the product and everybody would know what this meant. A static class structure was matched by a static view of consumption: you bought things to last.

Images of men were generic – often conveyed by an idealized drawing – and if there was any detail, it was to do with the war: there was little inherent sex or glamour in being a man. As the American model of consumption – a high turnover operation with a design aesthetic as a key selling point – came over to Britain from the mid-1950s on, advertising to men slowly changed (although not as quickly as advertising to women). It began to take on the American hallmarks of travel, power, money and sex appeal.

The central idea here was cool. There was no one unifying image for the mass male market in the 1950s: most people didn't think like that and if they did, the male icons were either too weedy (Dirk Bogarde), too tweedy (film stars like Kenneth More) or too rough (Elvis and the consequent rock 'n' roll stars). James Dean hadn't yet made the transition to youth cultural archetype. A sequence of extremely successful films translated youth-oriented mod styles and ideologies into an acceptable, older masculine image. Starting with *Dr No* (1962), the Sean Connery James Bond films mixed in American ad values – travel, money, power and sex – with a new style of technology consumption and an age-old patriotism.

This new cool became an instant adland hit: it marked an increasing democratization of consumption by apparently eradicating old class boundaries. A big mid-1960s fantasy that had originated in the teenage market was the idea of classlessness: the new leisure wear would be modelled by carefully rough-looking models to signify a *déclassé* butchness. Just in case, these ads added an adoring female or two for the sake of clarity – as they continue to today, if ironically.

Such 1960s ads reflect a confidence in a unity of taste that had disappeared by the end of the decade, as advertising unwillingly mirrored the social and gender changes of the period. Ads from the early 1970s reflect a deep social fragmentation rather than a marked segmentation: main themes reflect a post-hippie idea of 'escape' mixed in with a bit of ethnicity and you see the beginnings of nostalgia creeping into the most basic adverts. Interiors that in the 1960s had been sharp and modern were now countrified and muted: the man was still pictured with woman and children – recognizing his perceived economic dominance. *He* paid for the G-Plan, even if he didn't actually choose it.

Advertising is most effective when there is a consensus of taste and values which can be translated into a visual or verbal shorthand; the 1970s crisis of representation, which reflected a wider crisis of values, resulted at the end of the decade in an absence of all but the most basic male images. Just as in the 1950s, ads often displayed nothing but the product, or else, like Brut or Denim, they were butch in the most crass way possible.

Several factors have contributed to the change that has occurred in male ad imagery. As Lucy Purdy explains: 'There is the change in men's lifestyles and the change in the way in which advertisers now target their advertising, both of which date from the early 1980s. There is now a much more sophisticated planning approach within agencies: people employed as researchers have a much more sophisticated methodological tool – the group discussion – and they are much more involved creatively in ads than before. There is much more understanding about how consumers behave and feel.'

Purdy's research with Mantrack, a report produced by RBL since 1985 'as a targeting tool for clients who sell products or services to men', has turned up influential findings about changes in men's lifestyles: 'It is fair to say that traditional male/female roles *have* changed a little. Men are more likely to be on their own, are more likely to buy products that appeal to themselves. There is also a range of domestic tasks, like shopping, which it is socially acceptable for men to do.'

This research has coincided with, and fuelled, a more concerted attack on the problematic male market. The advertisers' task has been made easier both by the return to a social consensus (of the well-off) under the Conservatives – epitomized by the media term 'Yuppie' that came in here after 1985 – and the massive input of fresh male imagery provided by the style press.

It is now possible to conceive of a consensus, shorthand male image in a way that was not possible even five years ago: you can see these images – short-haired, polo-necked, be-501'ed, derived straight from Soho 1985 – all over current lager ads. These images from Style Culture, first disseminated through the pages of magazines such as *i-D* and *The Face*, are an acceptable solution to the problems of male representation.

The image is up-to-date yet aspirant, recognizably male yet admitting a certain vulnerability, still able to be interested, obsessed even, with clothes, male toiletries and new gadgets. They are an index of how the concerns of Style Culture have helped to fuel a new type of consumption and have codified a fresh marketplace: the 19–25-year-old male.

But how new is our New Man? Beneath the compulsively attractive surface, the same old angst gnaws. There is still the same old problem involved in putting the male body on display, now exacerbated by renewed social sanctions on homosexuality. Despite the fact that the gay milieu informs male representation at every level, it is now doubly necessary to disavow this. One solution is the use of women as props. But, even ironicized, this does not accord with social or commercial expectations: men are being asked to consume in areas that are traditionally 'female'.

'You can't assume that the New Man is a feminist man,' says Lucy Purdy, 'he's just more narcissistic.' The dominant New Man advert is remarkable indeed for its *absence* of women: women are either a threat (at their most extreme, personifying the AIDS virus in *Fatal Attraction*) or simply irrelevant to the new, self-enclosing world of male pressure and vanity. Far from marking a real change in gender roles, the New Man is yet another example of masculinity's privileged status in our society – the same old wolf in designer clothing.

Thanks to Kathy Myers.

Style People Out of Date

One current lager ad – no doubt helping to fuel the current spate of 'lager violence' so deplored by our rulers – sums up the dominant vision of pop culture. Very grainily shot, it carries the sales pitch within a storyline: young busker with a saxophone, Levi's and hair gel attempts to get close to his – black, balding, bit unstylish – sax hero.

Our absolute beginner has to pay his dues but it's *all right*: his hero is at hand. They interact: amid the trappings of male buddy buddy ritual – sales point: a can thrown lightly from the fridge; lifestyle point: 'hip', jive gestures – the younger man gets soul. The final, freeze-framed image is of him on stage. He is punching his full arm in the archetypal gesture of triumph: all you've got to do is win.

Symbols like the saxophone, hair gel, the grainy texture are the lingua franca of the new 'new youth': they show the impact of magazines like *The Face* and *i-D* at their moment of furthest outreach. You can only escape this world-view with difficulty: it's all over both Fleet Street and the High Street, all over advertising on the new breed of television programmes like *Network 7* or *Night Network*. The titles give them away: you're supposed to be part of the network, you're meant to *belong*.

How did it happen that the current pinnacle of pop culture should be found in adverts? A Radio 4 documentary broadcast this Thursday attempts to answer this and other questions. Called *The Stylographers*, it attempts to trace how the world-view of 'style' – a word which nobody likes but which has stuck – became a powerful motor of the integrated circuit of media industries in the late 1980s.

Both producer Peter Everett and presenter Nigel Fountain have considerable expertise in this area: Everett produced the 1985 BBC radio documentary about the history of post-war pop culture, *You'll Never be 16 Again*, and Fountain has written a recently published book, *Underground*, about the alternative press of the late 1960s. Relying on the voices of editors and journalists, the programme works well as a social history of a pop and publishing moment which has passed its sell-by date.

This is reflected by the proportion of airtime given to *The Face* and *i-D* magazines. It may be difficult to remember a time before Soho was swinging, but in 1980, when *The Face* and *i-D* appeared, there were very few glossy general-interest magazines. If Nick Logan founded *The Face* on a music base, Terry Jones founded *i-D* on a fashion: through instinct and talent, they both captured a new, socialized pop attitude that seemed refreshing after the extravagant alienation of punk.

Central to this was the idea of Club Culture: of group social activity and of black music used as a metaphor for cool, community and that indefinable, 'soul'. *i-D* had the brilliant idea of snapping the glitterati of this new mood as they stepped blinking into the daylight: many of these 'passers-by' – Boy George to name one – became pop stars, models, the 1980s equivalent of Mod 'faces'. And after a few issues, *The Face* found its cause: the new clubland life and attitude epitomized by groups like Spandau Ballet.

Although arresting at first, much of this message was quite simple: looking good and being aspirant were highly prized qualities. 'It was pre-Yuppie,' says former *Face* features editor Lesley White. 'That dreadful word.'

'It was a means for young, mainly middle-class people living in London to establish a voice,' says *Face* contributor Marek Kohn in the programme.

Areas *The Face* opened up included examinations of the way things looked, the style media themselves, the revolving past of youth culture.

This approach was both a creative and a commercial success – particularly within the mainstream media. '*The Face* was a useful passbook to the young idea for 40-year-old advertising executives,' says Marek Kohn. *The Face, i-D* and *Blitz* became more than magazines: they became, as *i-D* sort of ironically noted, 'style bibles', models for emulation rather than communication, as the celebrants of the clubland myth became the new aristocracy.

The media that these magazines were serving were on the point of expanding almost exponentially: a process which is still continuing in Fleet Street. An important part of this process was the increasing interaction of parts of the media that had previously been separate: the film, TV, pop and publishing industries become more enmeshed both structurally and creatively. The perfect expression of this was the pop video: eclectic, cut-up, postmodern, it was made by film-makers, funded by record companies and broadcast by TV.

Despite having considerable creative input, the pop video is promotion: it relies on the tired old arrogance of fashion photography. At its cross-over peak in 1984, the pop video blurred advertising and editorial: this blurring can now be seen all over baby-boomer TV shows like *Night Network* or *01 for London*, which exist to fill the space opened up by the deregulation which has resulted in many extra hours of broadcast-time. And indeed, despite, as Kohn says, 'anti-Thatcher, left-of-centre editorial values', *The Face* and *i-D* were classic models of modern entrepreneurialism.

If early copies of *The Face* suggested that people might see things in a new way, the lack of any coherent politics eventually meant that the magazine failed to resist wider pressures. The recent hundredth issue makes this clear: in between articles extolling the virtues of Mrs Thatcher or rerunning the tired old club culture myth just one more time is a running editorial which rewrites both the history of the magazine and the history of the 1980s. On the front, all 100 *Face* covers are run small: they look like an ad for Beck's beer.

'Youth' culture, if you can still call it that, no longer holds to its post-war democratic promise but is now to be made and consumed by the privileged. Much of this medium is now run on envy and exclusion, qualities which all too well reflect this Government's social policy. You can see the result in the new media district, Soho, as academic and author Judith Williamson says in *The Stylographers*:

> It's the supercilious way that people look, in places like the Soho Brasserie: they look like that in magazine photos and they look like that in the street. They don't look as though they could be open to any experience. It's as though the whole world is a market.

Style culture's current ubiquity is nothing less than a massive failure of intellectual nerve. Its insistence on *belonging*, derived from the club culture which forms the basis for Robert Elms's flimsy novel *Searching for the Crack*, is now restrictively defined to the ideal now seen in ads. This excludes the dissenting voices – whether the disenfranchised, the alienated or those simply dissatisfied with materialism – that will give the future shape. People will look back at this time and say: 'They fiddled while the world was burning. Then they put the burning world into car adverts. Couldn't they see what was really going on?'

Observer, 25 September 1988

There is the theory of the Moebius
A twist in the fabric of space
Where time becomes a loop

Orbital, "The Moebius", 1991

Speed

1988–95

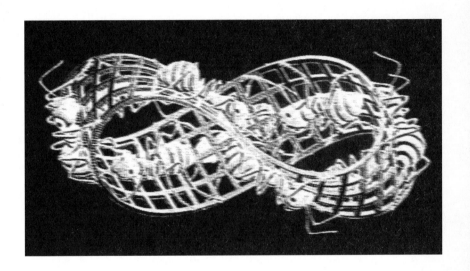

Let's Play House

A 1986 12-inch single perfectly mapped out the parameters and possibilities of House. One side began with a 'dub mix' of 'Donnie' by the It; built around a skeletal drum machine pattern, it was filled out over eight minutes by minor variations in percussion and synthesizer textures. The effect of the brief vocal line – almost doo-wop in its reliance not on meaning but on vocal tics built into pure sound – was startling.

Fingers Inc: 'Mystery of Love', early 1986

The other side, 'Mystery of Love' by Fingers Inc, was more of a song, although constructed in the same, totally synthetic manner. Its highly in-authentic, almost tinny sound was set off against quiet gospel vocals that repeated the title phrase. And then, after a tiny boogie piano-riff, the anony-mous vocalists started to *burn*, without histrionics, as they testified to their vision of sexual or religious transcendence. For, if rap is about frantically asserting the ego to master a hostile world, then House is about surrendering control to attain a fragile, fleeting state of grace.

These two tracks appear on *The Westside House of Hits*, an ambitious 14-album boxed set which aims to give a definitive history of House from 1983 on. This summer's media furore about Acid House has tended to obscure House's trajectory so far: from tiny Chicago clubs in the early 1980s to its cur-rent state as pop's lingua franca.

When it began in Chicago gay clubs, House fused the sexual insistence of late 1970s gay disco with the synthetic tones of European electronic pop like Kraftwerk. Original gay/electronic disco songs like Donna Summer's 'I Feel Love' worked as pick-up preludes: there, desire was out front. The new 1980s dance records concentrated on melodrama, or, post-AIDS, abstracted this sexuality into the mystery of love.

House's origins also lie in Philadelphia disco or the gospel androgyny of Sylvester, whose 'You Make Me Feel Mighty Real' caught the crossover between the sexual and the spiritual impulse.

All these ingredients contribute towards House's current ubiquity: because of both what is implicit in its history and the way in which it translates contemporary sexual tensions into an irresistible beat.

Like much black dance music, it is produced in a dog-eat-dog fashion: the lack of any House star system is not deliberate but a product of these chaotic conditions. But this effective loss of self is carried through in songs like 'Distant Planet', whose distant idealism has struck a chord in English pop audiences.

For House has it both ways. You can lose yourself in its textures. Or you can listen to the lyrics – and this is where House parts company with the rigid codes of London's Style Culture. Songs like Sterling Void's 'It's All Right' actually dare to look out into the world, mentioning politics, oppression, the positive power of music, even the future – all those things that the apologists for the New Right would like us to forget.

Observer, 23 October 1988

Brand New Dusty

If every interview is a performance, then Dusty Springfield's entrance is a bravura curtain-raiser. The theatre is a duplex hotel suite with a spiral staircase linking the two floors. Upon arrival you are greeted by a friendly but protective assistant who remains present. Springfield herself is upstairs, a disembodied voice offering greetings. When all is ready, she comes down the stairs with a flourish, all shoulders and hair, in a rush of words. Her initial impact is confident and glamorous; you're instantly won over.

'I always take the scenic route toward anything,' she says a little later. By then you have noticed the dark brown eyes beneath her fringe, humour alternating with doubt cloaked by her dark eyeshadow. There is the sense that the wrong word – 'tabloid', for example – will send her out of the room. Despite the fact that it is part of her persona, Dusty Springfield's volatility is not a ploy, but something with which she visibly grapples, her emotions appearing to switch with the rapidity of a shorting circuit.

'I've only just caught up with myself in terms of accepting a compliment,' she says. 'It's taken me 25 years to get to that point. Before, I never thought I deserved it. I never believed anyone who said they liked what I did because I knew different. It was a kind of arrogance which I didn't show. I was so convinced that I wasn't good enough, how could anyone tell me I was? I had so much praise: I could see it and I could hear it. But it was as though there was this bubble around my feelings and my body entirely.'

After 30 years in the music industry – her first record, 'Chimes of Arcady', was with the Lana Sisters in 1958 – Springfield remains an iconic performer, recognized as one of the finest, if not the finest, female singer that Britain has produced. Like all pop icons, her life *appears* to be reflected in and shaped by the material she sings: one of the qualities that we choose in stars (and Springfield expects and receives the fealty that all stars do) is an ability to act out a heightened sequence of contradictions and emotions on our behalf.

'Being good isn't always so easy – no matter how hard I try,' she sings on 'Son of a Preacher Man'. Springfield was melodramatic and contradictory: vulnerable yet strong, dignified yet wild, charming yet prone to tantrums. Sixties titles like 'I Just Don't Know What to Do with Myself', 'In the Middle of Nowhere' and 'Losing You' were given an emotional authenticity by her bruised voice: she had international success as the tragic torch singer, but the picture that these songs give of a confused, often self-destructive soul isn't without foundation in her own life.

This self-referential process continues with her new single, 'Nothing Has Been Proved', written by and recorded with the Pet Shop Boys. As far as the press is concerned, the jury has been out on various aspects of Springfield's life – her appearance, her sexuality, her professionalism, even her sanity – for more than 20 years. This process reached a new low with a particularly unpleasant *News of the World* spread last year, but it would be wrong to present Springfield as an innocent victim. She has enjoyed a flirtatious relationship with the press: often issuing outrageous quotes and seeming unable to establish the limits of self-revelation.

'It's a scandal,' whispers Neil Tennant on 'Nothing Has Been Proved'. The song is the theme for the forthcoming film, *Scandal*, about the infamous events of summer 1963, when the sexual revelations linking Stephen Ward, Christine Keeler, Mandy Rice-Davies, Russian spy Eugene Ivanov and John Profumo (then Minister for War) became a national soap opera. In the video for the single, the Pet Shop Boys play a journalist/photographer pair doorstepping Springfield while she lip-syncs through a whirlwind of newsreels from the time: there's Christine, there's Profumo, there's dashing Stephen Ward.

You almost expect to see Springfield herself, complete with beehive and kohl, singing 'I Only Want to be with You'. For if 'Nothing Has Been Proved' could be seen to refer to her own difficulties with her public persona and her private life, then it also loops her back to the year of her first solo success.

'I remember seeing it all in the papers,' she muses of the 1963 scandal. 'I wasn't really aware of what it was all about. I understood that people got very excited about it all, but that was a period of my life when I was extremely innocent. I literally didn't know the words.' She laughs. 'Because I am now associated with this, I've had to go back – not to look up the words, incidentally – to look up the dates. It was actually going on prior to the Springfields's split, wasn't it? But we never followed it through to see what happened.'

Dusty Springfield was well placed to benefit from the pop boom in the autumn of 1963: a boon to Fleet Street news desks anxious for a new story after months of scandal. She was born Mary Isobel Catherine Bernadette O'Brien in April 1939, to Irish parents. The family were comfortably off but troubled – her father was a tax consultant with frustrated artistic ambitions – and domestic dramas were commonplace. She has described her family in terms of 'intense bitterness'. 'I was brought up in Bucks and then we moved to Ealing, which is a step up,' she now says. She hated her convent school there.

In 1958 she went from a part-time job at Bentalls to join the Lana Sisters, an early attempt at a female rock 'n' roll group. 'I was the only one who had walked on a stage. I did my training at American air bases and Clacton Pier.' In 1960 she formed the Springfields with Mike Field (later replaced by Mike Hurst) and her brother Tom: 'We started out and got booked sight unseen at Butlin's; we did 16 weeks running round the country in this very old Volkswagen bus. We got TV exposure very quickly, and we discovered that if you sang loud and fast everybody got terribly impressed. We were terribly cheerful. It was extremely important to be cheerful.'

The Springfields bridged folk and pop with hits like 'Island of Dreams', a wistful, melodic song with a strong sense of possibility. They now sound anodyne, but then the songs, with a rare hint of emotions in the blend of voices, were something new in British pop. This was enough to take the Springfields to America, where 'Silver Threads and Golden Needles' was a top 20 hit in 1962. Flying to record in Nashville, Springfield found the country, and the music, of her dreams. 'It was love at first sight,' she says.

This was the start of a love affair with soul which brought her closer than any other British singer to the music aristocracy of the time: in 1968 she received the supreme accolade when she recorded *Dusty in Memphis* with the cream of Stax session men. 'I remember this quite clearly, because you always remember the first time. I was sitting in the Capitol Motel in Nashville when I heard "Don't Make Me Over" by Dionne Warwick. I actually had to sit down very hard because it was different to anything I had heard before. The other record was "Tell Him" by the Exciters, which I heard in New York. The pure power of it! "Tell Him" was more important because I could approximate the voices. I wanted that crispness, the ballsiness in the voice, which we hadn't had in England.'

Soon after that first US trip the Springfields broke up – 'we were all smart enough to see what was coming. That was the Beatles. We'd seen them play at the Cavern' – and Springfield had a chance to put her ideas into practice. 'I was going for the Phil Spector sound,' she says of her first solo record. The result was 'I Only Want to be with You', a song so endurable that it was again successful when revived in the late 1970s by the Tourists.

Suddenly, Springfield was a huge star, and, by her own admission, that's when the problems started. 'I was very sheltered, and suddenly we were taken out to these little clubs off Sloane Street. I didn't know what the people I was with were talking about, and I didn't know about the food they were serving. I was raised on meat and potatoes. So I developed this front so they wouldn't know. Because if they knew the real me, they wouldn't like me.' The stories of her wild and unpredictable behaviour – collapsing through worry, or driving her Italian sports car wearing sunglasses after dark – began.

So Mary O'Brien became fixed as 'Dusty Springfield'. Vital to this transformation was her appearance, which was artificial in the extreme: huge piled-up beehives and whole bottles of black eye make-up. 'I overdid a lot of things,' she says. Was it not a mask? 'It was a good thing to hide behind. Without the face I was a quivering wreck. I was terribly shy. So the more eyes I put on, the less shy I had to feel. And once you put that eye make-up on, it was such a hell getting it off that you would just leave it on. And put more on top after it for three weeks at a time. By that time it was really solid.'

She was also a perfectionist about her records, which didn't endear her to a music industry unused to assertive female performers. 'There was a lot of angst involved because it was not easy to get those sounds on to four-track. I was breaking a lot of new ground and asking musicians to play things they had never heard before because I had been in the States and they hadn't. So I scowled a lot: I got a reputation as a great scowler.'

The result was a sequence of 17 hits that spanned the rest of the decade. Because they fit the perceived melodrama of her life, she is best known for big ballads like 'I Close My Eyes and Count to Ten' or 'You Don't Have to Say You Love Me'. But she also recorded underrated stompers like 'Little by Little', as well as establishing herself as a fine interpreter of songs by composers like Jacques Brel, Randy Newman, Burt Bacharach and Hal David. Her own favourites remain Goffin and King's 'Going Back' and their 'Some of Your Lovin'', the song that 'sold her to Atlantic Records' and took her to Memphis.

These records established her as one of the most charismatic of a new generation of stars who embodied a new femininity: Cilla Black and Lulu with their rough voices and the cool Sandie Shaw. Springfield's appeal was the broadest: at once showbiz and existential, with a sophistication that appealed to men and women alike. The rapid passing of the 1960s, however, hit these stars hard: until the singer/songwriter, the only option available to female pop stars was the classic cabaret route. Dusty found being at the cutting edge exhausting; she fled to California.

'I had to leave for my own sanity,' she says. 'I knew that my career was disintegrating. If it was going to happen, I didn't want it to be here. I didn't want to land up playing northern clubs all the time.' Her 1970s are best glossed over: she had learned 'how to party' in the late 1960s and in Los Angeles went at it with a vengeance. 'I did the whole lazy self-destructive California bit,' she says, 'and thoroughly enjoyed most of it.' Her image in the press took a downturn: the storyline became melodramatic failure, a soap opera of pills and tantrums.

If in 1968 Dusty had sung 'Don't Forget About Me', by 1978 one of her many attempted relaunches was headlined 'It Begins Again'. A hard-core, mainly gay audience stuck by her. She went through several managers and record companies: one bizarre partnership linked her with Peter Stringfellow. But, in 1987, the storyline had a new twist; her collaboration with the Pet Shop Boys on 'What Have I Done to Deserve This?' resulted in an international hit and changed her life.

Her new career comes within a music industry both rediscovering and cannibalizing its past. Her own past has been revived, in last year's Britvic ad which used 'I Only Want to be with You' and in her best-selling *Silver Collection*. Nostalgia sweetens and trivializes, but her reappearance reminds us of her achievements and that the energy of the 1960s remains as an elusive touchstone. Like Morrissey with Sandie Shaw, Marc Almond with Gene Pitney, the Pet Shop Boys have reframed her in a way that uses melodrama to get to the reality of a misunderstood decade. As she says, 'Those songs were dark. I always tried to find a place somewhere on a record that was dark because that's an important side to me.'

So 1989 brings another 'Brand New Me'. In 1967 Dusty Springfield hoped that in 20 years she would 'have a settled mind'. If nothing else, she now says she has come to terms with her 'job' as a singer and the requirements that it makes of her life. She plans to move from Amsterdam – where she lives with several cats – back to Britain to work. She remains highly private, but intensely aware of her own myth. She has already turned her renewed confidence into a simile. 'I'm like a seagull over the marine bed, the way they swoop when you throw down food. Except I'm a very selective seagull. I just don't sweep down for crumbs: it's got to be a whole loaf of bread in the water.'

Observer, 12 February 1989

R.E.M.: Post-Yuppie Pop

'I could turn you inside out! But I choose not to do!' R.E.M.'s singer Michael Stipe, back arched, is bellowing into a megaphone. Five songs into their concert at the Onondoga War Memorial Hall in Syracuse, an upstate New York college town, R.E.M. are into their stride. The Memorial Hall, built in 1951 and running to seed, is filled with over 8,000 bodies and this mixed audience of students, rock fans and teens attracted by R.E.M.'s current hit single is going gently berserk amid purple and orange light.

Syracuse is the twenty-ninth American date out of R.E.M.'s eight-month 1989 world tour. The night before, they played to 17,000 people (85 per cent capacity) in New York's Madison Square Garden; the next day, they go to Toronto. Their current single, 'Stand', is at number six. Their first album for Warner's, *Green*, is in the top 20; released last November, it has sold over a million copies. The week of the concert, R.E.M. are dubbed 'America's hippest band' on the cover of that bastion of American rock values, *Rolling Stone*. It seems like business as usual.

Except that singer Michael Stipe is wearing make-up and a dress. It's a nice dress: a knee-length affair in red tartan. It covers the trousers of a baggy, dark brown 1950s suit and is covered by the suit jacket, which is held together with a safty pin. It is not the standard attire of a serious American rock group. R.E.M. are generally upbeat and often didactic, but 'I Could Turn You Inside Out' is designed as an all-out assault on the senses. The transformation that is the hallmark of any powerful pop event is beginning to take place. R.E.M. use surrealist backdrops throughout their performance: here, the screen is filled by murky film of fish shoals moving with a hypnotic slowness. The lyric examines the power of the performer, whether a pop star, 'a preacher or a TV anchorman', to manipulate a mass audience. Within this context, the dress has a particular significance: it marks R.E.M.'s passing from their cult rock-band status to the blurred, warping world of pop stardom.

The week of the concert, a big item in the US news is a rally in Washington attended by 300,000 people protesting against possible anti-abortion legislation. 'Causes are fashionable now,' says a friend in the centre of magazine Manhattan: R.E.M. are well placed to catch this post-Yuppie mood: they espouse green and specific issue politics. They are idealistic and forward-looking to a degree that might seem naïve.

'I'm over-simplifying,' says Michael Stipe, 'but I think as a motivating force for change pop culture is still at the forefront. Events like the Amnesty tour brought a lot of attention. Pop culture is still the one way in which someone who is without power can attain it and bring about a change.'

Aged between 29 and 32, R.E.M. represent the coming to power of a particular musical generation in America. As much as the Beastie Boys, they are the final products of English punk, which has taken 10 years to filter into the American mainstream. Like their nearest English equivalent the Smiths, they mark industry and public acceptance of the 1980s independent label sector. They are also a product of a new force in the American music industry, college radio, the success of which has made a dent in the programming policy of the notoriously conservative American radio networks.

'It's been a gradual build-up all the time,' says Stipe. The group, named after the rapid eye movement of the first, deep sleep, was formed by four college drop-outs in Athens, Georgia, in 1980. All had lived in the South for some while: only Stipe had lived outside America, in Germany. If bassist Mike Mills, guitarist Peter Buck and drummer Bill Berry appear straightforward and friendly, Stipe is the changeling of the group, hinting on occasions at the twisted dandyism of other Southern exports like Capote and Wolfe.

'"Think global, act local" is one of our catch-phrases,' says the spry Mills. A town of about 70,000 inhabitants, Athens remains important to the group: they all live there and are involved with local politics. The way Buck tells it, the town offered a sheltering bohemia: 'It has all the parochial small-town behaviour, but then you've got all the college kids who roar through. Its got the best art school in Georgia, maybe one of the better ones in the South.'

Like countless others, R.E.M. were inspired by the punk style and attitude that spread from New York and London in the mid- to late 1970s. 'I heard Patti Smith and Television when I was 15,' says Stipe. 'I'd found something that was dirty and exciting and sexy and *smart*. I realized that I was an outsider and I felt separated from most people. This music made the separation worse but it gave me an ace in the hole because I had something they didn't have.'

'I was the manager of a rare and used record store in Athens,' adds Buck, 'so I used to play records I hadn't heard all day long. In Georgia you were so far away from everything that the Sex Pistols were just the same as Ultravox. For two years, 1977 and 1978, we'd buy everything that came out. Punk filtered down to us; it meant that you didn't have to follow the rules.'

R.E.M. began playing in an Athens bar called Tyrone's – 'a good mixture of preppies and hippies,' says Berry – and soon began an extended period of playing any possible venue, from top 40 bars to gay bars. After their first single, they were picked by Miles Copeland's IRS Records for the first of seven albums, *Murmur*. Marrying the utopian jangle of the 1960s Byrds with the new forms of song construction illumined by punk, R.E.M. soon emerged as the best of an often mundane pack of new American rock bands.

'We've played with each other so long that we can intuit chord changes,' says Peter Buck, and this musical closeness has been an R.E.M. hallmark: songwriting credits are equally split between all four. 'The neat thing about us is our harmonies,' says Mills. 'We have three people who can not only sing but make up their own ideas about what to sing, instead of building a song on 1-3-5 harmonies.' The group have also been marked by Stipe's buried vocals and cryptic, allusive lyrics. 'In Television's songs, I never knew what they were singing about half the time,' Stipe says. 'But it doesn't matter because it sounds great. Some songs are written in one stroke but others are prepared for months and months. I have these notebooks: I'll pick a topic and run through the notes and say: "This applies and this applies." The moment of inspiration is extemporaneous but it's all been prepared before.'

This approach resulted in a sequence of songs that defines a new Americana. Albums like *Life's Rich Pageant* and the remarkable *Fables of the Reconstruction* captured the sense of space and possibility that lies within America, allied to a strong sense of loss and dreams betrayed. 'There's a lack of history here which would be the American version of Catholic guilt,' Stipe says. 'I think that's a big flaw in the American dream. You're not taught about the annihilation of the entire culture of the Indians whose land this was.'

Boosted by constant touring, superior material and college radio support, R.E.M. finally broke through to the US mass market with 1987's top 10 single, 'The One I Love'. During the 16-month lay-off that ended with the start of this tour, the group signed to Warner Brothers Records for several million dollars. The group's belief in the mass market inevitably required mass distribution. Says Stipe, 'IRS's distribution had gone as far as it could and it was time to move on to someone who could get the records out world-wide.

'Touring is a great pressure,' he adds, and at Syracuse fatigue is beginning to set in. The normally good-humoured Berry runs out of the photo session, while the night before Buck had broken a toe in a fit of frustration. All are coming to terms with the alienation of the mainstream music industry while attempting to retain their ideals and closeness.

Outside the hall in Syracuse, a strong wind blows gusts of snow through streets that to a European seem empty even when peopled. What R.E.M. offer their audience inside is a sense of community: their performance is a careful balance between raw feeling and downtown rigour, between outright didacticism and the dream state implied by their name. Their reward is a crossover appeal to intellectuals, rock fans and regular high-school students.

Perhaps what they react to is the transformation implicit in Michael Stipe's androgynous performance. As he dances, dervish-like, in the flicker of a strobe, or throws his head back and roars like a preacher, he dramatizes R.E.M.'s triumph at finding their own power. Unlike other groups, who use this power to dazzle, R.E.M. deliberately seek to draw the audience in, to offer a positive approach. 'Hope is important,' says Stipe. 'It's an intrinsic human emotion, to think there is some kind of light at the end of the tunnel.'

Observer, 21 May 1989

Woodstock:
Trip Down Memory Lane

'It's a financial disaster. We just opened the gates and let everybody in. It's a free festival paid for by the people who put up the money, who are going to have to pay it back, hey hey hey.' In the *Woodstock* film, promoter Artie Kornfeld squints at a television reporter. It's hardly fair: there he is, in front of 300,000 people, blasted on psychedelics, and here's this person asking him ... questions. What's a question? What's an answer? Focusing with difficulty (like the cameraman), Kornfeld grabs the joystick and attempts to land.

'This was paid for by all the people that took it to the point to get us in the position to just be a tool, a vehicle like everybody else just to get it to this point.' 'You're in the red,' the reporter interjects abruptly. Secure in the knowledge that he has just secured a film deal with Warner Brothers that will cover any costs, Kornfeld grins, beatifically: 'If you choose to talk in those terms when you're talking about something like this.'

Kornfeld then pushes the hyperspace button. 'This is really beautiful. It has nothing to do with money, with tangible things. You have to realize the turn-about that I've gone through in the last three days, the last 3 million years, you know that I, meaning us, all of us ...' A record company man of many years' standing, Kornfeld attempts to hype like crazy, yet the combination of atmosphere and drugs causes him to lose the plot. He really believes all of this.

The *Woodstock* film makes plentiful use of the split-screen device. While Kornfeld is delivering his rap on one side, a young man and woman are wandering off together into the long grass. Unknowingly caught by a telephoto lens, they disappear in a jumble of limbs just as Kornfeld reaches his zenith of incoherence. On the film's release, the young man sued. Working as a hairdresser in Montreal, his customers expected him to be gay: this heterosexual act, caught on camera and watched by many thousands, he claimed, ruined the illusion on which his business depended.

Twenty years on, Woodstock is full of more ironies and contradictions that its monolithic media image suggests. On the weekend of 15–17 August 1969, between 300,000 and 500,000 people (estimates vary) crammed into a field in upstate New York to see the cream of the day's pop stars: by the time the film was released nine months later, Woodstock was already a legend. For some, the film turned this legend into cliché; it gave others a condensed primer to the hippie lifestyle: the famed trinity of sex and drugs and rock 'n' roll.

'There are two sides to Woodstock,' says Grace Slick, acerbic lead singer in Jefferson Airplane, one of the festival's headliners. (Slick still sings with this group, long since renamed Jefferson Starship; when we spoke she was on her way to her daughter's high-school graduation in San Francisco.) 'It's become a symbol, but the event itself was ridiculous. It was so fucked up it was criminal. It was rainy. It was muddy. We were supposed to go on at 9 p.m. and we went on at 6 a.m. We couldn't leave the stage, but we were numb on drugs, like the audience. If you're numb and 21, you can take a lot.'

We think we know about Woodstock, an event synonymous with hippies and the indulgences of the late 1960s. Yet, just as the film divides into split screens, there is a contrast between the actual events of that weekend and the way these were propagandized and marketed. Woodstock was a huge hype, pretending, in the style of the time, not to be. Yet like all great pop events, it gained its power from an interaction between hype and audience belief.

The whole event began as a gleam in the eye of a charismatic hustler called Michael Lang. Together with his partner Artie Kornfeld, he was planning a studio complex in rural Woodstock, in upstate New York. In the late 1960s, the country was fashionable. After the success of 1967's Monterey Pop, festivals were fashionable also: the pair decided that this would be the perfect way to announce this development, and in early 1969 they got the backing of two rich young men looking for adventure, John Roberts and Joel Rosenman.

Although Lang now claims that things were a 'little less helter-skelter than they might appear', the festival was a mess. The final site was only found at the last minute, on land owned by a farmer called Max Yasgur in White Lake, near the town of Bethel, miles away from Woodstock. The partners got on badly; at times it seemed as though the event was only held together by Michael Lang's unshakeable belief. When the audience finally arrived, its size shocked everybody. Pop festivals were plentiful that summer – up to 40 were planned all over the country – but Woodstock was the one. Why?

'1969 was a bad year in America,' says Country Joe McDonald, whose 'I Feel Like I'm Fixin' to Die Rag', a Vietnam protest song, is one of the film's few references to the world outside. 'It was full of death: riots, assassinations, Vietnam.' The country was polarized: 800 American servicemen were killed in Vietnam each month in 1969, and in that year anti-war demonstrations mobilized half a million people nation-wide, with 9,000 campus actions. A whole generation was politicized and harassed. 'Woodstock was a breath of fresh air,' says McDonald, 'a respite from the struggle.'

Everybody who was at Woodstock talks about an extraordinary atmosphere which was not captured in the film. 'There was an energy you could tangibly feel miles and miles from the site,' says Michael Lang, 'a positive electricity in the air that is not reproducible.' This atmosphere was not created by the groups or the promoters, but from a need within the audience itself. 'People don't know how to live,' says a young man in the film. 'They don't know what to do. They think that if they come here they can find out.'

'Woodstock was important as an indication that there was a mass movement,' says Greil Marcus, a writer for *Rolling Stone*, 'that *we* existed in infinitely greater numbers than anyone had thought.' A principal mood of Woodstock is self-affirmation in the face of adversity; in this, politics were implicit rather than explicit. In 1969, it was enough for most to be there, to be able to smoke marijuana freely – as countless people do on camera – and to be able to have fun. 'It was party time with an eyeball out for disturbances,' says Grace Slick.

The chaos of the festival – declared a disaster area by Governor Rockefeller on Sunday – spilt over on to the stage. Despite a line-up including Janis Joplin, the Band, the Grateful Dead and many others, very few people played their best. Groups came on hours late, lots of people were spiked with LSD, and it was hard work getting feedback from such a huge crowd. 'Most people showed up, did their show and collected the money,' says Greil Marcus. 'The highlight was Crosby, Stills, Nash and Young: they were noisy and explosive. More than any other performers, they were inspired by the crowd.'

The festival itself closed on Monday morning with Jimi Hendrix's elegiac 'A Star-Spangled Banner'. Within days, the mythologizing started, through 'hip' TV shows like Dick Cavett and the underground media. 'The immediate news coverage concentrated on the disaster,' says Marcus, who wrote the *Rolling Stone* cover story. 'Then we and the *Village Voice* chimed in. I got carried away and wrote that the greatest possibilities of rock 'n' roll were to be contained in great gatherings because of their reciprocal intensity. That was utter horseshit because these gatherings became ritualized very fast.'

Woodstock had an immediate effect on the music industry. 'The financial side changed instantly,' says Michael Lang, 'as everybody realized the power these groups had to draw.' Many of the groups who had performed had their greatest period of success over the next year, as the Woodstock campaign mounted. Beneficiaries included Santana, Joe Cocker, Sly and the Family Stone, all with albums in the top three. Crosby, Stills, Nash and Young embodied the spirit of the event: their first LP, *Déjà Vu*, with its countrified 'Wild West' cover, shot to number one in America in May 1970.

This group also had the theme song to the *Woodstock* movie, a romanticized version of events written by Joni Mitchell, who'd been stuck in a New York apartment during the weekend. The film had begun when Michael Lang hired a documentary director called Michael Wadleigh to shoot on spec. After the ticketing débâcle, the film became the four partners' major asset. Unable to agree among themselves, they split: after buying out Kornfeld and Lang for $31,250 apiece, Roberts and Rosenman sold the film to Warners for $1 million flat. By 1979, *Woodstock* had grossed $50 million .

Seen today, the film is in parts moving, in parts irritating. There are an awful lot of very stoned people, like John Sebastian, wearing bad clothes and stumbling about. There is too much performance and not enough atmosphere: the best moments are vox pop interviews with the young crowd or with a benign, middle-aged toilet cleaner. Outside the context of America in 1969, the constant self-affirmation on the part of performers and audience comes off as smug: 'Woodstock was almost racist,' says Greil Marcus. 'It emphasized a rich, white definition of what rock music was.' The film soundtrack, which went to number one in America in June 1970 (and into the English top 10 in July), completed this first lifestyle package. The unprecedented success of both film and album began as a boon, but ended up as a millstone to many of those involved. 'Even today, people still come up and say, "Hey, Joe Woodstock!"' sighs Joe Cocker. The festival initiated a personal and musical self-indulgence which blighted much of the 1970s. 'It was the beginning of the end,' says Joe McDonald. 'The music industry learned how to market million upon millions of records, with uncontroversial people like Peter Frampton and the Eagles.'

During the last 10 years, the Woodstock industry has repeatedly been taken as a symbol of everything that is wrong with pop culture: in the UK, the punks gave it a mauling from which it has not yet recovered. Much of the twentieth anniversary celebration seems set to continue this: scads of interviews with participants and hours of footage from the film – including the outtakes of groups like the Band and Janis Joplin – on music industry outlets like MTV and its 'adult' channel, VH1. More baby-boomer marketing, you could be forgiven for thinking.

Yet what comes out of the memories of some participants and a new book, *Woodstock: The Oral History*, is a sense of possibility that is a refreshing antidote to today's cynicism. 'People went away with an idea about how things should be,' says Michael Lang (who today still manages Joe Cocker's still-thriving career). The 'free festival' ideal remains potent, even threatening. Last month an estimated 100,000 people attended the nineteenth Glastonbury rock festival, many of them without tickets. A few days later the police forced a four-mile exclusion zone on hippies attempting to celebrate the summer solstice at Stonehenge, an event that started in 1974 out of the 'free festival' movement.

'Woodstock is a powerful symbol,' says Country Joe McDonald. 'You can't think of it without the Vietnam War, and Vietnam is still a very hot potato in America. There's a lot of unfinished business. And the recent events in Tiananmen Square started me thinking about the differences between the old and the young. Faced with the holocaust, young people sensibly want change, or even just a chance to do something as trivial as getting together and having a good time, and the old use truncheons and bullets on them. It's a global gap.'

Observer, 9 July 1989

Morrissey: The Escape Artist

Last November, Morrissey played his first and only solo date in Wolverhampton. Arranged to promote his new single, the concert was free, first come, first served. From the middle of the day, the torpor of this depressed working town was disturbed by thousands of young fans from all over Britain. The streets of this repressive city were populated by youths in T-shirts displaying mythic, androgynous figures like Joe Dallesandro or famed drag queen Jackie Curtis.

The crush worsened. By the evening, the near-riot conditions outside the venue were reproduced inside, where Morrissey played a nine-song set on a stage swamped with youths of both sexes. This reception was as enthusiastic as any for current teen sensation Jason Donovan. Although you wouldn't have read much about it in the mass media, it clearly demonstrated Morrissey's ability to inspire a very physical fanaticism from an audience of the repressed and the dispossessed.

(I pick up the Smiths' Louder Than Bombs *CD and press that 1980s equivalent of the jukebox, the shuttle button. It alights on 'William, It Was Really Nothing': 'The rain falls hard on a humdrum town/This town has dragged you down …')*

With the Smiths, Morrissey had 15 top 30 hits and seven top 10 albums. After Morrissey and Smiths' guitarist Johnny Marr parted ways in 1987, Morrissey released his first solo album, *Viva Hate*, which entered the charts at number one. His most recent singles, 'Interesting Drug' and 'The Last of the Famous International Playboys', went straight into England's top 10. Morrissey has an importance that outweighs even these statistics, as he can discern from his mail. 'People seem to have an urge to say something terribly, terribly, terribly strong to me,' he says.

(The shuttle switches to 'Shoplifters of the World Unite': 'I tried living in the real world/Instead of a shell/But I was bored before I even began …')

Last night on TV, Channel 4 screened John Schlesinger's 1963 film *Billy Liar*. Shot in a black-and-white, energetic New Wave style, the plot tells of a young man, Billy Fisher, who lives in a generic northern city. Imprisoned by his total circumstances, he fantasizes about machine-gunning authority figures like his father or his boss. His one escape is to lie on his bed and lapse into reverie.

Driven to distraction by his shitty job in an undertaker's, Billy fiddles with petty cash. He gets engaged to two girls at once and still can't touch them. He writes a twist tune – this is just pre-Beatles – called 'Twisterella'. He even attempts to write a novel, but before he pens a word he stops at merely fantasizing the billboards of fame. The one chance he has of transforming his life is moving to London: the film's tension comes from us wanting him to take that leap into the dark. At the very last moment, he misses the midnight train. Billy's lie is that he just can't translate his dreams into reality.

(Flick. 'Is It Really So Strange?': 'I left the North/I travelled South/I found a tiny house/I couldn't help the way I feel …')

In his life, Morrissey offers an alternative ending. He is the Billy Liar who caught the train. As an adolescent, his sense of alienation was only matched by the depth of his obsession for pop music, and in particular the dreams of Manhattan that, through the New York Dolls, haunted punk Britain. In the Manchester punk milieu, he was obviously talented but apparently unable to cross that disabling gap between fantasy and practice. His unexpected triumph has led him to act out stardom as revenge, throwing in the public's face the rage of an outsider who has found his voice.

You will rarely find such a strong sense of place in the products of the London media industry – videos or pop stars like George Michael – that flooded America after 1982. These upbeat, Thatcherite celebrations of the metropolis's success drowned out the rest of the country. In 1986, however, the Smiths reached a peak on *The Queen is Dead*, which fused existential desolation and the hardest, most spacious rock possible into what remains the most accurate emotional map of non-metropolitan life in the 1980s. Despite the socialized, club-oriented diktats of the London media, the song carried the pleasures of alienation with the anger of an outsider who sees, in the words of the Buzzcocks, 'all these livid things that you never get to touch'.

(Flick. 'Rubber Ring': 'Don't forget the songs/That made you cry/ When you lay in awe/On the bedroom floor/And said: "Oh smother me mother …"')

Over a score or so records, the Smiths' sleeves featured a series of, in Morrissey's description, 'cover stars'. The references form not the usual pastiche, but a deep wish-fulfilment and an implicit engagement in a social and cultural struggle that still continues. Drawn from the history of pop culture, these images concentrate on the damned, the androgynous and the beautiful: Jean Marais in *Orphée*, James Dean, Candy Darling, Joe Dallesandro, Truman Capote. 'They make the best people,' Morrissey says.

The most concentrated set of references comes from a particular moment in British culture, in the late 1950s and early 1960s: early rockers like the scouser Billy Fury, teenage playwrights such as Shelagh Delaney, or northern working-class icons like Pat Phoenix, star of the longest-running, most popular soap ever, *Coronation Street*. TV shows such as these and plays like Delaney's *A Taste of Honey* focused attention on industrial conurbations like Manchester and imbued them with glamour.

This is the moment caught by *Billy Liar*. Deliberately filmed in 'anytown' in the north of England – in fact Manchester and Leeds – Billy drifts through a cityscape in transition. In 1845, Friedrich Engels used Manchester as a laboratory sample for his impassioned plea against the dehumanization caused by the Industrial Revolution. 'Everywhere half or totally ruined buildings, some of them actually uninhabited,' he wrote in *Condition of the Working Classes in England*; 'rarely a wooden or stone floor to be seen in the houses, almost uniformly broken, ill-fitting windows and doors, and a state of filth!'

(Time for a change: insert the soundtrack to Scandal, *tracks 18 and 19. Billy J. Kramer first, Brian Epstein's boy. His number-two version of the Beatles' 'Do You Want to Know a Secret': 'Closer! Let me whisper in your ear/Say the words I love to hear.')*

By 1962, the utterly degraded environment that Engels describes was being swept away, as consumer capitalism – an idea imported from America – began finally to take hold in Britain. *Billy Liar* is a whirl of the new, brassy confidence that the Beatles would synthesize the next year: a whirl of supermarkets, record stores, cheeseburgers, TVs and Bar-B-Qs. This was the British version of what had happened in America throughout the 1950s. Consumption was available to all, ran the propaganda, and it offered a degree of real emancipation.

Billy Liar itself symbolized this new confidence. It was a part of the wave of 'kitchen-sink' films that, after *Room at the Top* in 1958, became fashionable. If the previous wave of Ealing films had been cripplingly cute and middle-class, then, according to Raymond Durgnat in *A Mirror for England*, films like *Room at the Top* found: 'the appropriate strategy, which is to assert scandal and revolt against passivity and puritanism. By appealing to the working class, the young, and a new cosmopolitanism, it spreads its roots far more deeply and widely into human nature.'

Out of this moment of confidence came the British cinema of the 1960s that travelled the world: James Bond, the Beatles films, and the *Billy Liar* team of John Schlesinger and Julie Christie that went on to make the successful *Darling*. Yet these new films, in relying both on cosmopolitanism and on outside finance, tilted the balance of power. Just as countless Liverpool groups were forced to leave for London, the locale for popular films changed from the North to Swinging London. The centre of gravity quickly shifted back to the capital, the hub of the media industry.

(*Next: John Leyton, an early English facsimile rocker, packaged by famous gay producer Joe Meek. 'Johnny Remember Me', number one in 1960: 'When the mists are rising/And the rain is falling/And the wind is blowing cold across the moor …'*)

Hindsight has clearly shown the price to be paid for this brief moment of optimism. Much of Morrissey's iconography concentrates on the tragedy behind the illusion of consumer democracy, just as much mid-1960s pop carries a buried but persistent sense of loss. The headlong rush to consumer equality drew its subjects into a strange new world, which for all its allure was often no better than the one they had left behind. The story of a favourite Morrissey icon, Viv Nicholson, is exemplary. A young Yorkshire housewife, she won £152,000 on the pools in 1961 and promptly went mad. Within a decade she had gone through the money – blown it all, as she reveals in her autobiography, *Spend, Spend, Spend*, on Jaguars, racehorses, trips abroad and endless shopping – and five husbands.

(*Switch CDs to* The Smiths. *Shuttle. Mmmm … 'Suffer Little Children'. There are penalties to this random approach. 'Oh, Manchester, so much to answer for/Oh, Manchester, so much to answer for/Over the moor, I'm on the moor/The child is on the moor …'*)

In 1965, Manchester's Hollies had four UK top 20 hits, including their great, gleeful number-one version of Hank Ballard's 'I'm Alive'. That same year, the true story of what would be called the Moors Murders began to unfold: as Jeff Nuttall wrote in *Bomb Culture*, 'We were eaten up by repressed violence … and from this we strugglingly produced a culture. It's possible to get hysterical about the obvious connection between that culture and the Moors Murders. I did.'

'Working-class libertines' of the sort eulogized by the culture, Ian Brady and Myra Hindley tortured and killed upward of five adolescents between 1963 and 1965.

As the news of the murders unfolded, something in Manchester died. It was a moment Morrissey makes explicit reference to on 'Suffer Little Children', a title taken from Emlyn Williams's book on the murders, *Beyond Belief*. It was the end of that moment of optimism. By the mid-1970s, New Brutalist architecture, an oppressively religious police chief and the concentration of wealth had turned Manchester's centre into a night-time ghost town. So had the bustle of *Billy Liar* evaporated.

(Time for 45. The Buzzcocks' perfect 'Orgasm Addict': 'So well you're asking in an alley/And your voice ain't steady/The sex mechanic's rough/You're more than ready …')

Manchester was central to the spread of punk rock in Britain. The June and July 1976 Sex Pistols performances began a local scene that is still the most creative in the country. In the audience for those concerts were future members of the Buzzcocks, New Order and the Smiths. By 1977, Manchester punk was, in many ways, more interesting than that in London. 'You had to risk something to join in,' says Morrissey. The scene there was gender-friendly, a direct result of punk's finding a home in gay bars like the Ranch. For all those who, whatever their predilections, wanted to copy David Bowie's exaggerated, artificial appearance (itself stolen from early 1970s Warhol), gay bars were the only places where you didn't get beaten up.

'The Buzzcocks were *the* group,' says Morrissey. 'I liked their intellectual edge. I really despised the idea that in order to be in a group and to play hard music you had to be covered in your own vomit.' Along with the New York Dolls – 'Johansen was so witty' – the Buzzcocks are the blueprint for what Morrissey did with the Smiths. Over hyperactive, guitar-driven punk rock, Pete Shelley wrote deliberately gender-unspecific hits like 'Ever Fallen in Love', which he now says had a gay object. Morrissey has also used crunching hard rock to deliver a series of lyrics that trash male/female boundaries and masculinity in general.

(Another quick 45. The Smiths' 'How Soon Is Now?': 'There's a club, if you'd like to go/You could meet somebody who really loves you/So you go, and you stand on your own/You leave on your own/And you go home, and you cry/And you want to die.')

Between 1979 and 1982 I lived in Manchester; as a Londoner, I found the transition hard. The punk scene around the Buzzcocks that had, in part, brought me there had become a shadow of its former self. The hottest thing in town was Joy Division, with their uncanny, unsettling invocations of inner space. This post-punk scene, in which I fleetingly encountered Morrissey, was alienated and curiously numb. Mrs Thatcher had just been elected. Like much of the music of the time – Joy Division's *Unknown Pleasures* or PIL's 'Home is Where the Heart is' – I felt that I was slowly suffocating. 'It's this town,' says Julie Christie in *Billy Liar*. 'I don't like everybody knowing what I do. I want to be invisible.'

There was also the problem of sex. For a year I'd go to gay clubs like Heroes or Napoleons to try to get something started. With its mocking evocation of Bo Diddley's sex beat, 'How Soon is Now?' captures that experience exactly – except the last line: home and sleep were not upsetting but a relief. An important part, however, of Morrissey's teen appeal has been his extravagant self-pity.

> (*Joy Division:* Unknown Pleasures CD. *Track five: 'Interzone'. 'Down the dark streets, the houses looked the same/Getting darker now, faces look the same/I walked round and round/stomach, torn apart/Trying to find a clue/trying to find a way/To get out !'*)

Just as Engels had found Manchester a paradigmatic example of the chaos caused by the Industrial Revolution, now, in its manufacturing decline, it is a mecca for industrial archaeologists. There you can find the world's first railway station and, in Trafford, the world's first planned industrial estate. In the 1970s, the contraction of Britain's manufacturing industry – which created Manchester's second wave of wealth – and Brutalist town planning gave the city the look of a mouth whose teeth had been punched out. Yet these gaping holes – unlike London's tightly parcelled space – allowed new flowers room to breathe and grow; they also made Manchester a great place to take drugs.

But these spaces were an index of hardship and inequality. Within just over two years at the turn of the decade, unemployment more than doubled in Britain, with the North and the young (19 to 24) particularly hard hit. One result was a subcultural mixture that combined a deep wish for transcendence with a frustration a hair's breadth away from violence. You can hear the tension throughout Joy Division's music, or in A Certain Ratio's hypnotic deadpan 'All Night Party'. In the late 1970s, the club to go to was the Russell, also known as the Factory, situated on the outskirts of Hulme, a notoriously brutal 1960s development. There you were always half aware of the gang wars that might suddenly erupt, or the possibility that Perry Boys - *the* working-class subculture of the time – might be lurking in the skyways.

> (The Queen is Dead CD. *Play, not shuttle. First, the title out:* 'Passed the pub that saps your body/And the church who'll snatch your money/The Queen is dead, boys/And it's so lonely on a limb …')

Morrissey's work is shot through with the petty, peculiarly vicious nature of British violence. 'It's just my life,' he says, 'and my vision of the media. I do see a great deal of violence.' This knowledge tempers the extravagances of his self-pity and lends his androgyny – should it need it – some backbone. It is also not just a subject but an obsession. More recently, he has written a series of songs eulogizing 'The Ordinary Boys', the 'Suedehead', or the 'Sweet and

Tender Hooligan', whose impulses he envies: 'I don't feel natural when I'm fast asleep,' he says.

> (*Switch to the 'Interesting Drug' 45. 'There are some bad people on the rise/They're saving their own skins/By ruining people's lives …'*)

The 1980s have seen a concerted effort by Britain's Conservative government to roll back the gains of the last 40 years, whether in minority rights of every description or, in the wider sense, of the post-war democratic consumer ideal. One feels this most keenly outside London, where pop music is still a way of fighting back. Every non-London hit still resonates politically, in that it provides – to use Dave Marsh's description – 'a voice and face for the dispossessed'.

Manchester's music scene still thrives on the tension between oblivion and transcendence. In Britain, pop music still can be a way of fighting back, and you hear this struggle – for a voice, for some space – in Acid tracks like New Order's 'Fine Time' and A Guy Called Gerald's perfect 'Voodoo Ray', or in a clutch of rock groups who have gone nation-wide to explore the emotional map of Britain first traced by the Smiths.

Last month, I saw a group called the Stone Roses, who play a psychedelic dance-inflected rock infused with a spiky bitterness. Their first LP sports a Jackson Pollock version of the Union Jack. Inside is a vicious attack on that symbol of the English class system, the Queen. 'It's curtains for you, Elizabeth my dear', they sing to the tune of 'Scarborough Fair'. Even more directly, King of the Slums, on their new *Barbarous English Fayre*, delineate a whole gallimaufry of characters drawn from England's dispossessed: 'I'm going *nowhere*,' they sing on 'Fanciable Headcase'. 'I've got what it takes to take what you've got.'

Morrissey himself has become more insular. His most recent single, 'Interesting Drug', looked at last summer's Acid House scene from the outside, while *Viva Hate*'s 'Margaret on the Guillotine' was merely an adolescent fantasy of the prime minister's summary dispatch. These were the gestures of an isolate. Despite the quality of his recent work, Morrissey is finding it hard to remake the musical and social links that previously enabled him to engage with the world.

During the last two years, he has oscillated between Manchester and London, living alone and often incommunicado, even when there is promotional work to be done. His relationship to the export imperatives of his new record company, EMI, is arm's-length: 'I can't have a manager,' he says. 'I cannot plan my diary for the next 12 months and have things filled up. I just can't live that way. Also, I have a larger audience now but I have no inclination to be a massive global face. I have little interest in the world as a whole. That might be a dim thing to say, but I don't feel limited.'

More damagingly, Morrissey has struggled, since the break-up of the Smiths in 1987, to find a musical partner to match Smiths' guitarist Johnny Marr. For his solo work, he has used arranger Stephen Street – who engineered many Smiths' songs – to write the music, but this collaboration recently foundered after Street went public with a dispute over money. For his last two 45s, Morrissey has reconstituted the 1986–7 five-piece Smiths minus Marr, deliberately posing the question of a reunion. At present, he is becalmed.

The ending of *Billy Liar* contains an ambiguity. Is it cowardice that prevents Billy from getting on his train or the knowledge that leaving brings loss? For all its oppressiveness, his northern town offers a community of need and the security of the familiar. The act of transformation may bring release, revenge, fame and money, but the space it brings can reveal a warping waste. As Morrissey himself sang on 'Frankly, Mr Shankly': 'Fame, fame, fatal fame/It can play hideous tricks on the brain.'

One of the ironies of pop success is that it lends not only general recognition to the marginal but also the alienation of celebrity to the individual who embodies the many. For Morrissey, whose main project is his own life, this has meant a new isolation – from collaborators as well as the outside world – that cruelly mocks everything he first escaped. Unlike Billy Liar, he might have caught the train, but his journey has taken him back into another station.

(*Back to track one*, Louder Than Bombs. *'Is It Really So Strange?'*: 'My throat was dry, with the sun in my eyes/And I realized, I realized/That I could never/I could never, never go back home again …')

Village Voice: Rock & Roll Quarterly, Summer 1989

Flaring Up: The Stone Roses at Spike Island

You can see them all over the North-West, drifting through Manchester's arcades, doing the swim-dance in the high-tech Hacienda, travelling *en masse* to tribal events like the recent Stone Roses' concert at Spike Island on the Mersey estuary: groups of young men and women wearing floppy fringes and baggy, Day-glo clothes; 24-hour party people chasing the Great God Now in a millennial hedonism straight out of Prince's infamous '1999'.

The 1990s have begun, and with the new decade comes a new pop genera- tion. No one garment symbolizes this sea change more than the flared jean, that most proscribed of past styles. Flared, not from the knee but from the waist, these jeans are more than a simple generational challenge worn by peo- ple young enough not to remember the horrors of the early 1970s: they denote the looseness that is the hallmark of the new mood.

If recent pop culture, chasing the power politics of the time, has been largely exclusive and uptight, then the new mood is inclusive, relaxed, like a holiday at home. 'The 1980s were cynical,' says Ian Brown, singer for the Stone Roses, whose new single 'One Love', is poised to enter the top 10. 'People didn't want to participate. You can wake up in the morning and feel negative or positive. Why feel negative? It's better to be optimistic, to con- nect.'

Along with fellow Mancunians Happy Mondays, Inspiral Carpets and, most recently, the Charlatans, the Stone Roses have succeeded in codifying, for the pop charts and the media, a new pop bohemianism. This takes in equal parts from the social organization of the football terrace, from the holiday atmosphere of Ibizan clubs and from the mass transcendence of Acid House parties, updating the free spirit of the original hippies with a strong working- class flavour.

All this activity has been conducted away from the capital, in a city now careering into over-exposure in the rock media. Today, Manchester is unrec- ognizable from its grey, depressed state of the late 1970s. A combination of local boosterism, civic investment and shrewd exploitation of its road and air communications has resulted in a city that, if not booming, is more solvent than gridlocked, over-mortgaged London. In the centre's shopping precincts, the retail trade is buzzing. There is the smell of money in the air and the con- fidence that that brings.

In the boutiques of Afflecks Palace and the Royal Exchange, the predomi- nant look is a mix of terrace fashion and psychedelia: the Merseybeat fringe, the cagoule and 24-inch baggies of the Northern Soul fan, are put together with wild Day-glo and ethnic styles. Op-Art fashions, which would not have looked out of place on the King's Road in 1966, jostle with Brazilian shirts and old trainers: as House music blares from 20 different nooks and crannies, crowds flick through football shirts with the slogan, 'No Alla Violenza'.

Just past the domed public library, the rotting hulk of Queen Street station, derelict for decades, has been converted into gleaming G-Mex, a 7,000-seater exhibition and concert hall. A further 200 yards down Albion Street is Manchester's most celebrated night club, the Hacienda, eight years old this May. With its pin-sharp sound system and cutting-edge DJs like Mike Pickering and Graeme Park, the club has hosted two years of Acid House nights which, in their intensity, have redrawn the city's musical map.

If one group can embody Manchester's changes, then it is New Order, the group responsible for the recent chart-topping English World Cup theme, 'World in Motion'. In their late 1970s incarnation as Joy Division, they captured the alienated, terrible glee of a decayed city; in 1983, they recorded one of the most important singles of the 1980s, 'Blue Monday', which fused a rock concentration on inner space with the pure pleasure of electronic American disco music.

In 1982, New Order and their record company – the Didsbury-based Factory Records – decided to reinvest in the culture which had created them and put money into a new, specially designed nightclub. An oasis of high-tech in the middle of a redbrick industrial archaeology barely changed since Friedrich Engels wrote *Condition of the Working Classes in England*, the Hacienda has created a space where things can happen. Inside its grey grilles, a young, multiracial crowd has sought an almost spiritual transcendence in the hallucinatory, high-energy sound of American House music.

'Manchester is more spacious than London, where you can't see more than 200 yards in front of you,' Ian Brown says. 'And the city centre's compact.' This freedom of movement is all-important to the Manchester groups: you can hear it in the slowed-down James Brown backbeat which, in hits by Candy Flip, the Stone Roses and the Happy Mondays, has become the year's dominant rhythm. Large enough to support a cultural infrastructure, yet small enough to form a community, the city has fused styles from Ibiza, Chicago, Detroit and London into something recognizably, tantalizingly, new.

And that high profile is now part of the problem. If, according to insiders like the Hacienda's Paul Cons, the atmosphere of this movement reached its peak a year or so ago, then what we are now seeing is the commercial take-up. The hedonistic politics of Acid House parties are being packaged in a more conventional pop format: not the confusing detail of anonymous, fast-moving black dance records, but identifiable white, guitar-based groups with a message.

If the Happy Mondays are the most archetypal with their gang-like, couldn't give a damn attitude, then the Stone Roses are the most obviously groomed for stardom. After years of scuffling, the group broke through with the release of their first album, *The Stone Roses*, in May 1989. Like Joy Division's *Unknown Pleasures* or The Smiths' *The Queen is Dead*, the album interacted with last year's hot summer to capture the emotional experience of a new audience. With their mix of gentle tunes and violent lyrics, the eleven songs on the album helped map out a new romanticism: of travel rather than restriction, of sun rather than rain, of togetherness rather than atomized individualism – topped with a healthy dose of class revenge.

This is a volatile mixture, and the day before Spike Island, the group give a press conference which degenerates into a graceless shouting match: the world has come to Manchester and the city's subculture is buckling under the strain. Reacting to the prevalence of drug gang violence and a well-publicized death from Ecstasy within the club, the police are applying to have the Hacienda's licence revoked. Elsewhere in the North-West, there is a concerted effort to close down Acid clubs. If it is true that all the celebrants want to do is to party 24 hours a day, then simple good times are becoming harder to find.

'Time! Time! Time! The time is now,' Ian Brown shouts as the Stone Roses come out on to the massive stage at Spike Island. And, despite the extremely laid-back appearance of the concert crowd, there is a palpable sense of urgency in the air. The Manchester groups have succeeded in capturing and stimulating an ambience which is delicately balanced between ambition and solidarity, between radicalism and conservatism, between hedonism and idealism, between androgyny and laddishness, between gentleness and violence. Now all this looks to be on the point of shattering under the alienating effects of media attention and individual success.

There is something else stirring. Spike Island occurs during the week that there is world recognition that global warming, caused by the greenhouse effect, is more advanced than previously thought. In the unseasonal heat, everybody can feel it. And the sheer size of the crowd had fresh implications after the events of 1989, as the group notes: 'Anything is possible,' says Ian Brown. 'If Eastern Europe can change so quickly, so can England. It's strength of numbers. People coming together can always change things.'

Observer, 8 July 1990

And the Banned Played On: Censorship in the US

Last month, 43-year-old Dennis Barrie was in an Ohio court facing a year in prison and a $2,000 fine. His crime? Being responsible, as director of the Cincinnati Contemporary Arts Center, for an exhibition of 175 Robert Mapplethorpe photographs, seven of which were allegedly obscene. It was the first time that a museum and its director had been arraigned on criminal charges over work displayed.

Like all famous trials, the case dramatized a deeper social conflict. As Barrie explains, 'National forces – politicians like [Republican Senator] Jesse Helms and fundamentalist Christians like Donald Wildmon – were looking for a test case on obscenity. They saw Cincinnati as a logical place because it is the base for two powerful anti-pornography groups and we have a sheriff and a county prosecutor who have made political hay out of pornography for more than 20 years.'

After some preliminary legal skirmishing, the case came to court in late September. According to Barrie, 'Every ruling went against us from day one. In trying to get us convicted, the judge ruled that five very explicit photos – most depicting S&M – were to be considered in isolation from the rest of the exhibition. With the context removed, it's hard to understand why they were done.'

In American precedent law, the relevant ruling on obscenity is the 1973 Supreme Court decision in Miller v. the State of California, which established the 'three-prong test' – whether the object at hand is prurient to an average person in the community, whether it is patently offensive and whether it is without substantial literary, artistic, political or scientific value. During the 10-day trial, the Cincinnati Contemporary Arts Center argued 'artistic' value.

'The jury were all ordinary working-class people,' Barrie says, 'not usually interested in art; they found the pictures lewd and difficult, but our experts convinced them that the photographs were art, and that art doesn't have to be pretty. The jury acquitted us in two hours.'

The same week in which Barrie and his museum staff celebrated their acquittal, a black record-shop owner in Fort Lauderdale, Florida, Charles Freeman of E-C Records, was unable to invoke art as a defence. Charged with obscenity for selling the As Nasty As They Wanna Be album by rappers 2 Live Crew, Freeman was convicted by an all-white jury who felt that the record lacked artistic merit. But at the end of last month 2 Live Crew themselves were acquitted in Miami by a jury which said that they found the album's lyrics tasteless, but not obscene.

The United States is at war with itself. These three recent obscenity trials are only the tip of an iceberg that is chilling the cultural and social climate of the nation: the implementation by pressure groups, politicians and fundamentalist Christian leaders of a concerted attack on a wide range of cultural and social activities, from heavy metal, rap and rock to performance art and photography. American culture has always swung between the extremes of secular excess and religious puritanism. In 1930, the Hays Code put an end to the bacchanalia of Hollywood with a long list of on-screen restrictions. In the 1950s, Senator McCarthy presided over a climate of fear based, not on sex, but on politics. In 1966, religious groups burned Beatles records in response to John Lennon's comment that the group was 'more popular than Jesus'.

After the hedonism of the 1970s, America has swung back to its censorious, witch-hunting pole. Helped by a shift in power politics and the incidence of AIDS, the New Right has explicitly targeted and rolled back gains made in the 1960s, such as the liberalization of laws on abortion, homosexuality and women's rights. It is difficult not to see the current spate of anti-obscenity agitation as the cultural equivalent of this 'post-liberal' shift.

This impulse is usually denied by those involved in monitoring music or art. 'We are not trying to get groups banned,' says Jennifer Norwood, director of the influential watchdog organization the Parents Music Resource Center (PMRC), which was founded in 1985 by Tipper Gore, wife of the Democratic presidential hopeful Albert Gore. 'We seek to create an awareness about music marketed to young children that promotes destructive behaviour or is graphically violent.'

Both the PMRC and Jesse Helms, who has introduced an amendment to tighten up obscenity laws, have tapped middle- and working-class fears that America's culture is being swamped by violence – that it is, to quote *Time* magazine's May coverline, 'foul-mouthed'. 'My amendment does not prevent the production of vulgar works,' Jesse Helms has said. 'It simply provides some common-sense restrictions on what is and what is not an appropriate level of federal funding on the arts.' According to a July Gallup poll for *Newsweek*, 71 per cent of Americans believe obscenity has increased in the arts and 78 per cent think parents should do more to protect their children from it. Both are issues targeted by Helms and the PMRC.

As a result of the PMRC's campaign, the American music industry routinely labels albums by leading artists like Prince with the legend: 'Parental Advisory: Explicit Lyrics'. These records cannot be sold to minors. But the labelling has not prevented further attacks on pop music, whether in a spate of trials of heavy metal bands, local ordinances preventing minors from wearing clothing associated with rap or heavy metal, or outright legal bannings.

After a year of escalating legal action, both sides have become polarized. If the PMRC and Helms disavow illiberal intent, they are confronted by scores of voices shouting the Constitution's First Amendment. As displayed on 2 Live Crew member Luther Campbell's new album, *Banned in the USA*, this reads: 'Congress shall make no law respecting an establishment of religion, or prohibiting the free exercise thereof; or abridging the freedom of speech or of the press.'

The terms of the debate have been widened from control of excess to freedom of speech. This reflects the underlying seriousness of the struggle, which is not so much cultural as social and political: a struggle between opposing world views based on the lines of true power in US society. In one corner are senators, presidential hopefuls and wealthy fundamentalist leaders; in the other are music stars, outsider artists and anyone who is culturally, racially or sexually different – but most of all blacks. At stake is the governance of an America in crisis.

'The whole thing reminds me of a *Twilight Zone* episode where aliens invade a small town,' says artist Andres Serrano. 'Except you never saw the aliens. What happened was that the town destroyed itself with paranoia, each person pointing an accusing finger at their neighbour. The enemy was within.'

In May 1989, this quietly spoken New Yorker of Afro-Cuban, Chinese and Honduran descent made national news when he was denounced by Senator Alphonse D'Amato in the US Senate as producing 'shocking and abhorrent work'. The ensuing furore kick-started Helms's campaign against the National Endowment for the Arts, the body that distributes public funds for the arts. Two weeks later, a part NEA-funded Robert Mapplethorpe exhibition was cancelled in Washington. Earlier this year, feminist performance artist Karen Finlay was also refused an NEA grant.

Taken in isolation, the photograph that inspired this fury, 'Piss Christ', by Andres Serrano, can seem shocking. But as ever, the truth behind it is more complex. A five-foot by three-foot print of a wood and plastic crucifix submerged in yellow liquid, 'Piss Christ' would be a beautiful, if unsettling, image if Serrano had not broadcast his materials in his title. Once you know that urine is involved, the image changes instantly into an object of disgust or a paradigm of the soul/body split so common to religion.

Serrano's own motivation is complex. 'I have a Catholic background,' he says, 'so I am obsessed by symbols. As an artist and a former Catholic, I thought nothing of using that symbol. It's so ingrained, you really don't question why you're doing the work you're doing. "Piss Christ" reflects my own ambivalent feelings about the Catholic Church: feeling drawn to Christ, but resisting organized religion.'

'Piss Christ' is a serious work, but to some it is still offensive. Should it be funded out of the public purse? In fact, Serrano was one of 10 artists partici-pating in a show towards which the NEA contributed a quarter of the cost. The NEA's function is not to fund the hopelessly avant-garde, but to enable the potentially profitable to reach a wider public. This is how 'Piss Christ' was seen. The NEA is also prohibited by law 'from exercising any direction, super-vision or control' over the conduct of recipients of its assistance.

More importantly, the Serrano incident is a good example of how pressure groups have succeeded in making national issues out of small cases. 'Donald Wildmon's AFA [American Family Association] started a campaign two months after the show came down,' Serrano says. 'It consisted of over 500,000 letters. According to the records of the IRS, they spent another $2 million on a separate mailing which urged people to write to the NEA, to their Congressman and to the organizations which had co-funded the show, like the Rockefeller Foundation.'

Fundamentalist Christian organizations like the AFA and Focus on the Family (FOF) are trying, in Wildmon's words, 'to influence our society to adopt values that are the norm'. During the past three years, their success in influencing America – through censoring magazines, records, TV shows and adverts like Madonna's 'Like a Prayer' Pepsi campaign – has been dispropor-tionate to their constituency in the population. Despite their fears, 75 per cent of those polled by *Newsweek* 'don't want anyone imposing new laws on what they can see or hear'.

The new McCarthyites, as they have been dubbed, succeed because they are well organized and well off, and can therefore afford to exploit direct mail-ing and the fax machine, and because they are prepared to be covert. Dennis Barrie says, 'The only non-police prosecution witness in our trials was a "media consultant" called Dr Judith Reisman. When our attorneys asked her whether she worked for the AFA, she denied it. But our attorneys had the AFA's IRS statement which showed a payment to her of $24,000. She was paid by them to testify and had to back down in court, and the jury saw through her.'

Affiliations within the PMRC are more shadowy. 'We are not a member-ship organization,' says Jennifer Norwood. 'All we have is a board of 22 peo-ple, and I don't know what groups, if any, anybody on the board belongs to. We are not dictated to by the AFA or the FOF.' But a closer look at the PMRC board reveals some interesting power lines. According to researcher Dave Marsh, Susan Baker of the PMRC – who happens to be the wife of the current US Secretary of the State, James Baker – is also on the board of the FOF. And PMRC co-founder Sally Nevius also endorses ministers who stage record burnings around the country.

The fundamentalist agenda has been prominent in the censors' concentra-tion on heavy metal, which often delights in brandishing occult imagery. The fundamentalist view that heavy metal is part of a deliberate and carefully orchestrated plot to seduce teenagers to the Devil – as expressed in comics and books such as the controversial *Stairway to Hell* – is much quoted by politicians like former Senator for Missouri, Jean Dixon, who said in an interview that if rock and roll is about rebellion, then 'rebellion is witchcraft'.

Central to the fundamentalist case against heavy metal is the concept of backwards masking, the insertion of subliminal Satanistic messages into rock records. 'Induction into the world-wide Church of Satan is predicated on the ability to say the Lord's Prayer backwards,' states *Stairway to Hell* author Rick Jones. This was the crux of a trial in Nevada in September, when the group Judas Priest was accused of causing the suicides of two high-school drop-outs, Jay Vance and Ray Belknap, who were listening to the group's *Stained Class* album on the day, in late 1985, when they blew themselves to smithereens.

The judge's acquittal of Judas Priest hasn't stopped future cases – former Black Sabbath singer Ozzy Osbourne goes on trial on a similar charge in January – nor did it totally reject the concept of backwards masking. It also failed to address the teenagers' suicides in the context of which they occurred: the blasted lives of these two poor boys to which heavy metal – with its fantasies of power and control – was the soundtrack.

In the second half of 1990, America's recession has deepened into depression. Even before the full economic effect of the military operation in the Gulf is felt, unemployment figures are showing a rise of 5.7 per cent, with half a million jobs going in the past three months alone. Pop music speaks out about this and is therefore targeted because it brings to light a whole range of experiences which are often denied in more mainstream media.

Nowhere is this truer than in rap music, which is now bearing the brunt of hard-line censorship. Many think 2 Live Crew were targeted precisely because their record could not fulfil the three-prong test against obscenity. The case itself was another fundamentalist baby: prosecution lawyer Jack Thompson first heard of the group as a result of an AFA mailing.

As Nasty As They Wanna Be is firmly within a tradition that lies between National Lampoon's *Animal House* and the comedy of Redd Foxx or Eddie Murphy. Although there is a call and response section where female voices 'dis' (disrespect) the male rappers' smut, the 2 Live Crew's general view of women is demeaning. Yet *As Nasty As They Wanna Be* is nowhere near as bad as the comedy performed by the white American Andrew Dice Clay, who, instead of a ban, was given a Hollywood contract.

Despite their legal notoriety, 2 Live Crew are nowhere near rap's cutting edge. In response to appalling social conditions in the inner cities, rap is now moving into a stage like British punk in the late 1970s. 'We tell the truth,' says rap artist Dr Dre of the group NWA (Niggaz With Attitude). 'And the truth is bitter.' Coming out of the South-Central Los Angeles district of

Compton, where two-thirds of all young black men have been arrested, NWA present a heightened ghetto reality, both to make money and to inform the world about inequities that are usually swept under the carpet.

Their style of music is called 'gangsta rap', and shock is its initial aesthetic. NWA's first album, *Straight Outta Compton*, is a furious mixture of hard, compressed music and raps that express frustration, black pride and activism, and the social Darwinism of the black dispossessed. This is a potent mixture: apart from sundry bannings, NWA have also attracted the attention of the FBI with their rap, 'F — k tha Police', a protest about the LAPD's infamous 1988 drug sweeps.

The more they are demonized, the more these groups throw it in your face. In September the most brutal rap album yet, by Def American's Geto Boys, was so offensive to the workers at the CD plant that they refused to press the record. When the group's American distributor, David Geffen, heard the record, he terminated his agreement.

Listening to the Geto Boys seems to bear all this out: 'F — 'em' they roar incessantly. But further listening reveals not only a critique of white media demonization but positive messages: wear a condom, don't let white America tell you to fail. 'When people were offended by our first record,' says Geto Boys writer Bushwick Bill, 'we then made a radio record, "You Gona Be Down", which said: "If you see a crook in action, be down and call the police." The people used to hearing our stuff didn't like it, so we have to write about the way it is where we live, in Houston's fifth ward – it's violent.'

It may be true that gangsta rap plays on white fears and prejudices, but it is often the only way these groups can get heard. And the records sell, in hundreds of thousands, to whites as well as blacks. 'When it comes to blacks in the media,' says Bushwick Bill, who has studied sociology for three years at night school, 'we're all rapists, murders, pushers. Whenever a productive black person who is off drugs and is intelligent speaks out, they are a threat to white America.'

If the rappers speak of realities many white Americans do not wish to hear about, they do so in a language that very few take time to understand. This intensifies the cycle of frustration which now permeates America's culture: any understanding of plurality is in short supply in a country racked by racial and class conflict. In this, the fundamentalists offer certainty cloaked in intolerance.

The response of the industries under attack reflects their cultural atomization. While the art world has reacted quickly, the record industry has been much slower to act. Many companies have thought it better to work with the PMRC, an approach rejected by Virgin America's president, Jeff Ayeroff. 'The PMRC speak from a bully pulpit, with their husbands in the background, and the music industry has acted like wimps on this issue. For us to be corrupted by this reactionary agenda betrays a basic lack of trust in what we are making our money from.'

Ayeroff labels Virgin's products, not with the PMRC-approved text, but an anti-censorship tract. He was also involved with Rock the Vote, an industry pressure group which was collecting funds to bring out thousands of new voters in North Carolina for last week's Senate race between Jesse Helms and a black moderate Harvey Gantt.

'One of the reasons we're being pushed around is that we have no constituency,' Ayeroff says. 'The minute these politicians realize that there is a significant margin in the 18–25 vote and the black vote, they'll have to call a different tune.'

Censorship and freedom of expression will be key issues of the 1990s. As Western societies go into recession, dissonant opinions will become more common and less tolerated. Recently, Britain has seen a whole range of restrictions on the media and on sexual activities, whether in Clause 28 of the 1989 Local Government Bill or the latest broadcasting bill demanding political 'balance'. Gay bookshops are regularly raided and last month several record chains refused to stock NWA's oral sex rap 'Just Don't Bite It'. The power struggle in the US between fundamentalism, hedonism and secular beliefs is already being enacted here.

Observer, 11 November 1990

City of Quartz: Mike Davis's LA

'Bloody red sun of fantastic LA,' the Doors sang in 1970, and we think we know all about Los Angeles's apocalyptic allure. We think we should, so thoroughly have images from LA's several fantasy factories penetrated our subconscious. For 80 years now, this city at the edge of the world has been the setting for myriad dreamscapes: to name but a few, the paranoid odysseys of Chandler; the Beach Boys' endless summer; the futuristic techno-playground of the Byrds; the stinking cesspit of James Ellroy's hyper-*noir*; the Darwinist ghetto reality of NWA.

Ever since the Sennett comedies used the building sites of Hollywood as an idealized location that advertised Sun! Space! Glamour! to the world, we have seen the world through Los Angeles. It is not so much a city as a way of perception which we now share, thanks to Hollywood's status as the cultural vanguard of American consumer capitalism. As Michael Sorkin writes: 'LA is probably the most mediated town in America, nearly unviewable save through the fiction scrim of its mythologizers.'

When you first visit Los Angeles, the persistent sensation – half pleasurable, half queasy – is the buzzing hallucination that you are living in someone else's script. Yet, as Mike Davis notes in *City of Quartz*, these half-formed impressions dominate discourse about the city, which is described principally by expatriates, superior New Yorkers or Europeans, or by tourists: French philosophers who trip out on the city's surface unreality, or pop fans trying to find out about their culture.

'LA is usually seen through the lens of West Hollywood and the Hollywood Hills,' Davis told me. 'It seems to be a strange phenomenon that happens in Hollywood: the critical faculty put to sleep by the swimming pool. Even heroes of mine like Brecht, when they ventured out into the city, were remarkably unadventurous. The absence of autobiography from intelligentsia born and raised here has been unusual. It's due to hyper-mobility, and it's also a fault of the institutions. Nobody teaches South Californian history locally.'

The result is to clear the mediation that envelopes the city like an intellectual smog. The whole point of Davis's dense, brilliant book is to point out that LA is not just a dream but a carefully planned and patrolled city which – contrary to the postmodern gloss which it has acquired – enshrines class and racial inequality for the millions of people who live or exist there. The reality of Los Angeles is far stranger and harsher than we could have ever imagined.

Take MacArthur Park. In 1968, Richard Harris broadcast the name of this small green space throughout the world with an unsettling seven minutes of super-schlock. Ten years later, Donna Summer did it again. Before the Second World War the park had been a middle-class area, situated between the business district of Downtown and the consumer zone of Wilshire Boulevard, but by the time Jim Webb wrote about it you would have been more likely to find John Rechy's hustlers there rather than any cakes left out in the rain.

If in 1978 MacArthur Park was the site of the first major LA punk riot, today theatrical violence is replaced by class violence. Mike Davis explains, 'The park is a battlefield between the poor section of the Salvadorean community and the Los Angeles Police Department [LAPD]. It's a teeming tenement area: there are so many kids living feral off the streets. There's crack and teenage prostitution. The LAPD used the excuse that there were street sellers to have a crackdown: if you're in any Latin city, there are always street sellers, so a whole section of the community has been criminalized.'

In *City of Quartz* Davis deep-mines the various strata of Los Angeles to draw a new graph of the city he both loves and hates – a graph based not on tourist trips but on power lines. In his seven chapters and prologue, almost all the bases are covered: Hollywood; the Catholic church; the art and architecture industries; heavy industries; international banking; the police force; middle-class pressure groups; LA's starring appearance in its own myth.

Davis begins with an end – the ruined 1910s socialist utopia at Llano, in the Mojave valley – and ends with a beginning: a heartfelt history of faded steel town Fontana, 60 miles east of LA, where he was born. 'Writing as a socialist, and as someone who comes from the absolute outer periphery, the last suburb,' he explains, 'I have a decentred vision. I was very influenced by having lived in Glasgow and Belfast: my ambition was to write something as rooted in Los Angeles.'

To counterbalance the postmodernists' nihilist *jouissance*, Davis fuses more traditional Marxist analyses – class and economics – with a *noir* sense of terrain. If LA today shows 'an unprecedented tendency to merge urban design, architecture and the police apparatus into a single, comprehensive security effort', then Davis's history shows how fantasy and climate merely serve to massage – for the world and the city's residents, who buy into California's dream if they can – the steely exercise of oligarchical power.

The LA that we know was created by a tiny group of white men like Otis Chandler, who promoted the foundling SoCal storyline. As Kevin Starr writes, 'a mélange of mission myth, obsession with climate, political conservatism (symbolized in open shop) and a thinly veiled racialism.' According to Davis, this 'ideology of Los Angeles as the utopia of Aryan supremacism – the sunny refuge of White Protestant America' has informed LA's film, scientific and business communities since the turn of this century. For proof that it is still vigorous, you only need to listen to the music of Guns 'N' Roses.

The result is a city where there is no such thing as society. 'From the 1940s to the 1960s,' Davis comments, 'California invested very heavily in the social infrastructure for growth. LA's schools are among the worst in the country, as are the emergency services. The levels of infant mortality are astronomic. No social infrastructure has been created. In that sense, the Californian dream has collapsed.'

Enforcing these social conditions is the LAPD, historically a state within a state, and a body which has now become completely autonomous. Naturally, this 50-year process has been fictionalized: it's the difference between *Dragnet* and *Robocop*. 'Police chief Gates is more powerful than the mayor,' says Davis. 'He now says repeatedly that casual drug users should be shot and all the petty criminals of the city should be put in concentration camps and surrounded by minefields. Is this hyperbole? No.'

Today, the LAPD have transformed the poor suburbs of South-Central LA into a domestic Vietnam. This is an extreme response to 30 years of inequality: in districts such as Watts and NWA's Compton, black youth 'experienced the 1959–65 period – the white kids' endless summer – as a winter of discontent – black unemployment skyrocketed from 12 to 20 per cent (30 per cent in Watts). In August 1965, Watts exploded in a "festival of the oppressed" that the black community called a rebellion and the white media a riot.'

The transition from the pride that you can see in 'Eyes on the Prize' to the 'Social Darwinism' of today's ghetto can be told through the story of the Black Panthers. In February 1969, two Panther leaders were murdered on the campus of UCLA, and the following year an armed SWAT team surrounded their HQ. 'Their' decimation led directly to a recrudescence of gangs during the early 1970s, when the Crips and the Bloods took over the Panthers' social organization, but without the politics. Politics had failed, it was now me first, or as Eazy-E raps, 'It ain't about colour; it's about the colour of money.'

By 1983, it has been estimated that black unemployment in LA was four times higher than in 1960. As Davis states, quoting UCLA economist Paul Bullock, 'the last rational option open to Watts youth – at least in the neo-classical sense of utility-maximizing behaviour – was to sell drugs.' In a direct imitation of the business behaviour valorized today by white society, black gangs sell crack as the ultimate freelance activity, as the ultimate enslaving product.

With this trade as an excuse, the LAPD have turned South-Central into an exclusion zone. With increased powers, thanks to a media demonstration of drugs gangs, the police can now carry out sweeps like 1988's 'Operation Hammer', during which innocent civilians were arrested, and very few real criminals apprehended.

The impact on civil rights has been incredible. According to Davis, since 1974 two-thirds of all younger black males in the state of California have been arrested, and they are virtually unable to move out of the ghetto without arrest. 'In Los Angeles, once upon a time a demi-paradise of free beaches, luxurious parks and 'cruising strips', genuinely democratic space is all but extinct. The Oz-like archipelago of Westside pleasure domes is dependent on the social imprisonment of the Third World service proletariat who lives in increasingly repressive ghettos and barrios. In a city of several million yearning immigrants, parks are becoming derelict and beaches more segregated, libraries and playgrounds are closing, youth congregations of ordinary kinds are banned, and the streets are becoming more desolate and dangerous.'

Today, even a cursory swerve off the beaten track can reveal hell. I found it, by accident, in LA's Skid Row, which, even from a car, is the worst thing I have ever seen. Even more disturbingly, Skid Row is only a few blocks away from Downtown's financial district where, *inter alia*, most of Japan's $3.05 billion investment in LA real estate is turned into steel and glass. No attempt is made to ameliorate the situation; indeed, the local strategy is 'containment' and petty harassment in what is seen as an 'urban civil war'.

'What's now really turning Skid Row into an inferno is crack,' says Davis. 'There are also increasing amounts of poor Central Americans living there alongside the blacks. The fear now in the city is that we are facing the first regional recession since 1938, and the city's problems have always been perceived as knowing what to do with growth. There will soon be tens of thousands of the newest, poorest immigrants swelling the homeless. We are on the edge of the real social catastrophe which even a mild recession will bring about.'

'The ultimate world-historical significance of Los Angeles,' Davis writes in *City of Quartz*, 'is that it has come to play the double role of utopia and dystopia for advanced capitalism.' The city's importance to us is emphasized by its science fiction mythology: LA's future is, in part, ours.

Teetering on the edge of recession, Los Angeles is facing a massive struggle between a definite Anglo minority which nevertheless holds – along with the Japanese – all the sites of power, and *tsunami* wave of immigration, not only from Mexico, but from all around the world.

'The cultural definition of the poly-ethnic Los Angeles of the 2000 has barely begun,' writes Davis. 'The energy and idealism of these people is just like New York in 1910,' he told me. 'The most important reality in LA has gone unnoticed, and that is a whole raft of new writing which is coming onstream from Mexicans, Salvadoreans and blacks. This is a new, multicultural intelligentsia, based on the streets and barrios, which will transform the vision of LA. I feel like John the Baptist in my book, paving the way for them.'

20/20, Autumn 1990

I'm All Right, Jack

Adverts are always a good repository for received values – that's their job – and the current Ford Orion clip is a good case in point. The storyline goes: against a wash of urban chaos, late-thirties beefcake begins to get into car with younger, leggy blonde. The camera zooms in on a pair of 'designer punks', an over-made-up compendium of new romantic, punk and more recent club styles. They look stupid. The beefcake rolls his eyes upwards. The blonde, with a look of infinite distaste, buzzes up the electronic window and slots in a CD: the hubbub is replaced with the soothing strains of classical music.

The car moves off, on a dual carriageway so empty that it could be one of Cecil Parkinson's fantasy toll roads. Up comes the Ford logo with the advertising copy, selling the Orion as 'a classic'. Standard aspirational stuff, you might think, but in its baldness, this Ford ad contains the trickling down of the New Right values and practices: fear of the outside world, distaste for the pluralism of popular culture, social and economic privatization, concentration on presentation to mask overt ideology.

It also prompts the question not fully addressed by the comments on Mrs Thatcher's resignation: to what extent has her personal style influenced the way we see and think about things. It's useful to remember that Mrs Thatcher began to find her idiosyncratic voice at the same time as punk, and that the two were to some degree symbiotic: the negative and positive solutions to the same problem – the collapse of the post-war consensus. Both shared a tough, absolute style which was explicitly anti-liberal; both used an apocalyptic rhetoric and appealed to the sado-masochistic instincts of the English.

As Peter York wrote: 'The Thatcher government was definitely an act of faith in the hard style. Might work. Try it.' There was also a clash between the punk and the Thatcher vision of Britain, and, when the Tories won in 1979, cynicism swept through post-punk pop culture. With the failure of the left to sustain any links with punk, the New Right, which, with Mrs Thatcher's support, was assuming the intellectual high ground, moved in and slowly usurped the rhetoric, the style, even the cast list of cultures which had opposed it, whether punk or feminism.

It is in the media that you can still see punk's deepest influence: two former punk commentators, Julie Burchill and Garry Bushell, attempt to perpetuate punk's class fantasies in terms of bigotry and bile. Their world is simple: might is right. This is a degraded, Little Englander version of a social Darwinism apparently endorsed by Mrs Thatcher and New Right intellectuals such as Roger Scruton; as the PM herself said in 1987: 'There is no such thing as society.' No doubt Thatcher was attempting to say something more complex, but when this statement interacted with the 'Loadsamoney' archetype it threw into stark relief the inequality promoted by Government policies.

There were so many who did not benefit from the Thatcher project, who indeed were specifically excluded: the old, the poor, the disabled, the unionized, and the more vocal 'minorities' such as blacks and gays. Thatcher's achievement, as Ian Jack wrote in the *Independent* last Saturday, was to make 'moralists – the people who droned and drivelled that they cared – seem like humbugs'. The charge of hypocrisy – 'You're doing OK so you can't complain' – is still the standard New Right rebuttal of any social criticism. This is a vision now rejected by the majority of the population.

The seventh British Social Attitudes report, published the week before Thatcher's resignation, showed 56 per cent of the electorate in favour of increasing taxes to pay for better social welfare. Areas particularly targeted were the NHS, education and state pensions. Also indicated was a growing awareness of green issues: 88 per cent said that industry should be prevented from causing damage to the countryside.

The British still want a collectivist, welfarist society: a desire which Mrs Thatcher has failed to eradicate. It's also interesting to note that most of the entertainment media have remained uncommitted towards Mrs Thatcher. This is quite correct, since the New Right has attacked them all in different ways. Pop music has been written off in a return to bourgeois 'high art' values, while the film and, in particular, the TV industries have been censored, and then destabilized by financial restructuring.

In the face of these attacks, there has been little organized intellectual opposition. Stemming from a much-needed early 1980s overhaul of traditional left attitudes towards consumption and pleasure, the 'new times' project by *Marxism Today*, and later the *New Statesman*, produced an intellectual collapse.

One of the few consistent beneficiaries of her three terms has been advertising: between 1980 and 1988 spending leapt from £43 billion to £88 billion a year. This is not surprising, considering the amount of Treasury money, £22 million, pumped into campaigns such as that for water privatization: just compare this to the £2 million on ads for teacher recruitment.

This is it, isn't it, the metaphor for the Thatcher years: the promise of a much-needed change which, despite the rhetoric, delivers a short-term, cosmetic solution. Like the advert, Mrs Thatcher's vision has been expensively and attractively dressed, but is ultimately transient: service industries rather than manufacturing, PR rather than genuine interaction, style rather thansubstance. Even her famed moral authority has become just another style, to be superseded by this year's more sombre model.

Observer, 2 December 1990

The Criminal Justice Bill: Clause 25

Within the next two weeks, the House of Commons will be debating a bill which, if passed, could exacerbate the combination of law and enforcement which has made Britain the sick man of Europe as far as gay rights and freedoms are concerned.

Clause 25 of the proposed Criminal Justice Bill will empower the judiciary to impose stiffer sentences on 11 offences, some of which relate to child sexual abuse and prostitution. The three offences which involve gay men come under the 1956 and 1967 Sexual Offences Acts and include: indecency between men (S. 13, 1956: this includes any sexual or even affectionate contact by gay men outside the privacy of their own homes), solicitation by a man (S. 32, 1956) and procuring others to commit homosexual acts (S. 4, 1967).

These sections have been widely interpreted by the courts. Section 13 has been primarily used to prosecute outdoor sexual activity like 'cottaging' – having sex in public toilets. But the 1967 act, which decriminalized homosexuality, failed to repeal, or made serious crimes, other activities like procuring or soliciting (introducing or arranging a meeting) which might result in otherwise perfectly legal behaviour: i.e. sex between two men over the age of 21 in private.

Like most legalese, this is complicated and dry, but the effects are potentially devastating. At its most extreme, Clause 25 means that any contact made by one gay man to another in a public place – like chatting up, passing addresses, even smiling – to arrange sex has been ranked with serious sex crimes like rape and child abuse, and can be punished on first offence with the maximum sentence of two years in prison.

The Home Office state that it is necessary 'to protect the public from serious harm', but there is no serious harm to the public in the facilitation of otherwise legal acts, or even, although it may well not be to many people's taste, outdoor sex.

This is absurd and unjust, and has brought forward a storm of protest from gay pressure groups like OutRage and Stonewall. On the surface, the Government has been making encouragingly ameliorative noises: two weeks ago, Sir John Wheeler, Chairman of the Select Committee on Home Affairs, told *Capital Gay* that it was not intended 'to discriminate against the lifestyles of those people in the community who form relationships of a consenting nature where the law should not be concerned'.

Despite these words of comfort, this Government's good intentions about the gay community have been in considerable doubt since Clause 28 of 1988's Local Government Bill, which made it illegal for local councils to 'promote homosexuality'. Informed by evangelical organizations like the Christian Family Concern, there is a very strong Moral Right lobby within the Tory party – some estimates put it at about 90 MPs.

Any statement by the Government about gay rights has thus to be taken with a pinch of salt. The Government made ameliorative noises at about the same stage of Clause 28: this was just a marker and would not be put through the courts. Although there has been no test case, a recent study by the Law Department at the University of Wales, Cardiff, found examples of six local authorities which had refused applications for research grants or to allow the showing of plays because of the homosexual nature of the subject matter.

But the Government is just part of the story. Laws are powerful only in so far as they are applied, and this task is handled by the police and the judiciary. The police are very gung-ho about arresting gay men: according to researcher

Peter Tatchell, more than 2,500 men were convicted or cautioned for the offences of indecency and soliciting in 1988. This is the highest level of convictions since the mid-1950s, when male homosexuality was still illegal.

This high incidence is due to several factors, of which the ease of fulfilling arrest quotas by targeting gay men is not the only one. As pertinent is the moral agenda assumed by the police as an integral part of their social policy, as recently expressed by top-ranking policemen like David Owen, the president of the Association of Chief Police Officers, who blamed the 'anything goes ideas of the 1960s' for a 'decline in public and private morality'.

In practice, this translates into an obsessive interest in gay sexual offences: four-fifths of indecency offences result in a conviction or a caution, but only one-fifth of burglaries are solved. In certain parts of England, the rate of legal activity around gay lifestyles is so great that it amounts to a near recriminalization. So much so, that organizations like Stonewall are arguing the Criminal Justice Bill the other way: for less serious sentencing under the 1956 and 1967 acts, with the eventual aim being the repeal of the 1967 act in full parity with the laws governing heterosexual sex.

The Government's nervousness about Clause 25 stems partly from ignorance and partly from the fact that it is pursuing an agenda of its own without a strong mandate from the public. Despite the damage done to the public's perception of gays by the onset of AIDS and the consequent tabloid propaganda, there is little demand for the law enforcement which has made Britain the most punitive country – as far as homosexual matters are concerned – of Europe.

This national scandal is one result of the New Right agenda being fully worked out within the institutions of law and order. Like the moves against abortion and gay fostering, the attacks against the 1967 act are part of a general New Right attack against the liberal pluralism and genuine social gains of the 1960s – stimulated by the constant anti-1960s rhetoric pumped out by senior politicians like Mrs Thatcher and Norman Tebbit.

Despite the fact that the New Right is almost played out intellectually, their influence and ability to set an agenda are still barely challenged. The only papers to report the scandal of Clause 25 and the increased rate of convictions have been the gay weeklies like the *Pink Paper* or *Capital Gay*; what used to be called the 'liberal media' have been almost silent so far. There's no reason why they should be: the time for the liberal backlash has never been riper.

Observer, 10 February 1991

My Bloody Valentine: Feedback to the Future

There may be a half-formed thought in your head, buzzing vaguely like low-level background noise, that the language we use to talk about music – of whatever kind, although here we're talking about pop – is inadequate.

So many people want pop to be some*thing* – meaningless fun, baby-boomer nostalgia, age-cohort marketing, a political statement – that they will not let it be indistinct. Like the images you see on the television screen, pop is almost always over-lit. Yet it is exactly in the right-brain (intuitive) area of human activity that music can work so well: away from genre definitions, away from the cross-media apparatus of consumption, away even from musicians' intentions.

That age-old pop cliché – 'We don't want to be defined' – has a basis in truth. Today, there is a gulf between the music that is being produced and the way that it is being perceived. The dominant perception of pop music – in upmarket newspapers, at least – is of nostalgia, boredom, the heavy sighs of people grown old before their time. This is an unthinking, arrogant reflex. I'd much rather talk about the materiality of sound, or, as Dimensional Holofonic Sound state so eloquently on their House cut-up of the Beatles' 'Revolution Number Nine', 'the difference between noise and music'.

My Bloody Valentine's latest EP, *Tremolo*, begins with a song called 'To Here Knows When', which 'samples' a low-level catastrophe. You wouldn't listen for it unless it was pointed out to you, but you would hear it as a vague, percussive rumble that fades in and out, stimulating, perhaps, a sense of unease, at odds with the sweet female voice and the reassuringly looped – if distorted – guitar melodies.

'That's the main thing that throws people,' says MBV writer and guitarist Kevin Shields. 'The sound of a disaster sampled and looped all the way through. It doesn't get past a certain frequency, so it has the sound of a bad cassette, and yet further up there's a tambourine on more like a hi-fi frequency, which you wouldn't get with a bad cassette.

'The idea was to make you feel that the rhythm had gone off one way or another, while it stayed perfectly in time. Like a train noise which is rhythm and rumble at the same time. But this nice little 10-minute idea took six weeks of work on the rhythms, all of which was scrapped in the end and replaced by this rumble. I find it fascinating the noises that people will accept. Have you ever heard the tube at Old Street? It's the most extraordinary screeching sound. All these City people stand around every day, blotting it

1out. If you put them in a room with this noise they'd cover their ears, but because it's escalators they accept it.'

Through a sequence of three EPs and one LP – 1988's *Isn't Anything* – My Bloody Valentine have explored that crossover between what we think of as noise and as music, testing the limits of studio technology and their own inexperience. Working instinctively, using trial and error, they have produced a body of work that tests our perception of music and the world around us with an experimental glee rarely heard outside the most extreme House or Techno records.

The title of their new EP, *Tremolo*, refers to their first breakthrough. Beginning, as Shields put it, as 'a kind of version of the surrealness of the Ramones,' My Bloody Valentine had by 1987 evolved into a 'number one Jesus and Mary Chain rip-off band. We had the same noisy guitar.' But there were scores of independent groups attempting to marry Beach Boys melodies with squalls of feedback; what marked My Bloody Valentine out, on 1988's 'Slow', was a swooning intensity, conveyed both by new arrival Bilinda Butcher's voice and the liberal use of the tremolo arm.

'We use it all the time, in a way where you don't think about what you're doing,' says Shields. 'You just do it, get into it, listen to it when you play it, in the same way that happens when you do anything a lot – you don't think about it any more. Most people who use tremolo arms think about it; no one has made the tremolo just an extension of strumming.'

Allied to this self-induced hypnosis is a whole array of perceptual effects. Most obvious is the mark of 'the devil's music' – songs played backwards. 'We don't actually use backward guitars,' says Shields. 'That sound is to do with the relation between the attack and the decay on the sound envelope. In most people's minds, "backwards" is attack and delay being the wrong way round: we're doing that constantly, but we don't run the tape backwards.'

A good example of this is 'Glider', the title track from their most successful 1990 EP (which also contains 'Soon', the most convincing fusion of white pop and House music to date). This is an instrumental so psychoactive that it near makes you faint. 'It sounds as though there are loops but there are no loops at all. I actually played it manually. The sampled noises are samples of feedback. We sat for hours listening to feedback, waiting for pieces of melody to come through and then sampled them, let it build. Feedback has some great qualities, it twists itself into tunes, it changes frequencies as it goes along.

'"Soon" could have been made in the 1960s,' Shields adds. 'Every piece of equipment we use was available then: Jaguar guitars, Vox and Marshall amps. Yet it doesn't sound 1960s, because we approach old equipment with a completely new attitude. The guitar sound on 'Slow' comes from a kind of reverb that reverses the attack, so you get a backwards feel; if you play it a certain way you don't hear that backwards effect, you get a smooth, constant sound, as if it joins everything up. That was only invented in the mid-1980s. You have people like Phil Collins who have the most advanced equipment possible, yet the music they make doesn't reflect that. Only the Beatles used the most advanced equipment of their day and moved everything forward. In a way we've been given the sort of freedom they had,' says Shields.

My Bloody Valentine have benefited from the space opened up within music production by the increased commercial viability of independent labels. Thanks to distribution systems like Pinnacle, independents now regularly get chart records. Even 'Glider' went to number 41.

In person, Kevin Shields – a Dubliner brought up in New York – engages in enthusiastic conversation about the properties of sound but defies any analysis of the overt meaning of their lyrics or the politics of the group's severe androgyny: 'I hate saying what the songs are about.'

So what you hear is what stimulates your imagination. We've all heard that before, but like all great pop MBV convince you to suspend belief, to see the world in a new way. Much of this is due to their thorough immersion in their own subconscious: 'We just follow our instincts,' says Shields. 'There's a point at the verge of unconsciousness when you're much more able to do something than you ever would otherwise.'

It's this moony, hermetic quality that gives My Bloody Valentine their power. Unlike most modern pop, they offer not a palliative but an aural solution to contemporary problems. Both in their constitution – two women and two men – and their records – both noise and music – they suggest not only a fusion of apparent opposites but a way through the chaos that is today's emotional and physical reality. Through their profound, almost environmental, acceptance of confusion they make a different future conceivable.

20/20, Spring 1991

Tasteful Tales:
Todd Haynes's *Poison*

The first thing you see is a caption: 'The whole world is dying of panicky fright.' Is this to locate us within the film's plot, or is it a general statement about our underlying emotional condition? The first shot is a black and white POV, of police battering down your door. The POV rushes through a nondescript, 1950s/1980s apartment – utility furniture, shot in harsh black and white – before rushing to a window and dissolving into light.

Jump cut to a 1980s news report, all flat US tape colour, about a Long Island boy who has flown away. But the voice-over – 'Who was Richie Beacon and where is he now?' – is too rhetorical for the pawky certainties of most news reporters. Another cut takes you immediately into the title sequence, where another POV travels with a hand around a lush set design. It then unfolds into the more familiar territory of period costume and Genet homage.

Poison is written, directed and co-edited by Todd Haynes, a Brown University art/semiotics graduate. Costing $255,000, it is his first feature proper; previous credits include *The Suicide* (1978), an examination of contemporary teenage life, and *Superstar: Karen Carpenter Story* (1987), where the story of the Carpenters is told through a dazzling variety of media techniques: vox pops, POVs, voice-overs, captions and reconstructions.

Superstar is infamous for its central device: the reconstruction of Karen Carpenter's life and death (from anorexia) is told not through actors, but mutated Barbie dolls. As Karen's desperate fight for control over her life unfolds, the narrative reveals the emotions raging beneath the Carpenters' blank façade. The formal shock posed by the dolls' appearance forces you to concentrate on the film's overt message: 'As we investigate the story of Karen Carpenter's life and death, we are provided with an extremely graphic picture of the internal experience of contemporary femininity.'

As you might expect from Haynes's previous work, *Poison* is a film full of ideas, aphorisms and deliberate perceptual tricks, all spinning around the montage of three separate storylines. As with all good montage, this placing together of disparate elements sets up a vortex of ideas and feelings, not all of which may have been consciously intended by the assembler.

Containing explicitly male homosexuality and funded by the NEA, the film fell foul of the American Family Association and was, as usual, boosted by the attendant publicity. This slots it into several easy matrices (America's current moral spasm; homophobia; the public funding of the arts; the symbiosis between the censor and censored). Haynes is nothing if not a didactic filmmaker, but *Poison* is most interesting, and moving, when seen as a film about perception and limits.

For instance, you could jump right in and say that the first sequence, 'Horror', is obviously 'about' AIDS. Shot in the apocalyptic style of 1950s science fiction films – with a voice-over that has the luxurious menace of a Vincent Price and tight close-ups on the lights, fans, grilles and vortices that were the staples of *noir* – the storyline follows a scientist, Thomas Graves, who synthesizes a distillation of the sex drive into liquid form. He drinks it by accident and develops a disfiguring, easily contagious disease characterized in the film as leprosy.

'Do I look like the pitiful, decrepit result of some indulgence,' Graves hisses, parroting the Moral Right's dismissal (both rhetorical and financial) of gay PWAs. There is much anger here, but as the sequence moves in the last third of the film into pure *Invasion of the Body Snatchers* territory, with a vengeful crowd of 'ordinary people' in pursuit of the outcast, a wider alienation presents itself. As David Wojnarovicz says in his searing account of seropositivity, *Close to the Knives*, 'My rage is really about the fact that when I was told that I'd contracted this virus it didn't take me long to realize that I'd contracted a diseased society as well.' This is a magical conversion necessary for 'deviants' – that is those described not by themselves but by society as abnormal – to function with any sense of self-worth. This treacherous odyssey is undergone by any of us whose sexual proclivities define us, both from within and without, as homosexuals.

The second storyline, 'Hero', concerns the agents of naming: the media, and in particular current affairs. With its intrusive, normative style – voice-overs, vox pops, name and status captions – 'Hero', like the worst of news journalists, bludgeons the story out of its protagonists; but then the story itself starts to slide. Most documentaries set up contradictions. In 'Hero', wildly varying opinions on the Bad Seed, Richie Beacon – the 'perfect child' who 'liked controlling people', who had an infectious anal discharge, who killed his father – unfold to the point where the story is 'understood', but then this understanding is cut away by Richie's magical disappearance – like Peter Pan, he flies away. Here, the intensity of his mother's awe cuts through the 'objective' framing of the documentary.

The power of naming

Naming is crucial here. As *Poison*'s voice-over says, 'A child is born and he is given a name. Suddenly, he recognizes his position in the world. For many, this experience, like that of being born, is one of horror.' The consequences of naming can be disastrous, and have been at times for the homosexual communities. For instance, once the word 'homosexuality' was coined by the Viennese writer Karoly Maria Benkert in 1869, it quickly came to replace all previous epithets and, as David F. Greenberg points out in *The Construction of Homosexuality*, marks a mid-nineteenth-century change in attitudes, the net effect of which was to 'strengthen antihomosexual beliefs'. Homosexuality was soon legislated against in the UK by the Labouchère amendment of 1885. And 10 years later, this legal naming resulted in the archetypal scandal of Oscar Wilde.

Wilde's downfall gave a flippant piece of law-making the sanction of legitimized social prejudice: a current in British and American public life that continues today. Like many who have been the subject of scandals after him, Wilde was punished for being 'public' (or 'out') with activities which were regarded as only conductible in private, if at all.

Yet for members of a putative 'deviant minority', what is in the 'public' sphere – the language of law, government, police, market research, media – almost always fails to reflect life as it is lived. The disparity between your individual sense of self-worth and the pejorative estimation placed on you by society is often vast. The psychic consequences are considerable, and still ill-understood: they can be productive or they can be annihilating.

The disparity between the public and the private is the area in which pop culture and subculture work so effectively – and is why they have been targeted so heavily by the New Right. As Richard Davenport-Hines states in *Sex, Death and Punishment*: 'Gay men and lesbians cause doubt, confirm uncertainty, emphasize differentiation, symbolize contradiction, and confront normality. The vagrant impulse which characterizes homosexual desire, the insolent language and camp gestures, and above all *doubtfulness* are all horribly threatening to authority and authoritarian personalities.'

In the conversion of values necessary for empowerment, gestures and objects become talismanic. In his crystallization of subcultural theory, Dick Hebdige focuses on Jean Genet's tube of Vaseline: 'We are intrigued by the most mundane objects – a safety-pin, a pointed shoe, a motor cycle – which, none the less, like the tube of Vaseline, take on a symbolic dimension, becoming a form of stigmata, tokens of a self-imposed exile.'

Genet's writing centres on this magical conversion. His is a world of pure insult, a world in which society's values are turned on their head. And this is the subject of *Poison's* third narrative, 'Homo', an adaptation of *The Thief's Journal*. If Richie Beacon 'was always doing these private things', then John Broom, the hero of 'Homo', flourishes in his own, private world within prison: a micro-society in which privacy is well nigh impossible.

Broom is named 'homosexual' on his entry into prison, but this public definition is so far away from his experience that he shrugs it off. His love affair with Jack Bolton is played out within the normative public/private dichotomy. Whatever their 'private' feelings for each other, when others are present, Broom and Bolton bark at each other and collude in the ritual humiliation of the one out, femme homosexual prisoner. Although set in prison, 'Homo' plays down any sense of public activity. All the encounters are set within enclosed spaces, yet even here, volcanic feelings are barely to be admitted in a private language.

Broom and Bolton have internalized society's hostility, and the raunchiness of the climactic penetration sequence is not enough to dispel their implicit, and self-hating, equation of sex with violence. This is how oppression works: not just overtly, but covertly; in the subconscious as well as in the everyday. As Broom says in voice-over, 'All night long I built an imaginary life of which he was the centre. And I always gave that life, which was begun over and over, a violent end.'

Back to the 1950s

Since the onset of a 'gay cinema', gay skinheads and prison toughs have appeared *ad nauseam* in the films of Fassbinder and Derek Jarman. The relief of seeing confident imagery cannot be over-emphasized, but the very success of these and other directors has resulted in a new erotic shorthand, and it is in this hand that 'Homo' writes. Admitting a particular kind of sexiness is a good idea, but in the case of 'Homo' it also places a third of this ground-breaking statement into more familiar definitions. The attempt here by the *Village Voice* to equate *Poison* and *Salo* is unfortunate; in place of Pasolini's fundamental smash-up, *Poison* seems like an end-of-term paper.

Part of the problem lies in the film's sourcing of 1950s imagery; although both 'Homo' and 'Hero' source different periods (an idealized 1940s; a flat 1980s), it is 'Horror' that provides the dominant look. *Poison* uses the blurred 1950s/1980s aesthetic that in the hands of directors like David Lynch has defined pop postmodernism during the last five years. But it then transgresses it: unlike Lynch, Haynes has a sharp sense of progressive politics and the interaction between his narratives throws up fundamental psychic questions.

It has been the project of the New Right of the 1980s in both the US and the UK to erase the wildness and progressivism of the 1960s and restate the 1950s ideal of suburban, materialistic conformity. Pop culture has played its part in this: beginning with punk/new wave fashions, which re-emphasized the 1950s shape, that decade's style has proved dominant. Films like Lynch's *Blue Velvet* and Demme's *Something Wild* define this blur of 1980s values with the stylistic trappings of the 1950s. Yet the return of the decade in which pop culture was young and vigorous is a mockery: the 1950s flawed but powerful ideal of democratic consumption has been reneged upon; now it is enough, as Barbara Ehrenreich has pointed out, for fewer to consume more.

The result is the exclusion of those 'minorities' for whom pop culture brought visibility and power. It is impossible to speak about pop culture without the minority subcultures which have given individual artists, groups, stylists and writers their strength. Yet in the 1980s, subcultures have become style and style has become the preserve of the rich or right wing. In this, as Hanif Kureishi has noted in the last issue of *Sight and Sound*, *Blue Velvet* and *Something Wild* portray the deviant roots of pop culture as something dangerous, other. In the terrifying abyss that suddenly opens up beneath the suburban ideal, nonconformists are demonized in order to be excluded.

Poison speaks defiantly from the other side. The paranoid, apocalyptic air of 'Horror' reminds you that consumerism was as much a product of the Cold War as it was a liberation. It also rams home the point that we are now living in the future predicted by those paranoid 1950s science fiction films: the vortices that once suggested a near-terminal loss of control are now the staple of pop videos.

Queer planet

'The world is dying of panicky fright' and *Poison* makes all too clear the disastrous impact of AIDS and exclusion on the psyches of queer men: horror, loss, pain, paranoia, self-hatred, and an all-pervasive sense of the world closing in. *Poison* concentrates on periods inimical to queer culture – both the 1950s and the 1980s – and while it makes the point that queer life comes out of these periods, it presents this life against an environment that is uniformly hostile; despite moments of transcendence, it internalizes much of the decade which it pastiches. One way through this would be to move the time frame forward. To source that brief period between 1966 and 1982 that was more supportive of homosexuality would be to dispel the internalized message of self-hatred and doom which, in the hands of the New Right, AIDS has appeared to reinforce.

Unlike *Salo*, or another film from the 1960s widest outreach, *Performance*, *Poison* is a film of *limits*. In this respect, it is a film of its time. This is not a period when the public discourse of media, law and government is tolerant of wildness and experimentation. After 10 years of the assiduously promoted ascendancy of the New Right, the freedoms of the late 1960s now have to be regained, and then extended.

Like Jarman's *Queer Edward II* – the book of the film – *Poison* is an explicit engagement with the onset of AIDS and the bigotry which has grown in its wake. In this climate, a scene as, let's face it, *tasteful* as the climax of 'Homo' becomes scandalous. 'Sex is the only way to infuriate them,' wrote another master of insult, Joe Orton. 'Much more fucking and they'll be screaming hysterics in no time.'

Poison taps into the new, hostile homosexual politics, the very name of which is a conversion: Queer Nation, Queer Planet, Queer Queer Queer. Just like the captions in *Queer Edward II*, which turn around the slogans routinely used to chastise homosexuals – 'If you must be heterosexual, please try to be discreet' – *Poison* is a deliberate, stinging slap in the face.

Sight and Sound, October 1991

Sex and Martyrdom:
The Spanner Appeal

Next Tuesday, five middle-aged men will enter the House of Lords Committee Rooms in a final attempt to clear their names, for what they hope will be the end of a five-year nightmare. Within the Lords' ambience of panelled walls and leather chairs, they will look nondescript, like the local government officers, computer operators and engineers that they are. Or were, before the police investigation that changed their lives and sent them to prison for a crime they didn't know they had committed.

On 19 December 1990, in a judgement described variously as 'illiberal nonsense' (*The Times*), as setting 'disturbing precedents' (*Independent*) and as 'wide and worrying' (Liberty), Anthony Brown, Roland Jaggard, Colin Lasky, Saxon Lucas and Christopher Carter were convicted at the Old Bailey under the 1861 Offences Against the Person Act, for assault and aiding and abetting assault. They were part of an alleged 15-man 'homosexual porn ring' then undergoing two trials: one by law, the other by the tabloids.

What they actually did was to engage in homosexual, sado-masochistic sexual practices which, although on occasions extreme and certainly not to many people's taste, were consensual – i.e. there was no coercion involved – and, until the judgement, not thought to be illegal. However, in what is now recognized as a test case, Judge James Rant ruled that consent was no defence to the charge of assault, and that the acts were injurious to the public good.

This judgement was upheld in the Court of Appeal last February by Lord Lane – the man who had previously refused the appeal for the Birmingham Six and the Guildford Four. The principal precedent was an obscure 1934 case, where a man was found guilty of assault when he caned a woman 'for the purposes of sexual gratification'. Lane also rejected the defence of consent, quoting a 1980 Court of Appeal case about two youths who received minor injuries when they agreed to fight each other. Fights between consenting adults in public or private were ruled illegal, because there was 'no good reason' for the assaults to be allowed.

'The satisfying of the sado-masochistic libido does not come within the category of good reason,' Lane has added. This ruling has effectively tightened legal control over the body: S&M sex which involves the breaking of the skin is now effectively illegal, as is body-piercing where sexual pleasure and intent are involved. An ultimate application of the judgement would make other sexual activities like spanking and even the giving of love bites illegal, although a prosecution on this basis is highly unlikely.

The dry, abstruse language of the law often masks severe clashes of world-view. The Spanner case (named after the police's original operation) has been well chosen as a test case: the activities under question would be revolting to many people – especially when reported in lurid tabloid headlines like 'Castrate Me Plea of a Homo Ring Perv', or summarized, say, as 'branding with hot metal, beating with stinging nettles, sandpapering each other's testicles, and passing wires and safety pins into each other's penises'.

But should something be made illegal – and that law enforced with condign penalties – just because it offends taste? In this, the courts seem out of step with public opinion. After the Spanner verdict in 1990, editorials in the *Independent* and *The Times* put forward a similar view to that of Mrs Clare Brakspear, JP: 'Most of us are sickened by many of the sexual activities to which we are not personally inclined; but that such behaviour should come within the ambit of criminal law without reference to Parliament is cause for alarm to this law-abiding citizen.'

'I find it extraordinary that in 1992 a case like this could ever be prosecuted,' says Ann Mallalieu, the QC acting for two of the appellants and a Labour speaker on Home Office affairs in the House of Lords. 'It does not have public support; in my experience, people think it is a complete waste of public money. What on earth are we doing wasting police time and manpower on this case when so many more serious crimes are being committed and are not, apparently, being dealt with by the police force?'

To what extent, indeed, should the law intrude into people's sexuality and their private lives? England is already the most sexually policed country in Europe. There are many grounds for disquiet in the Spanner case, not only about the wider application of the judgement, which potentially criminalizes thousands of people, but also about the legal process by which judgement was reached on the basis of an obscure precedent dating back to the 1930s.

There are other questions: about the way in which the case was reported so emotively by the tabloids that one of the defendants was attacked outside the court and hospitalized; about the role of the Obscene Publications Squad in bringing the prosecution; about the severe sentencing of the defendants. Six received terms of three years or over, more serious than some cases of rape.

In December 1990 I spent a few days at the Old Bailey witnessing the hearing. It was one of the most depressing experiences of my life. Despite the full weight of the English legal machine, it was clear that this was a show trial and that the defendants were scapegoats. As a result, the law has effectively extended its reach into areas where, it can be argued, it has no remit and where, furthermore, it is seen to be contrary to reasonable opinion. Is this yet another case of English injustice?

Operation Spanner was triggered by the seizure in 1987 of four home-made videos in Bolton, Lancashire – one of which showed the cutting of a penis.

The material was so extreme that the investigation quickly passed to the Obscene Publications Squad, which made it its flagship operation. During the next two years, about 100 people – all homosexual – were interviewed, 42 were arrested, and 16 eventually committed for trial at Lambeth Magistrates Court.

There, the charges of assault were widened to the charge of a conspiracy to corrupt public morals, which meant that the case had to be referred to the Old Bailey. The conspiracy charges were then dropped as soon as the trial began. In a televised interview for Channel 4's *Out* series, defendant Anthony Brown specifically denied the conspiracy charges and any notion of a closed circle of people: 'I'd never met more than half of the defendants until we first appeared in court.'

At the same time it was ruled that the defence of principle offered under the 1967 Sexual Offences Act, which states that homosexual acts in private between consenting adults are within the law, did not apply in this case. Despite the fact that the defendants had not realized they were breaking the law (and thus willingly cooperated with the police investigation), consent was no defence. The defendants were therefore *de facto* guilty, and the Old Bailey hearing was purely about sentencing.

In this, a hostile climate of opinion was fostered by the police and the prosecution, who capitalized on legitimate fears about serious crimes like paedophilia and any potential public distaste for the S&M sex practised by some of the defendants. The prosecution successfully drew a vivid picture of a wild smorgasbord of violent, criminal sexuality, lumping together forms of activity like piercing and sado-masochism which are often quite separate and which are, furthermore, practised by both hetero- and homosexuals.

Had the defendants actually broken the law? Apart from the question of assault – until the judgement not previously applied to S&M activities – there is some doubt as to whether any of the defendants had broken the law as they thought it stood, particularly in regard to the videos. It would have been illegal if they had distributed the videos after making them, but there is no hint that they were produced for anything other than private consumption.

There is no doubt that Judge James Rant took the video-tapes into account – as many of us would have done. In his summing up, he told the Old Bailey, that 'much has been said about individual liberty, but the courts must draw the line between what is acceptable in a civilized society and what is not.'

However, it is unique in English law that someone has been found guilty for consenting to what can legally be described as an assault on their person, or giving an 'assault' to which the other person assents, indeed fully agrees.

Twelve of the 15 defendants received prison sentences, four of them suspended. Their private lives were exposed throughout the press: some were vilified in public, some lost their jobs and were forced to sell their houses, a couple became ill as a result of the strain. 'They have been shellshocked by the whole experience,' says solicitor Angus Hamilton, who has been working on the appeal. 'They didn't, and don't, believe that what they were doing was wrong.'

The Spanner trial has all the hallmarks of a classic moral panic: hysterical (and often inaccurate) media coverage exploiting current anxieties; an imputed conspiracy subverting the fabric of society; scapegoats selected and punished. The police often think in terms of the tabloids, with whom they have a symbiotic relationship: here a new 'threat' was discovered, exposed and dealt with quickly. A nice, neat storyline with sequel possibilities.

Moral panics are a constant in English society. The phrase was coined by sociologist Stanley Cohen in reference to the mod 'riots' at Brighton in 1964. Cohen used it to describe a process where 'deviant' outgroups are suddenly discovered (when they might have been in existence for years), and put under the harsh glare of media attention and legal scrutiny. The interesting question to ask is why this English perennial should take the form that it did in the Spanner trial: why, now, sado-masochistic homosexuals?

In Spanner, the police appear to have blended the discovery of extreme homosexual material into two other moral panics current at the start of the investigation in 1987: the 'paedophile club' then under investigation for the revolting murder of Jason Swift, and the persistent but unproven stories about 'snuff movies', films where someone is actually killed on camera. No one has yet found a snuff movie, but the severity of some of the Spanner material was enough to send the police down what has been an expensive road: one estimate is up to £3 million.

Spanner is also part of a wider pattern. As S&M activist Kellan Farshea states, 'It's a direct part of the backlash against gay people which comes out of the AIDS crisis.' Homosexuality has had a foul press, in the tabloids at least, since the recognized onset of AIDS in 1983. In 1986–7, at the height of what now seems like a modern hysteria, phrases and headlines like 'Gay plague', 'Gay killer bug' and 'Gay menace' were frequently used.

As Richard Davenport-Hines writes in *Sex, Death and Punishment*, his magisterial survey of the English sexual maze since the Renaissance, 'contagion fears have been politicized in order to attempt to frighten people into accepting a regime in which sexual appetite is regulated, eroticism is repressed, social conformity equated with health, and conspicuous people of all sorts treated as undesirable.'

This 'vindictive sanctimony' has been actively pursued by some fundamentalist Christians, and was given a boost by the Government in 1988, when Section 28 of the Local Government Act forbade local councils to 'promote homosexuality'.

This restrictive morality – which, along with its contradictory twin, free-market economics, has been a major plank of Conservative policy over the past two governments – has been actively pursued by the Obscene Publications Squad itself. The Squad has had a chequered history during the past 20 years: a May 1977 corruption trial ended in six officers being found guilty of taking bribes, while the previous Squad chief, Leslie Bennett, was forced to resign after appearing on the front of the *Sunday Mirror* in a dress.

The current chief, Superintendent Michael Hames, has allied himself with the values espoused by Mary Whitehouse's National Viewers' and Listeners' Association. Whitehouse herself brought a successful prosecution for blasphemous libel against *Gay News* in 1977. The NVALA has had an influence on government and police disproportionate to its status as a pressure group, reinforced by a fringe meeting at each year's Tory party conference. In March 1991 Kenneth Baker, the then Home Secretary, was guest speaker at a NVALA conference.

The NVALA has long campaigned for more restrictive obscenity laws, a campaign which accords with the wishes of the Obscene Publications Squad: the head of the Squad has attended NVALA meetings at the past seven Tory conferences. In October 1991, Superintendent Hames spoke at a meeting advertised in this way: 'URGENT! CHILDREN and the FAMILY increasingly AT RISK from violence/pornography/obscenity. NEW LAW NEEDED NOW!'

David Webb, head of NCROPA (National Campaign for the Reform of the Obscene Publications Act) found Superintendent Hames's presence at the meeting disturbing. He felt that Hames had become too closely associated with the views of the NVALA. When Webb wrote to Sir Peter Imbert, the then Chief Commissioner of the Metropolitan Police about the matter, however, he was assured by Imbert's office that Hames addressed the meeting 'with the full knowledge and support of the Metropolitan Police'.

The actual remit of the OPS is, at times, as vague as its founding law, the Obscene Publications Act of 1959. In January 1991, the Squad gave 'its stated priorities in order' in answer to a question from MP Chris Smith: '(a) child porn; (b) bizarre material; (c) extreme bondage and flagellation: (d) British publication of obscene material.' The last three categories exploit the OPA's vagueness: 'It's so loosely drafted that there's nothing to stop the Squad from attempting to widen the definition of what is obscene.'

As part of a campaign to widen the remit of the OPS, Superintendent Hames has fed tabloid prurience. In a June 1991 feature in the *Daily Mail*, written by Geoffrey Levy and David Gardner, and entitled 'Open House for the Merchants of Porn', Hames gave his opinion on pornography: 'It is addictive, like a drug. And the addict needs ever stronger drugs to satisfy himself.' When, in 1990, the Home Office published the Cumberbatch Report – which stated that the casual links between pornography and sex crimes were quite weak, but that there was some 'correlation and association' – the report was publicly criticized by both Mrs Whitehouse and Superintendent Hames.

As for the conduct of the Spanner trial, it is worth noting that the accused were not charged in fact under the OPA, but under the quite separate Offences Against the Person Act. There is a sense in which the law was moulded to achieve the desired convictions. 'The trial came down to a prosecution for morality,' says Angus Hamilton, solicitor for some of the appellants. 'The OPS is supposed to apply the law as it stands; in this case, the investigation was based on speculation about the law.

'Spanner has given the police a green light for prosecution,' Hamilton adds. 'They are waiting for the outcome of the appeal, but there has already been an increased OPS interest in S&M activity.'

The sequels have already begun: one early casualty of the Spanner ruling was a book called *Modern Primitives*, a serious, 200-page investigation of tattooing, scarification and piercing. Again, some of the practices discussed in an anthropological format are shocking to some people, but does this then mean that the book should be banned? Bow Street Magistrates Court thought not, in a March 1991 judgement by JP Ian Baker.

Another, more recent, OPS investigation has, again, exploited an existing moral panic. Coinciding with a February 1992 programme by Channel 4's *Dispatches*, which used an extreme performance art video by the rock group and 'art terrorists' Psychic TV to hint at ritual murder and satanic abuse, Psychic TV's co-founder Genesis P-Orridge had his house raided and has been effectively forced to remain out of the country until the OPS makes its intentions clear.

The conclusions of the *Dispatches* programme have been brought into question – the testimony of the chief witness, 'Jennifer', was pronounced a 'fantasy' in two March 1992 *Mail on Sunday* features – and there were no ritual murders or satanic abuse of children. In the wake of the moral panic created by the programme, the OPS has exploited justified and unjustified fears to reinforce what is still a disputed judgement: it is doubtful that P-Orridge would be guilty of anything except the new areas of illegality opened up by the Spanner verdict.

The issues in the Spanner trial, quite apart from any injustice, are legal and ethical, if not constitutional. 'We should not have test cases to find people morally rather than criminally in the wrong,' says Angus Hamilton. In attempting to force through a moral agenda not necessarily shared by the general population, the OPS has used the media to bring the judiciary and the public in line with their way of thinking. In feeding fears and prejudices in what can only be described as a modern hysteria, they undermine that very 'civilized society' which they claim to protect.

Additional research by Chris Woods and Mayavision.

Observer, 29 November 1992

Swimming Pool Season:
The Regeneration of Liverpool

It's Thursday night and the Baa Baa is jammed. At 10.45 p.m., the floor area alongside the long, blue bar is full of eddying groups of people. I'm sitting with a group of Liverpool natives, who are regarding the scene with a finely tuned mix of cynicism and appreciation. There is a strong element of display about the crowd, taking their positions around the free-standing metal counters in the centre of the floor, or walking self-consciously down the stairs from the eating area. A new style is emerging: short hair, brushed forward, with long sideburns; tight, waisted 1970s jackets; straight-leg trousers – a touch of dressing-up after the casual sloppiness of the past three years.

Liverpool is a city that appreciates the show, and the post-industrial blue, silver and fibreboard environment of the Baa Baa is a good stage. Situated on the corner of Slater Street and Fleet Street, this former hemp warehouse is at the heart of Liverpool's Soho: the area just north of the city centre, bounded by Bold, Duke, Berry and Hanover Streets. Club flyers, wittily designed, adorn the tables: from here, the eddies sweep the crowds to the 051, the Mardi Gras, Lola's, the Academy. Outside, the night air is cool: you can smell the sea. The streets are punctuated with noisy groups of young clubbers; once they have passed, there is a curious stillness.

Despite the efforts of entrepreneurs such as Tom Bloxham, who owns the Baa Baa and its sister building, the Palace – a rabbit-warren of record and clothes stalls – this former warehouse area is spotted with dereliction. Many locals think of the 1960s as the last time the sun broke through on Liverpool: the Beatles and Merseybeat, Liverpool FC's ascent to the First Division and FA Cup victory in 1965 – the golden age of Bill Shankly. The city is untouched by the febrile boom of the 1980s: there are hardly any post-modernist office buildings, few Legoland housing developments. Parts of the city look much as they have done for the past 20 years, since decline started to set in. These empty spaces are thronged with ghosts.

Yet Liverpool has changed considerably since I first spent time there, in the late 1970s. The economic blows keep coming – the recent announcement that shipbuilders Cammell Laird are to shut down with a loss of 900 jobs was seen as the last nail in the region's industrial coffin. Still, there is a pervading sense that, after the ravages of the past decade, the city has nothing more to lose; practised in slump, there is nowhere to go but up. And if the old industries have gone, the city will find for itself other things to do. Always in performance, Liverpool's new show is confidence.

During the nineteenth century, Liverpool was at the centre of things; now it is isolated. Unlike its neighbour Manchester, at the hub of an intricate motorway network, Liverpool has not adapted well to the age of the car: the route from the capital runs along a spur of the M62 and ends in confusion four or five miles out of the centre. Once there, you're in an apparent dead-end: the original road plan leads to the now barely used dock terminals, where employment has shrunk from 35,000 in the 1950s to only 1,300.

My favourite way into the city offers another perspective: from the south-east, past the chemical works that dominate the Mersey narrows, over the Runcorn Bridge, past Spike Island and Widnes, through Halewood, Speke, Aigburth, up to the brow above the Dingle and the Anglican cathedral. From this angle, you follow the course of the River Mersey itself and see Liverpool not as a dead-end but connected to a landscape of variety and potential.

More than London, Liverpool is a city dominated by its river. It is this river which has brought Liverpool its past wealth, from the middle of the eighteenth century onwards, and the vast movements of people which make the city's character. In the mid-1840s, the city saw mass immigration by hundreds of thousands of Irish fleeing the potato famine: some 370,000 of them, more than doubling the population. At the turn of the century, Liverpool vied with London as the busiest port in the world.

Even as late as the 1950s, the city was exceptionally vibrant and active. 'There were masses of people in the street,' says Sir Desmond Pitcher, the Liverpool-born chairman of the Merseyside Development Corporation. 'There was a most extensive tram network, and the overhead railway down by the docks. Friday night was always special: you always had the liners – Cunard, Canadian Pacific – and frequently Royal Naval ships. It was an interesting place to be. 'When I was 22, I went abroad for 15 years, and when I returned, it had all changed. All the things I've described just went. Dirt and dereliction appeared on the scene; there was a sense of frustration, and an air of deprivation that was new.'

The Merseyside Development Corporation (MDC) was created by the Government in 1981, after the Toxteth riots, which made international headlines. Answerable to the Department of Trade and Industry, its original remit was to regenerate the south dockland area: roughly from Pier Head to Otterspool Promenade. Recognizing, in Pitcher's words, that 'the river is at the heart of Liverpool', the MDC began two major projects in 1981–2: the Garden Festival and the restoration of Albert Dock.

Today, Albert Dock houses a bewildering array of restaurants, bookshops, cafés, clothes and craft stalls. There is the Beatles Museum with its labyrinthine installation, bizarre but affecting. Past the stalls selling books like *Teach Yourself Scouse*, past the floating map of the UK which Granada uses to do the weather forecast, there is the Tate Gallery in the dock's south-west corner, rebuilt by James Stirling. Since its opening in 1988, 600,000 visitors a year have passed through to see exhibitions such as the current Stanley Spencer show.

Despite the bustle, however, there is a strong sense of the theme park about Albert Dock: a good place to visit, but curiously divorced from the city. There is a flimsiness there which makes it feel like a PR exercise, rather than a fully integrated part of the city. Indeed, the MDC is good on PR. Pitcher has cogent reasons why industry should be attracted to Merseyside: 'low cost of operation, good quality of life and a high standard of labour'. Yet the question remains: can a city be refloated through PR? It is a good idea – nobody has any better – but the next few years will see whether this confidence will attract and sustain new industry.

These questions underlay the summer's big event in the city, the MDC supported Fanfare for a New World, a gala concert at Kings Dock to celebrate the return of the tall ships – at the end of a transatlantic race – to Liverpool. For one week, Liverpool was a port again: up to 1 million people travelled to Birkenhead's Victoria Dock to view the sailing ships from Colombia, Russia and America. In a city centre full of foreign sailors there was a tingle in the air and a light in people's eyes.

But despite the quality of the cast – and the music, sublime enough to make you forget the chill wind whipping off the Mersey – the concert failed to attract national publicity. It might have been successful in attracting foreign interest, but for locals, it was unrepresentative of the city. 'There's not many black people,' said one of our party as we walked through the crowd.

'In the analysis of race and racism in the city, people have failed to put into context Liverpool's involvement in slavery,' says Adam Hussein, a second-generation Liverpudlian of Somali origin. 'Because of the city's inability to look at its own past,' he says, 'it is impossible to ensure that all of its citizens are treated equally and fairly. We need to put the other side of the coin, to educate people in our hidden history.' Just as the city's success was built on this trade, so has its subsequent decline seemed as though, to use the title of James Pope-Hennessy's famous book, the 'sins of the fathers' have been visited on the sons and daughters.

After lunch, Hussein takes me down Granby Street, Liverpool's 'front line' and centre of the 1981 riots. Despite that protest, conditions have not improved much since then: the government-appointed task force has made improvements to the area, but it is still difficult for local blacks to get jobs, funding to start small businesses, let alone licences to run clubs. Hussein is proud, with a quiet anger; but, as he says, 'We have the skills and ability to participate fully in Liverpool's economic life. It is the city's loss that we are not given the opportunity to develop fully.'

Eight miles away from the MDC remit area, in Kirkby, councillor Frances

Clarke echoes these sentiments: 'We're not overbearing or overproud. We're hard-working people, and we'd rather have our own job, our own dignity.' A product of the great post-war migration out of the bomb-wrecked Scotland Road area, Clarke has lived in Kirkby since 1958. She has seen this dream suburb flower in the 1960s and decline in the late 1970s, when much of the local industry, attracted by grants, moved out once the grants finished.

In a recession, what industries will invest? One answer might lie in what is now, according to economists, one of the few holding areas of our economy, the culture industry. Clarke herself is writing, for Virago, a book about her life: *I Was Born Aged 29*. Of her children, daughters Angela and Margi are actresses: Margi is currently best known as Queenie in BBC's *Making Out*. Both have acted in films written by son Frank: *Letter to Brezhnev*, *Fruit Machine* and, most recently, *Blonde Fist*. All three were made, on minimal budgets, in Kirkby and Liverpool.

'The biggest stimulant for this city in the 1980s was *Letter to Brezhnev*,' says John Smith. 'It demonstrated that anyone could do it.' Another Scotland Road native, Smith now runs the 051 Club and Media Centre, located in a triangular 1970s 'white elephant' near the Adelphi Hotel in Mount Pleasant. Upstairs are film and video facilities and a lecture theatre, downstairs a night-club remodelled by designers Shed – Miles and Jonathan Falkingham, also responsible for Baa Baa. 'The way forward is through tourism,' Smith says. 'We aren't a manufacturing base; we need to export our talent by using the marketability of the city and making it international. Liverpool's future is as a centre for music.'

The big change that has happened in the Liverpool music industry is that there is a local music industry. In the Merseybeat days, groups would go down to London and disappear. Punk reversed that, with small labels such as Bill Drummond's Zoo and Peter Fulwell's Inevitable: this was the start of acts as diverse as the KLF, Dead or Alive, Julian Cope, the Christians and Ian McCullough, the last two of whom still live and work in the city.

Since the mid-1980s, music has spawned into a whole infrastructure: city centre studios such as Amazon; designers such as Shed and the Farm's Carl Hunter; management such as Andy Carroll and James Barton, whose K-Klass had a massive hit last year with 'Rhythm is a Mystery'; record labels such as Produce and 3 Beat, which have steered the careers of recent top artists such as the Farm and Oceanic. It is now possible to see through every stage of mak-ing a hit record without having to go to London.

There has been a concerted effort to promote Liverpool as an arts base in other ways. The council actively promotes Heritage Liverpool – more Georgian terraces than Bath, more listed buildings than any other city in the UK – to film and television companies as a fertile source of locations: recent films shot in the city include *Let Him Have It* and the latest Indiana Jones epic, its Easter uprising storyline shot in the old warehouse area of Henry Street. Another attention-getter – this year co-funded by the MDC – is the city's successful Festival of Comedy, which attracted 120,000 people.

Finally, there is an extraordinary expansion in higher education. The statistics are impressive. Liverpool's population currently stands at 475,000. With a planned 40 per cent increase in student numbers, there will be more than 50,000 higher education students in the city by the year 2000.

'There's a real sense of hope here now,' says Colin Fallows, senior lecturer in art and design at the John Moores University (previously Liverpool Polytechnic). 'There are many possibilities for the future.' Fallows is also involved as coordinator for a new venture, the Liverpool Institute for Performing Arts (LIPA), announced by Paul McCartney in 1990. LIPA has already received government funding to aid the restoration of the old Liverpool Institute building – McCartney's old school – situated to the west of the Anglican cathedral entrance. Still actively seeking donations and sponsorship, the institute is scheduled to open in 1995.

Fallows refutes criticism that LIPA will just be a cosmetic *Fame* school. 'We have a real opportunity here to offer a multi-disciplinary approach,' he says. 'Art, design and pop can work together: our model is a late twentieth-century version of the Bauhaus, where skills are transferable across disciplines.'

There are deeper questions. How are all these students going to support themselves, in an era of declining grants? Are they going to be able to get jobs when they graduate? Will the recession kill this renewed confidence before it has time to put down real roots? No one has any answers, except a determined confidence and a sense that the sheer injection of youth into an ancient city is a good start.

Outside the rowdy Flanagan's Apple in Mathew Street, there is a plaque with a legend from Carl Jung: 'Liverpool is the pool of life.' The city has strong underlying resources as a spiritual centre. For the religious there are scores of churches – Catholic, Greek Orthodox, Chinese, Muslim. The city's skyline is dominated by Gilbert Scott's Anglican cathedral, one of my favourite buildings in the world. For the pagan, there are many Druidic sites, including the standing stones in Calderstones Park, a stone's throw away from the Strawberry Fields immortalized by local resident John Lennon.

'There are musical ley-lines, and they run from Liverpool to New Orleans,' says Rogan Taylor, author of *The Life and Resurrection Show* – the first book about shamanism and pop music – and, until last year, president of the Football Supporters' Association. 'There's no doubt that Liverpool is one of the hot spots. And this is a very warm little speck; the bird life here is fabulous, we get very little snow. If you were God walking around with a forked-stick, you'd go twitch twitch. There is something going on here.'

The Times, 12 December 1992

Snapshots of the Sixties: Swinging London in Pop Films

In the 1960 film *Village of the Damned*, a small village is knocked out for several hours by an invisible, localized and highly potent force. Within a few months, 12 children are born (many to unmarried mothers). The children develop alarming characteristics: telepathy, a group mind and a group appearance, a certain glassy abstraction.

These children are aliens: they do not think like us, they do not want what we want, they are everywhere. The film's tension comes from the adults' reaction: should the children be penalized or understood? Like all paranoid science fiction films of the time, *Village of the Damned* is more about contemporary fears than fictional universes: the inevitable destruction of the children shows us an essentially pre-war, hierarchical society fearful of its children, fearful of the group, fearful of the future.

It's April 1964 and the aliens are here. To promote their first feature film, *A Hard Day's Night*, the Beatles – all of whom are over 21 – are placed in prams, two by two. To the uninitiated, they are difficult to tell apart, with their identical clothes and long hair. That same week, the group has five records in the US top 10. The film for which this clip is shot will be premièred in three months' time, July 1964: during its first week's release in the US, it will take more than six times its budget. Suddenly time has speeded up, as it will throughout the rest of the decade. Britain is living in the present.

It's obvious – to the point of cliché – that something happened between the years 1960 and 1964: the jump-cut into present and future time that those four years represented was so jarring that it is still a matter of furious debate. Much of this debate is carried on at a very low level: spectacular recantations from former 1960s libertines, the yelping envy of today's cultural conservatives, whose house god, Philip Larkin, wrote:

> Sexual intercourse began
> In nineteen sixty-three
> (which was rather late for me) –
> Between the end of the *Chatterley* ban
> and the Beatles' first LP.

The 1960s have long been a target for conservatives and Little Englanders, as the decade when everything went wrong in British life: when the verities were swamped by a rising tide of drugs, permissiveness and socialism, when traditional family values were under attack.

This hostility to the 1960s has become a kind of pathology – which is interesting. What happened during those years, between 1960 and 1969, that is so disturbing to so many people today? We think we know about the 1960s, but as the current exhibition 'The Sixties Art Scene in London' at the Barbican and its accompanying film season 'Blow-Up' make clear, we don't. A decade seen today either as the time when the fabric of society began to unravel or as a pop golden age reveals itself as something far more complex and disturbing.

My own experience of the 1960s was as a teenager in west London. My environment was saturated in pop, to the extent that it was the only thing through which I could make sense of my world: the blank, mobile suburbia of Ealing. In fact, as I now realize, that suburbia was one of the main breeding grounds for the British pop music that went global during the decade: the clubs of Richmond, Ealing and Twickenham threw up the Rolling Stones, the Yardbirds, the Who. In 1966, my favourite record was 'Substitute'; although I didn't know it, Pete Townshend then lived half a mile away, just off my daily journey to school.

Ealing even had its own Pop Art exploitation group, the Eyes. Their first record, 'When the Night Falls', came complete with auto-destructive feedback. Their second, 'The Immediate Pleasure', was heralded by larger than life-size advertisements on the platform of Ealing Broadway tube station. 'The Eyes are Smashed to Fragments', ran the copy, and to promote the idea, the five Eyes posed in rugby shirts emblazoned with a huge, unblinking eye – in the centre of which was the head of the relevant Eye.

The arrival of the Eyes, early 1966

The eye of the media was everywhere then, for the first time. 'The period was already framed by the TV screen and, to a lesser extent, the cinema screen,' writes David Mellor in the catalogue to the Barbican exhibition. 'Whether it was possible to gain access to an actual world beyond this media "spectacle", an access to the life-world and the ground of existence, concerned several artists. At times the metaphoric media frame became a real one.' Our dominant image of the decade, Swinging London, is in fact a journalistic conceit coined after the event, in April 1966, by *Time*'s Piri Halasz.

In the crossover between the 1950s and the 1960s, artists like Richard Smith and Robyn Denny were exploring the links between fine art and advertising. In 1963 the Beatles were photographed in front of Denny's *Austin Reed* mural: the first image you see as you enter the Barbican show. Everything was coming together: new technology resulted in the spread of media such as television (BBC2 in 1964, the first addition since ITV in 1955), pop music and advertising – all still comparatively new to Britain.

In 1957 critic Laurence Alloway talked about film scenographies that used 'the saturation of an environment with communication devices'. The London films of the mid-1960s take this for granted: nothing is pure, everything is mediated, even the idea of youth itself. 'That's why you chose her, wasn't it?' says an advertising man to his boss in John Boorman's first feature, *Catch Us If You Can* (1965). 'That's her image: rootless, classless, kooky, product of affluence, typical of modern youth.'

In *A Hard Day's Night*, as you might expect from Richard Lester, a director with experience in advertising and television, the Beatles are shown to be the victims of a relentless media onslaught: during the blindingly cut press conference scene, George Harrison is photographed and you see flash flash flash flash the still image of the photograph come up on screen. Elsewhere, Harrison is kidnapped by a cynical youth programmer: 'Now you'll like these,' says Simon, shoving a collection of shirts under his nose. 'You'll really "dig" them. They're "fab" and all those pimply hyperboles.' 'They're dead grotty,' Harrison replies, instantly creating slang.

The main location is a television studio and at least three numbers are shot from the multiple viewpoint of studio cameras, often in shot, then compounded with a shot of several studio monitors at once. We're always aware that this is mediated. The effect is to make everything more intense, capturing the real insanity the Beatles were then suffering (much of the budget was spent on avoiding fans).

In fact, it becomes clear that the media distort time: expanding it, fragmenting it, until the distinction between mediated and actual ceases to exist. Michelangelo Antonioni's *Blow-Up* – shot in summer 1966 – takes as its hero a fashionable documentary photographer, a mixture of the real-life Donald McCullin and David Bailey. Much of the film is taken up with the photo-

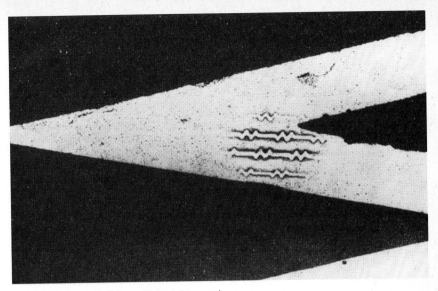

Roadmarking, 1961 (photo: Robert Freeman)

graphic process – from shooting to developing to cropping and framing to sell-ing the end result in the mass media.

The plot is a metaphor for media voyeurism: David Hemmings pho-tographs two lovers on one of many whims, realizes later, when the pho-tographs are blown up, that a murder has been committed, and goes back to find the body. At first cynical, he is profoundly shocked and becomes passion-ate – a no-no in this and today's media world. When both the body and the evidence of his photographs are removed, he is left wondering 'Was this real?' but in media, the imaginary becomes real.

The London of these films is almost unrecognizable from today's stasis: a city in transition, with its history becoming subsumed in a new cybernetic space, all speed and arrows as in Robert Freeman's 1961 photograph of Park Lane (then being redeveloped into a three-lane mini-motorway). The images of London that you take away from these films are of the demolition of St Luke's Church, Notting Hill, and the construction of new houses in Ealing (*A Hard Day's Night*), of advertising hoardings cloaking bomb-sites, of under-passes and arrowed traffic signs (*Catch Us If You Can*), of new modernist wastes like London Wall (*Blow-Up*).

If this is exciting, it is also temporary. Despite their official status as prod-ucts of youth culture's golden moment, both *A Hard Day's Night* and *Catch Us If You Can* have as their dominant theme a brief escape inevitably curtailed. In the Beatles' film, the infamous 'Can't Buy Me Love' sequence – where the group lets rip on a Gatwick helicopter pad: more speed – is a rare exterior sequence in a film dominated by interior locations. In *Catch Us If You Can*,

Dave Clark and Barbara Ferris take off on a whim, to be hunted down relentlessly until the final Felliniesque shot of a crowd of cameramen and journalists, caught where sea and sky meet in nothingness.

However crudely, both films make a connection between the sudden enfranchisement of youth – which may well be temporary – and the sudden shift in political power from Conservative to Labour in 1964. In *A Hard Day's Night*, the Beatles' group identity is seen to embody a gut northern socialism. (As Christopher Logue wrote then, 'I vote Labour … because Ringo does.') Hard though it may be to imagine today, after nearly 14 Conservative years, a change in government from right to left always stimulates the arts. But this is a fragile freedom. In *Catch Us If You Can*, an ad man says, 'We've flogged energy until it's tired,' before advocating a return to the idea of 'gracious living', while Simon admonishes a truculent George Harrison, 'The new thing is to care passionately, and be right wing.'

Both films show the insecurity of a new phenomenon (in 1964 and early 1965, British pop culture was still a novelty). By the time *Blow-Up* was shot, a year and a half later, the Beatles had their MBEs, London was established as a youth city and people had started taking marijuana instead of amphetamines. Antonioni could go further and deeper. He amplifies the point made by the two earlier films, that rapid change creates ellipses in communication. In *A Hard Day's Night* and *Catch Us If You Can*, people talk at and across each other all the way through, but they are sure of who they are, they are in the same time. This parallels both films' forward motion.

In *Blow-Up*, the process is slowed down. The certainties of black and white shift into washed-out colour. Hemmings's confrontation with Sarah Miles, where she is expecting him to tell her he loves her, is revealed in silences and awkward pauses, followed by spurts of intensity. This is the rhythm of the film itself. In the climactic sequence, Hemmings goes to tell his agent about his discovery. He finds him at a smart party, a fabulous *mise-en-scène* in a panelled house overlooking the Thames. His agent, Ron, is stoned out of his mind, but pot doesn't alter reality, it intensifies it; Ron's inability to understand Hemmings's need is not created by but is revealed by the drug. The chasm that has always been there between them opens up.

In similar fashion, the pop music used to express exuberance throughout the two earlier films is now seen as an agent of alienation. *Blow-Up* is not a musical film – the Herbie Hancock soundtrack is lousy – and its fame as such rests on another climactic sequence, this time featuring the Yardbirds (with twin lead guitarists Jeff Beck and Jimmy Page). If the Beatles play in an ornate theatre (the old Scala) to screaming teenagers who end up dominating the soundtrack, the Yardbirds play in a scruffy club, to a catatonic and silent if trendy audience of student age.

Only two people are dancing – a black man and a teenage Janet Street-

Porter as Hemmings scours the club for Vanessa Redgrave. The group's equipment starts to malfunction, and the moodier guitarist – Jeff Beck, not required to act – starts to attack the amplification system. Antonioni had wanted the Who – whose Pete Townshend would smash his guitar in a fit of auto-destruction (he had studied at Ealing Art College when Gustav Metzger was lecturing there) – but they were not available. So he persuaded the Yardbirds, not then known for such antics, to do the same.

When Beck throws the remains of his trashed guitar into the audience, they go mad, instantly. Hemmings struggles furiously to secure the neck and rushes out of the club; once on Oxford Street, he looks at the thing and suddenly discards it. What was desired one moment is now useless. Objects are being discarded willy-nilly through all these films (the autographed postcards in *A Hard Day's Night*, half-eaten food in *Catch Us If You Can*) – obvious enough in a period of overproduction, but Antonioni's heavy-handed comment on the fickleness of pop fame has a resonance.

Films take a long time in production, too long sometimes to come out at the right time to coincide with pop's fast-changing moods. If *A Hard Day's Night* succeeded because of its four-mouth turnaround from the start of shooting to the première, the many months needed for Antonioni to finish *Blow-Up* – from summer 1966 to spring 1967 – made his film out of time on its release. For one, the Yardbirds had stopped having hits and the Beatles had withdrawn from public performance, amid rumours that they had split up. For another, the whole fantasy of Swinging London had foundered in the first halt to the post-war boom: a wage freeze followed by the devaluation of the pound in 1967.

Its casual sexism aside, though, *Blow-Up* remains timeless as an examination of the warping effect that media can have on their participants. If *A Hard Day's Night* and *Catch Us If You Can* reveal their stars to be fundamentally unchanged by what has gone on during the 90 minutes or so we have spent watching them, we are left in no doubt at the end of both *Blow-Up* and the last, and perhaps greatest, of this quartet of London films, *Performance*, that something shattering has happened to the protagonists. In these last, pop/media culture is seen not as something transient, but as fundamentally transforming, both wonderful and malign.

With *Performance*, shot in summer 1968 but not released until 1970, we could be in any time until the late 1970s. As the young girl comments, staring at a late 1960s poster of Jagger/Turner, 'You should have seen him 10 years ago.' Jagger's burnt-out rock star, Turner, predates the better-known cinematic variant, David Essex in *Stardust* (1974), by a good six years. If time starts to slip in *Blow-Up*, then *Performance* blasts it into an everlasting present.

If there is an auteur of the film, it is generally regarded – after his subsequent success – as Nicolas Roeg. But in fact it is Donald Cammell's film: he

'September 16 1968: Rolling Stone Mick Jagger starts shooting his new film': James Fox and Jagger on the set of *Performance*. (Keystone)

wrote the script, co-directed and spent months in an edit, as he told me in 1984, which became ever more abstracted, ever more intellectual. With his *haut/demi-monde* experience as a fine artist, Cammell was well suited to show the underside of the 1960s explosion: his script casts James Fox as an abstraction of the Krays, who had become criminal stars in Swinging London (and in the Barbican show), while Turner takes on the precious, full-tilt persona more accurately associated with the Rolling Stone who would die between the film's completion and its release, Brian Jones.

Performance dazzles on several levels. If Lester in *A Hard Day's Night* used hand-held cameras and cut to the beat to emphasize the excitement of the moment, and Antonioni used a wide-angle lens to communicate dislocation, Roeg and Cammell let loose a whole box of effects, such as the many mirrors that run through the film. Fish-eye lenses, coloured stock, gauze over the lens produce images that are cut across or with the many rhythms that make up the brilliant score.

Former Phil Spector arranger Jack Nitzsche uses early synthesizers to conjure up psycho-active drones and pulses – some but not all of which are on the released soundtrack. These are matched subliminally by cuts which often cross time and, at first, logic. A QC pontificates to a jury about market forces; as he speaks, a drone intensifies and the jury become the spectators of flagellation movies in a seedy soho cinema. Chas (James Fox) is in a meeting with his boss, Harry Flowers; as Flower pontificates, the music distorts and changes phase. Chas appears shot through a blue filter with Flowers's words echoing as if from very far away.

Performance dramatizes stress, even breakdown, and shows us how good can come from a state that most of us find terrifying. It does so by going right into that state. Here, everything is seen either in instantaneous, hallucinatory close-up – as where Chas, stuck in Wandsworth and talking to his boss, realizes his days are numbered: his doodle of a man disappears under a hail of black ink from his pen (this within a few seconds) – or in the most profound of blurs. 'I wonder,' says Turner, 'if you were me, what would you do?' 'I don't know,' Chas replies. 'It depends, it depends on who you are.' 'Who am I? Do you know who you are?' 'Yes.'

What's great about *Performance* is the way it takes up every challenge thrown down by 1960s pop ebullience. The modernist environment under construction in the mid-decade is shown here, in Chas's flat, to be heartless, soulless, a mask for violence (fast forward to *A Clockwork Orange*). The conversational ellipses of a time when things are happening so fast that everyone talks at each other are turned into a lightning-fast, logic-crunching, deadly duel. (Just look at the three-hander in the kitchen between Fox, Jagger and Anita Pallenberg: I'm not transcribing it.)

The soundtrack flashes forwards and backwards, going into the past and future of the R&B so charmingly plundered by the mod groups. Jagger does a passable version of Robert Johnson's devilish invocation, 'Come on in My Kitchen', while for the first time in these films black voices are heard – Merry Clayton (later to find fame as the voice behind the Stones' *Gimme Shelter*) and the Last Poets, whose influence on 1980s rap would be considerable. Indeed, the film parallels only *A Hard Day's Night* in the way it has fed back into popular culture.

Most acutely, *Performance* goes straight for the promise of transformation held out by 1960s pop – and pop ever after, for that matter. The androgyny that is played up so much in *A Hard Day's Night* and *Catch Us If You Can* (although not without a certain resistance from the determinedly blokeish Dave Clark Five in the latter), is taken to its logical conclusion. In the climactic mushroom sequence, Chas is transformed from a gun-packing psycho into a bewigged, drugged, feminized man. In the morning he wakes up next to someone who appears to be Turner; he starts to make love, and the figure changes to Lucy, the more androgynous of the two women in Turner's household.

For the first time, he offers to do something for someone else – to fetch Lucy some shampoo – and it is while on this errand that he meets his erstwhile colleagues, now executioners. He has changed, but they haven't, and they stand appalled in the seedy glamour of the Powis Square location.

And now, Cammell goes as far as he possibly can. We have already seen Turner and Chas blending – quick facial superimpositions – and now they bond, irrevocably, with Turner's murder by Chas. As Chas himself goes to his death, driving through sunny Richmond Park in a white Rolls-Royce, he looks out of the car window, but the person staring out is not Chas but Turner. The blending is complete, an ending that is satisfying and curiously hopeful.

If it is hard today to imagine anyone backing a film as intense as *Performance*, that may tell us something about the cynicism of the 1990s and remind us of one of the vital strengths of the earlier decade. These films dramatize Paul Virilio's challenge in *Pure War*: 'Can't we envision, isn't it incumbent upon us to imagine what an *intensive* life could be? Being alive means to be lively, to be quick. There is a struggle between metabolic speed, the speed of the living, and technological speed, the speed of death which already exists in cars, telephones, the media.'

The tension in these 1960s films lies in this struggle for time, for life: their life mocks the cynicism of today's culture. In seeking to deny and denigrate the 1960s, the New Right merely reinforces that decade's mythic power. That power remains strong for a very good reason: a cool look at that time, 30 years on, reveals a Britain where people were not afraid to confront the present, were not afraid to think about how the future might be. This optimism, the will to survive even, is still a beacon.

Sight and Sound, May 1993

Machine Soul:
A History of Techno

Oooh oooh Techno city
Hope you enjoy your stay
Welcome to Techno city
You will never want to go away

Cybotron, 'Techno City' (1984)

The 'soul' of the machines has always been a part of our music. Trance always belongs to repetition, and everybody is looking for trance in life … in sex, in the emotional, in pleasure, in anything … so, the machines produce an absolutely perfect trance.

Ralf Hütter, 1991, quoted in *Kraftwerk: Man Machine and Music*,
Pascal Bussy

'It's like a cry for survival,' a panicked male voice calls out. The beat pauses, but the dancers do not. Then Orbital throw us back into the maelstrom: into a blasting Terry Riley sample, into the relentless machine rhythm, into a total environment of light and sound. We forget about the fact that we're tired, that the person in front of us is invading our space with his flailing arms. Then, suddenly, we're there: locked into the trance, the higher energy. It does happen, just like everybody always says: along with thousands of others, we lift off.

The Brixton Academy is a 3,500-capacity venue in south London. Built at the turn of the century in the style of a Moorish temple, it may look beautiful but it's hard to enliven: groups as diverse as the Beastie Boys and Pavement have disappeared into its dark, grimy corners. Tonight, however, it is full of white light and movement: the whole stage is a mass of projections, strobes and dry ice, in front of which a raised dance floor has been put in. Above us is stretched white cloth: at the sides of the building, the alcoves are lit up, and flanked by projections of pulsating globules.

The whole scene reminds me of the place I wanted to be when I was 18, the same age as most of this audience: the Avalon Ballroom. Never mind that most of the dancers were born long after the San Francisco scene had passed: they're busy chasing that everlasting present. The sound is Techno, but psychedelic references abound: in the light shows, the fashions (everything ranging from beatnik to short hair to late 1960s long hair), the T-shirts that read 'Feed Your Head' (that climactic line from Jefferson Airplane's 'White Rabbit'), the polydrug use that is going on all around us.

This event is called Midi Circus: an ambitious attempt by the London promoters Megadog to make dance music performance work. It's obvious from the lightness of the atmosphere that time and energy have been spent on the staging. The acts selected – the Orb, Orbital, the Aphex Twin – are the most interesting working in the Techno/psych crossover that has moved into areas formerly associated with rock: large public events, raves, festivals. It's here you will find the millenarian subculture of Techno primitives, half in electronic noise, half in earth-centred paganism.

Orbital's name is taken from the M25 orbital motorway that circles London: it comes from the period, three years ago, when huge raves were held around the capital's outer limits. They've had a couple of hits, have just released a fine second LP. Tonight, they stand behind their synths wearing helmets with two beams roughly where their eyes would be. When the dry ice and the strobes are in full effect, they look like trolls from *Star Wars*, or, perhaps more unsettling, coal miners. And then, as machine noise swirls around us, it hits me. This is industrial displacement. Now that Britain has lost most of its heavy industry, its children are simulating an industrial experience for their entertainment and transcendence.

At first the art of music sought and achieved purity, limpidity and sweetness of sound. Then different sounds were amalgamated, care being taken, however, to caress the ear with gentle harmonies. Today music, as it becomes continually more complicated, strives to amalgamate the most dissonant, strange and harsh sounds. In this way we come ever closer to *noise-sound*.

Luigi Russolo, 'The Art of Noises' (1913)

Punk rock, new wave, and soul
Pop music, salsa, rock & roll
Calypso, reggae, rhythm and blues
Master mix those number one blues.

G.L.O.B.E. and Whiz Kid, 'Play That Beat Mr DJ' (1983)

Techno is everywhere in Britain this year. Beginning as a term applying to a specific form of dance music – the minimal, electronic cuts that Detroiters like Derrick May, Juan Atkins and Kevin Saunderson were making in the mid-1980s – Techno has become a catch-all pop buzzword: this year's grunge. When an unabashed Europop record like 2 Unlimited's 'No Limit' – think Snap, think Black Box – blithely includes a rap that goes 'Techno Techno Techno Techno', you know that you're living within a major pop phenomenon.

My experience of it has been coloured by my recent circumstances: frequent travel, usually by car. Techno is the perfect travelling music, being all about speed: its repetitive rhythms, minimal melodies and textural modulations are perfect for the constantly shifting perspectives offered by high-speed travel. Alternatively, the fizzing electronic sounds all too accurately reproduce the snap of synapses forced to process a relentless, swelling flood of electronic information.

If there is one central idea in techno, it is of the harmony between man and machine. As Juan Atkins puts it,' You gotta look at it like, Techno is technological. It's an attitude to making music that sounds futuristic: something that hasn't been done before.' This idea is commonplace through much of avant-garde twentieth-century art – early musical examples include Russolo's 1913 'Art of Noises' manifesto and 1920s ballets by Erik Satie (*Relâche*) and George Antheil (*Ballet mécanique*). Many of Russolo's ideas prefigure today's Techno in everything but the available hardware, like the use of non-musical instruments in his 1914 composition *Awakening of a City*.

Post-war pop culture is predicated on technology, and its use in mass production and consumption. Today's music technology inevitably favours unlimited mass reproduction, which is one of the reasons why the music industry, using the weapon of copyright, is always fighting a rearguard battle against its free availability. Just think of those 'Home Taping is Killing Music' stickers, the restrictive prices placed on every new Playback/Record facility (the twin tape deck, the DAT), the legal battles between samplers and copy-right holders.

There are obviously ethical considerations here – it's easy to understand James Brown's outrage as his uncredited beats and screams underpin much of today's black music – but at its best, today's new digital, or integrated analog and digital, technology can encourage a free interplay of ideas, a real exchange of information. Most recording studios in the US and Europe will have a sampler and a rack of CDs: a basic electronic library of Kraftwerk, James Brown, Led Zeppelin – today's sound bank.

Rap is where you first heard it – Grandmaster Flash's 1981 'Wheels of Steel', which scratched together Queen, Blondie, the Sugarhill Gang, the Furious Five, Sequence and Spoonie Gee – but what is sampling if not digitalized scratching? If rap is more an American phenomenon, techno is where it all comes together in Europe as producers and musicians engage in a dialogue of dazzling speed.

> Synthetic electronic sounds
> Industrial rhythms all around
> Musique non-stop
> Techno pop
>
> Kraftwerk, 'Techno Pop' (1988)

Kraftwerk stand at the bridge between the old, European avant-garde and today's Euro-American pop culture. Like many others of their generation, Florian Schneider and Ralf Hütter were presented with a blank slate in post-war Germany: as Hütter explains, 'When we started, it was like shock, silence. Where do we stand? Nothing. We had no father figures, no continuous tradition of entertainment. Through the 1950s and 1960s, everything was Americanized, directed towards consumer behaviour. We were part of this 1968 movement, where suddenly there were possibilities, then we started to establish some form of German industrial sound.'

In the late 1960s, there was a concerted attempt to create a distinctively German popular music. Liberated by the influence of Fluxus (LaMonte Young and Tony Conrad were frequent visitors to Germany during this period) and Anglo-American psychedelia, groups like Can and Amon Düül began to sing in German – the first step in countering pop's Anglo-American centrism. Another element in the mix was particularly European: electronic composers like Pierre Schaeffer and Karlheinz Stockhausen, who, like Fluxus, continued Russolo's fascination with the use of non-musical instruments.

Classically trained, Hütter and Schneider avoided the excesses of their contemporaries, along with the guitar/bass/drums format. Their early records are full of long, moody electronic pieces, often with simple, classical melodies, using noise and industrial elements – music being indivisible from everyday sounds. Allied to this was a strong sense of presentation (the group logo for their first three records was a traffic cone) which was part of a general move towards control over every aspect of the music and image-making process: in 1973–4, the group built their own studio in Düsseldorf, Kling Klang.

At the same time, Kraftwerk bought a Moog synthesizer, which enabled them to harness their long electronic pieces to a drum machine. The first fruit of this was 'Autobahn', a 22-minute motorway journey, from the noises of a car starting up to the hum of cooling machinery. In 1975, an edited version of 'Autobahn' was a US top 10 hit: it wasn't the first synth hit – that honour falls to Gershon Kingsley's hissing 'Popcorn', performed by studio group Hot Butter – but it wasn't a pure novelty either.

The breakthrough came with 1977's *Trans-Europe Express*: again, the concentration on speed, travel, pan-Europeanism. The album's centre is the 13-minute sequence that simulates a rail journey: the click-clack of metal wheels on metal rails, the rise and fade of a whistle as the train passes, the creaking of coach bodies, the final screech of metal on metal as the train stops. If this wasn't astounding enough, 1978's *Man Machine* further developed ideas of an international language, of the synthesis between man and machine.

The influence of these two records – and 1981's *Computer World*, with its concentration on emerging computer technology – was immense. In England, a new generation of synth groups emerged from the entrails of punk: Throbbing Gristle, Cabaret Voltaire, the Normal all began as brutalist noise groups, for whom entropy and destruction were as important a part of technology as progress, but all of them were moving toward industrial dance rhythms by 1976–9.

The idea of electronic dance music was in the air from 1977 on. Released as disco 12-inch records in the US, cuts like 'Trans-Europe Express' and 'The Robots' coincided with Giorgio Moroder's electronic productions for Donna Summer, especially 'I Feel Love'. This in turn had a huge influence on Patrick Cowley's late 1970s productions for Sylvester: synth cuts like 'You Make Me Feel Might Real' and 'Stars' were the start of gay disco. Before he died in 1982, Cowley made his own synthetic disco record, the dystopian 'Mind Warp'.

More surprisingly, Kraftwerk had an immediate impact on black dance music: as Afrika Bambaataa says in David Toop's *Rap Attack*: 'I don't think they even knew how big they were among the black masses back in 1977 when they came out with "Trans-Europe Express". When that came out, I thought that was one of the best and weirdest records I ever heard in my life.'

In 1981, Bambaataa and the Soulsonic Force, together with producer

Arthur Baker, paid tribute with 'Planet Rock', which used the melody from 'Trans-Europe Express' over the rhythm from 'Numbers'. In the process, they created electro and moved rap out of the Sugarhill age.

> The Techno Rebels are, whether they recognize it or not, agents of the Third Wave. They will not vanish but multiply in the years ahead. For they are as much part of the advance to a new stage of civilization as our missions to Venus, our amazing computers, our biological discoveries, or our explorations of the oceanic depths.
>
> Alvin Toffler, *The Third Wave* (1980)

> Music is prophecy: its styles and economic organization are ahead of the rest of society because it explores, much faster than material reality can, the entire range of possibilities in a given code. It makes audible the new world that will gradually become visible.
>
> Jacques Atalli, *Noise* (1977)

In the inevitable movement of musical ideas from the avant-garde to pop, from black to white and back again, it's easy to forget that blacks – who to many people in England must be the repository of qualities like soul and authenticity – are equally as capable, if not more, of being technological and futuristic as whites. A veiled racism is at work here. If you want black concepts and black futurism, you need go no further than the mid-1970s Parliafunkadelicment Thang, with its P-Funk language and extraterrestrial visitations.

Derrick May once described Techno as 'just like Detroit, a complete mistake. It's like George Clinton and Kraftwerk stuck in an elevator.' 'I've always been a music lover,' says Juan Atkins. 'Everything has a subconscious effect on what I do. In the 1970s I was into Parliament, Funkadelic; as far back as 1969 they were making records like *Maggot Brain*, *America Eats Its Young*. But if you want the reason why that happened in Detroit, you have to look at a DJ called Electrifyin' Mojo: he had five hours every night, with no format restrictions. It was on his show that I first heard Kraftwerk.'

In 1981, Atkins teamed up with a fellow Washtenaw Community College student, Vietnam veteran Richard Davies, who had decided to simply call himself 3070. 'He was very isolated,' Atkins says. 'He had one of the first Roland sequencers, a Roland MSK 100. I was around when you had to get a bass player, a guitarist, a drummer to make records; you had all these egos flying around, it was hard to get a consistent thought. I wanted to make electronic music but thought you had to be a computer programmer to do it. I found out it wasn't as complicated as I thought. Our first record was "Alleys of

Your Mind". It sold about 15,000 locally.'

Atkins and 3070 called themselves Cybotron, a futuristic name in line with the ideas they had taken from science fiction, P-Funk, Kraftwerk and Alvin Toffler's *The Third Wave*. 'We had always been into futurism. We had a whole load of concepts for Cybotron: a whole Techno-speak dictionary, an overall idea which we called the Grid. It was like a video game, which you entered on different levels.' By 1984–5, they had racked up some of the finest electronic records ever, produced in their home studio in Ypsilanti: tough, otherworldly yet warm cuts like 'Clear', 'R-9' and, the song that launched the style, 'Techno City'.

Like Kraftwerk, Cybotron celebrated the romance of technology, of the city, of speed, using purely electronic instruments and sounds. One of their last records, 'Night Drive', features a disembodied voice whispering details of a rapid, nocturnal transit in an intimate, seductive tone – this set against a background of terminal industrial decay. After the riots of June 1967, Detroit went, as Ze'ev Chafets writes in *Devil's Night*, 'in one generation from a wealthy white industrial giant to a poverty-stricken black metropolis'. Starved of resources while the wealth remains in rich, white suburbs, the inner city has, largely, been left to rot.

Much has been made of Detroit's blasted state – and indeed, analogous environments can be found in England, in parts of London, Manchester, Sheffield, which may well account for Techno's popularity there – but Atkins remains optimistic. 'You can look at the state of Detroit as a plus,' says Atkins. 'All right, you only take 15 minutes to get from one side of the city centre to the other, and the main department store is boarded up, but we're at the forefront here. When the new technology came in, Detroit collapsed as an industrial city, but Detroit is Techno City. It's getting better, it's coming back around.'

By 1985, 3070 was gone, permanently damaged by Vietnam. Atkins hooked up with fellow Belleville High alumni Derrick May and Kevin Saunderson. The three of them began recording together and separately, under various names: Model 500 (Atkins), Reese (Saunderson), Mayday, R-Tyme and Rhythm is Rhythm (May). All shared an attitude towards making records – using the latest in computer technology without letting machines do everything – and a determination to overcome their environment: like May has said, 'We can do nothing but look forward.'

The trio put out a stream of records in the Detroit area on the Transmat and KMS labels: many of these, like 'No UFO's', 'Strings of Life', 'Rock to the Beat' and 'When He Used To Play', have the same tempo, about 120 bpm, and feature blank, otherworldly voices – which, paradoxically, communicate intense emotion. These records – now re-released in Europe on compilations like *Retro Techno Detroit Definitive* or *Model 500: Classics* – were as good, if not better, as anything coming out of New York or even Chicago, but because of Detroit's isolation few people in the US heard them at the time. It took British entrepreneurs to give them their correct place in the mainstream of dance culture.

Like many others, Neil Rushton was galvanized by the electronic music coming out of Chicago mid-decade, which was successfully codified in the British market under the trade name 'House'. A similar thing had happened in Chicago as in Detroit: away from the musical mainstream on both coasts, DJs like Frankie Knuckles and Marshall Jefferson had revived a forgotten musical form, disco, and adapted it to the environment of gay clubs like the Warehouse. The result was a spacey, electronic sound, released on local labels like Trax and DJ International: funkier and more soulful than Techno, but futuristic. As soon as it was marketed in the UK as House in early 1987, it became a national obsession with number one hits like 'Love Can't Turn Around' and 'Jack Your Body'.

House irrevocably turned around British pop music. After the successes of these early records by Steve 'Silk' Hurley and Farley 'Jackmaster' Funk (with disco diva Darryl Pandy), pop music was dance music, and, more often than not, futuristic black dance music at that. The apparent simplicity of these records coincided with the coming onstream of digital technology whereby, in Atkins's words, 'you have the capability of storing a vast amount of information in a smaller place'. The success of the original House records opened up more trends: Acid House – featuring the abrasive sounds of the Roland 303 – was followed by Italian House, and later, Belgian new beat's slower, more industrial dance rhythms.

'The UK likes discovering trends,' Rushton says. 'Because of the way that the media works, dance culture happens very quickly. It's not hard to hype something up.' House slotted right into the mainstream English pop taste for fast, four-on-the-floor black dance music that began with Tamla in the early 1960s (for many English people the first black music they heard). In the 1970s, obscure mid-1960s Detroit area records had been turned into a way of life, a religion even, in the style called 'Northern Soul' by dance writer Dave Godin.

'I was always a Northern Soul freak,' says Rushton. 'When the first Techno records came in, the early Model 500, Reese and Derrick May material, I wanted to follow up the Detroit connection. I took a flyer and called up Transmat; I got Derrick May and we started to release his records in Britain. At that time, Derrick was recording on very primitive analog equipment: "Nude Photo", for instance, was done straight on to cassette, and that was the master. When you're using that equipment, you must keep the mixes very simple. You can't overdub, or drop too many things in: that's why it's so sparse.

'Derrick came over with a bag of tapes, some of which didn't have any name: tracks which are now classics, like "Sinister" and "Strings of Life". Derrick then introduced us to Kevin Saunderson, and we quickly realized that there was a cohesive sound of these records, and that we could do a really good compilation album. We got backing from Virgin Records and flew to Detroit. We met Derrick, Kevin and Juan and went out to dinner, trying to think of a name.

'At the time, everything was House, House, House. We thought of Motor City House Music, that kind of thing, but Derrick, Kevin and Juan kept on using the word *Techno*. They had it in their heads without articulating it: it was already part of their language.' Rushton's team returned to England with 12 tracks, which were released on an album called *Techno! The New Dance School of Detroit*, with a picture of the Detroit waterfront at night. At the time, it seemed like just another hype, but within a couple of months Kevin Saunderson had a huge UK hit with Inner City's pop-oriented 'Big Fun', and Techno entered the language.

> In the future, all pop music will bring everyone a little closer together
> – gay or straight, black or white, one nation under a groove.
>
> LFO, 'Intro' (1991)

The sheer exponential expansion of dance music in Europe since House is attributable to several factors. First, the sheer quality of the records coming out of the US, whether swingbeat, rap, New York garage, House or Techno. Secondly, Acid House – acid being a Chicago term for the wobbly bassline and trancey sounds that started to come in from 1987 on – coincided with the widespread European use of the psychedelic Ecstasy. In Europe, Acid House meant Psychedelic House, and this drug-derived subculture has become the single largest fashion in Britain and across the Continent: gatherings of up to 5,000 people were common after 1988 and have become an important circuit for breaking hits.

Thirdly, the deceptively simple sound of the Detroit and Chicago records, together with the spread of digital technology like the Roland 808 sequencer, encouraged Europeans to make their own records cheaply, often in their own home studios, from the mid-decade. The long delay between Kraftwerk's 1981 *Computer World* and 1986 *Electric Café* occurred in part because the group was converting its Kling Klang studio from analog to digital. The result is greater flexibility, more storage space, and more sonic possibilities – vital in an area of music as fast-moving and competitive as the dance economy.

The big British breakthrough came in 1988 with S'Express's number one hit 'Theme for S'Express' – a playful reworking of that old travel motif, with Karen Finlay and hairspray samples for percussion. The acid sound development from the Roland 808 explorations of Phuture's 'Acid Tracks' – the sound of buzzing bees discovered by accident from a synthesizer straight out of the shop. Squeezed, bent, oscillated, this buzz became the staple of the 1988–9 acid boom: you can hear an early English version on Baby Ford's proto-hard-core 'Ooochy Koochy F*** You Baby Yeah Yeah'.

By 1990, the relentless demand for new dance music was such that, in Neil Rushton's words, 'The Detroit innovators couldn't take it to the next stage. What did was that kids in the UK and Europe started learning how to make those Techno records. They weren't as well made, but they had the same energy. And, by 1990–91, things became more interesting, because instead of three people in Detroit, you suddenly had 23 people making Techno, in Belgium, in Sheffield.'

Beltram's 'Energy Flash', released on the Belgian R&S Records in early 1991, defined the new mood. Inherent in the man/machine aesthetic is a certain brutality that goes right back to the macho posturings of the Futurist F. T. Marinetti: even in records as soulful as those made by Model 500, you'll find titles like 'Off to Battle'. With its in-your-face bass, speeded-up industrial rhythms and whispered chants of 'Ecstasy', 'Energy Flash' caught the transition from Detroit Techno to today's hard-core – the aesthetic laid out for all time on Human Resource's 'Dominator':

> I'm bigger and bolder and rougher and tougher
> In other words, sucker, there is no other
> I wanna kiss myself.

'In Belgium we had all the influences,' says R&S label owner Renaat Vandepapeliere. 'We had new beat, which was slowed-down industrial music: Cabaret Voltaire and Throbbing Gristle were very big in Belgium. Detroit Techno and Acid House came in and everything got mixed up together.' Other Beltram cuts, like 'Sub-Bass Experience', with its sensuous, psychedelic textures and rock samples, pointed the way forward to other R&S releases like the Aphex Twin's 'Analogue Bubblebath', which spun Techno off into yet another direction.

In Britain, the Techno take-up came not in London or Manchester (which by then was busy with rock/dance groups like the Happy Mondays), but in Sheffield, which in the late 1970s spawned its own electronic scene with Cabaret Voltaire and the Human League. 'There are no live venues here in Sheffield,' says WARP records co-owner Rob Mitchell. 'The only way to be in a band and be successful is to make dance records.

'All these industrial places influence the music that you make. Electronic music is relevant because of the subliminal influence of industrial sounds. You go around Sheffield and it's full of crap concrete architecture built in the

1960s: you go down into an area called the Canyon and you have these massive black factories belching out smoke, banging away. They don't sound a lot different from the music.' You can hear this in early industrial cuts by Cabaret Voltaire, like 1978's 'The Set Up', with its deep throbbing pulse.

In 1989, CV's Richard Kirk was looking for a new way to operate. 'Cabaret Voltaire had just finished a period on a major label, EMI, and we weren't working together. I spent a lot of time going to clubs, and working in the studio with Parrot, a DJ who ran the city's main club night, Jive Turkey. We made a record, as Sweet Exorcist, called "Test One", which we made to play in the club. It was very, very minimal: WARP was a shop where everyone bought American imports, and they put it out. We started to move seriously in that direction.'

WARP released 'Test One' in mid-1990: by the end of the year they had two top 20 hits with 'LFO' and 'Tricky Disco', both with eponymous dance cuts. The WARP material is less brutal than the Belgian Techno: still using crunch industrial sounds, but more minimal, more playful. And then another change occurred, as Techno went hard-core in 1991. 'I didn't like the hard-core stuff,' says Mitchell. 'It was too simplistic, crude and aggressive. We were getting sent lots of tracks that we couldn't sell on singles, so we thought, "Let's just do an LP." We got the title, *Artificial Intelligence*, and a concept: "Electronic music for the mind created by transglobal electronic innovators who prove music is the one true universal language."'

The cover of *Artificial Intelligence* is a computer-generated image: a robot lies back in an armchair, relaxing after a Sapporo and what looks like a joint. On the floor surrounding him are album sleeves: the first WARP compilation, featuring LFO and Sweet Exorcist among others, Kraftwerk's *Autobahn*, and Pink Floyd's *Dark Side of the Moon*. The music inside has slower beats, and is a ways away from the minimal funkiness of Detroit Techno: cuts by the Dice Man, the Orb and Musicology are nothing other than a modern, dance-oriented psychedelia.

Featured on the album was the then 17-year-old Richard James, who, under his most familiar pseudonym Aphex Twin, has become the star of what most people now call Ambient Techno – although it doesn't quite have a name yet. Coincidental to the *Artificial Intelligence* compilation, R&S released the Aphex Twin's *Selected Ambient Works 85–92*, which developed a huge underground reputation at the end of last year. With its minimal, archetypal graphics – a mutated boomerang shape on the sleeve – the *Ambient Works* album trashed the boundaries between Acid, Techno, Ambient and Psychedelic. It defined a new, Techno, primitive romanticism.

When Richard James was finally found and interviewed, he came up with a story that has already become myth: how the by-now 19-year-old student from Cornwall recorded under a bewildering variety of pseudonyms – the

Aphex Twin, Polygon Window, Dice Man and Caustic Window, to name but a few; how he built his own electronic machines to make the speaker-shredding noises you hear on his records; how he already has 20 albums recorded and ready to go. WARP plans to release his next ambient collection as a triple-CD set with a graphic novel.

The Aphex Twin's success comes at a moment when, in Britain and on the Continent, one wing of Techno is going towards Ambience. The slowing of pace is partly in response to the still-popular working-class fashion of hard-core, which regularly throws up generic chart hits like those by Altern-8 and the Prodigy. At the same time as the drug supply in clubland has changed from Ecstasy to amphetamines, hard-core has gone far beyond the linear brutalities of 'The Dominator' into a seamless dystopia of speeded up breakbeats, horror lyrics and ur-punk vocal chants. Like gangsta rap, it's scary, and it's meant to be.

'Ninety per cent of the Techno records you hear now are made for a fucked-up dance floor,' says Renaat Vandepapelier. 'That's what I see now in a lot of clubs: no vibe, no motivation, aggression – the drugs have taken over. The majority don't understand it yet, but most of the guys who are really good, like Derrick May, don't take drugs. Techno was a sound but it is now an attitude, and that's to make records for drug-oriented people. There is another category, where people are making music for you to pay attention with your full mind, and we're trying to make something now that will last.'

'I believe that the 1970s are the parallel for what is going on in the 1990s,' says WARP's Rob Mitchell. 'Musical moods tend to be a reaction against what has just gone on: we've just had very aggressive period. The original Detroit Techno is very sophisticated: what we're putting out now – Wild Planet, F.U.S.E. – has a similar level of sophistication. The real change for us since we started is the fact that this music is 99 per cent white, but the idea of raising Techno to an artier level is really exciting.'

If the 1970s are back, then it's the early part of the decade: you can see 1970–71 in the long hair and loose clothes of R&S/WARP acts like the Aphex Twin, Source, C. J. Bolland; you can read it in their titles ('Neuromancer', 'Aquadrive', 'Hedphelym'); you can hear the hints of Terry Riley, German romanticists Cluster and Klaus Schulze, even Jean-Michel Jarre. The very idea of boy keyboard wizards goes back to that moment in the early 1970s when Kraftwerk began their electronic experiments, when rock went progressive. Techno has moved into psychedelia with groups like Orbital; now it's gone prog.

It's hard to avoid the impression that Ambient has come as a godsend to the music industry. The very success of the dance-music economy has thrown up problems, as Rob Mitchell explains: 'There is virtually no artist loyalty in dance music; the record is more important than the artist. Dance is incredibly fast-moving; which is good, but very difficult to build careers in.' With ambient acts like the Aphex Twin, the music industry has something it recognizes and knows how to promote: the definable white rock artist, as opposed to the anonymous, often black, record. And Ambient Techno also slots directly into the music industry's most profitable form of hardware: the CD.

The term *ambient* was popularized by Brian Eno in the late 1970s. The percussionless, subtle tonalities of records like *Music for Airports* were perfect for the CD format when it came onstream in the mid-1980s. Ambient Techno and its kitsch associate, New Age, are the modern equivalent of the exotic sound experience that developed to fit the technologies of the 1950s. Just as mass distribution of the LP and the home hi-fi gave us film soundtracks and Martin Denny, the CD and the Discman have given us Ambient Techno.

Ambient could go horribly wrong, but hasn't yet. A cyberpunk/computer games aesthetic is always patched somewhere into the screen, but is not obtrusive. Inherent in the genre is a lightness of touch, and a rhythmic discipline that comes from its Detroit source. The best material, like Biosphere's *Microgravity* and Sandoz's *Digital Lifeforms*, also has a holistic spirituality that goes back to the Detroit records. As Sandoz's Richard Kirk says, 'I've been making music for a long time. Much of it has been very cold, very aggressive, very stark. It's time to do something that makes you feel good, that makes you feel warm.'

Recorded by a 27-year-old from Norway, Geir Jensen (a.k.a. Biosphere), *Microgravity* stands at the apex of ambient. Its nine cuts (sample title: 'Cloudwater II') forms a perfectly segued 45-minute whole that balances the utopian/dystopian pull inherent in the machine aesthetic. Their ebb and flow, between fast and slow, between playful and awful, between moon and sun, hold some of the queasy balance within which we live. At the end, a resolution: 'Biosphere' merges the sound of technology – the thrum of heavy industry, an electric alarm – into a bass pulse and atmospheric effects, warning but enclosing. The last sound is wind.

> There's something in the air called objectivity
> There's something in the air like electricity.
> There's something in the air, and it's in the air, the air.
> There's something in the air that's pure silliness.
> There's something in the air that you can't resist
> There's something in the air, and it's in the air,
> And you can't get it out of the air.
> > Theme song, Schiffer-Spoliansky revue, *Es Liegt in der Luft*
> > (*There's something in the air*) (1928)

Techno, how far can you go? 'A lot of it was kind of as we planned,' says Juan

Atkins, 'but nobody knew how far things would go. Nobody knew it would be a global thing as it is now, from little Detroit.' 'We have played and been understood in Detroit and Japan,' says Ralf Hütter. 'That's the most fascinating thing that could happen. Electronic music is a kind of world music. It may be a couple of generations yet, but I think that the global village is coming.'

The computer virus is loose. Right now, Techno presents itself as a paradox of possibilities (and limitations, the most glaring being gender: where are the women in this boys' world?). In its many forms, Techno shows that within technology there is emotion, that within information access there is overload, that within speed lies entropy, that within progress lies destruction, that within the materiality of inanimate objects can lie spirituality.

These tensions have been programmed into our art and culture since the turn of the century, and it is fitting that at the century's end, a form has come along which can synthesize the encroaching vortex of the millennium. You can do anything with Techno, and people will. As our past, present, and future start to spin before our eyes, and our feet start to slip, the positivism inherent in Techno remains a guide. Like Juan Atkins says, 'I'm very optimistic. This is a very good time to be alive right now.'

Village Voice: Rock & Roll Quarterly, Summer 1993

The Velvet Underground: You are What You Perceive

Sometime during 1991, I go on a pilgrimage. About 15 years before, I'd read John Rechy's *City of Night* in a kind of frenzy: what is this book telling me? What can it tell me? Written in the pace and rhythm of early rock 'n' roll, *City of Night* became central to my mental map of Los Angeles. Years later, I'd drive around MacArthur Park and Pershing Square looking for the ghosts of Chuck, Skipper, Miss Destiny.

Rechy lives in Los Feliz, at the foothills of his beloved Griffith Park. In 1955, Nicholas Ray shot the climactic scenes of *Rebel Without a Cause* at the park's Observatory: hidden in the film's love triangle – James Dean, Natalie Wood and Sal Mineo – was an explicit homosexual subtext, which Mineo lived and died out. In 1966, Rechy made the park the star of *Numbers*, as Johnny Rio cruises its hidden glades to the sound of 'Wild Thing', 'Summer in the City', 'Dirty Water' – punk nirvana.

Rechy has the physique of a body-builder and the carefully tended face of a middle-aged queen: deliberately artificial, as he is pleased to admit. He is cautious, but soon we begin to talk about how his work has fed back into popular culture – he cites the Doors' 'LA Woman' with its 'city of night' refrain, I tell him about Soft Cell's gleeful 'Numbers'. And we both realize: what was once taboo, hidden, right at the edge, is now at the heart of pop culture. What has happened?

When Rechy first wrote what would become *City of Night*, in the late 1950s, homosexuality (let alone hustling, drag queens and drug taking) was just not talked about. Much of the 1950s existed in order to edit out of history the freedoms of wartime: a renewed, McCarthyite puritanism drove homosexuality further underground with the inevitable psychic consequences. By the mid- to late 1960s, there were all sorts of *exposé!* books, but not then: just a few coded, discreet novels (like James Barr's *Quatrefoil*), which would usually end in suicide or death.

Before the full industrialization of media and of pop, there was an edge of the world, and Rechy was falling off it. Now that the leather queen is a pop video staple, now that trannies are high fashion, now that it may be possible – although still very rare – for the outcast to gain access to the mainstream and thus lose his outcast status, it's hard to imagine the sheer desperation of being cast out in the late 1950s/early 1960s, with no way in, no way never ever.

The question, what is pop? is an ideological battleground. If 'pop is dead', who is pronouncing the death sentence? Not those of us who still have much to gain from it. The 1980s insistence – post *The Great Rock 'n' Roll Swindle* – on pop as industry, pop as process, pop as hyper-capitalism is, of course, an important part of the picture, but not all of it, never ever. I'd rather talk about the idea of pop as an enfranchising process – imperfect but powerful: let us celebrate those who have contributed to this ongoing process, to these hard-won freedoms.

The Velvet Underground came in through the door marked *exposé!*: they are now in the museum. That's a long, nearly 30-year journey. Their reputation rests on a perfect quartet of albums: *The Velvet Underground and Nico* (1967), *White Light/White Heat* (1968), *The Velvet Underground* (1969) and *Loaded* (1970). All the other records released after the event, whether the dozens of bootlegs, or official releases like *The Velvet Underground Live at Max's Kansas City* (1972), *1969 The Velvet Underground Live* (1974), *VU* (1985) and *Another View* (1986), are after the event.

1965: the name comes from a paperback by Michael Leigh, ostensibly about S&M but, like all *exposé!s*, promising much more than it delivers. It is found on a New York street by Tony Conrad, an early associate. The members all live in New York, then on the cusp between its recent folksinging past and the electricity of Dylan-derived rock. Lou Reed teams up with Sterling

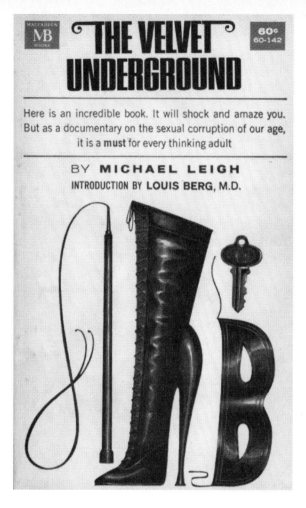

THE VELVET
UNDERGROUND

MB
60¢
60-142

Here is an incredible book. It will shock and amaze you.
But as a documentary on the sexual corruption of our age,
it is a **must** for every thinking adult

BY MICHAEL LEIGH
INTRODUCTION BY LOUIS BERG, M.D.

Michael Leigh: *The Velvet Underground* (1963)

Morrison to make hilarious exploitation Merseybeat records for a company called Pickwick. John Cale studies piano under John Cage and plays with Downtown minimalist LaMonte Young. Maureen Tucker lives on Long Island and plays garbage cans for drums.

Film-maker Barbara Rubin takes the Warhol crowd to see them in the Café Bizarre and the Velvet Underground become the Factory house band. Paul Morrissey brings Nico in to sing with them; Mary Woronov and Gerard Malanga do the dance of the whips – a variation on the 1965 *Vinyl* film – to the flicker of the same strobes that will trigger the suicide of their operator, Danny Williams. They take a residency at a Polish hall called the Dom, in St Mark's Place, attract a lot of interest, play in LA, where people hate them, record the first of those four albums.

That quartet: *The Velvet Underground and Nico* mixes blank, sharp pop songs with journalistic *exposé!s* of various taboo states: S&M, drugs, nihilism. Reed works at detachment but the band – especially John Cale – start to take you in there. Recorded in the red, *White Light/White Heat is* nihilism, with a certain black humour. John Cale is fired and replaced by Doug Yule. Distortion is replaced by quiet. *The Velvet Underground*: blind faith as a way out of a dead end. *Loaded*: pop culture as the ultimate redemption. Doug Yule squeezes Reed out of his own group.

As Lou Reed said: 'If you play the albums chronologically they cover the growth of us as people from here to there, and in there is a tale for everybody in case they want to know what they can do to survive the scenes. If you line the songs up and play them, you should be able to relate and not feel alone – I think it's important that people don't feel alone.' The perfect trajectory: from cult to pop, from nihilism to affirmation, and then nothing – for a while.

During the 1970s, there is a slow accretion of the legend. In 1971, Polydor UK re-release the first three Velvet Underground albums to critical acclaim and a few more sales. The next year, Lou Reed hits with 'Vicious' and 'Walk on the Wild Side'. David Bowie popularizes Warhol, bisexuality and the Velvets, who were there first, weren't they? The Reed-less Velvets tour, and nobody notices. In mid-decade, Mercury release the 1969 live set, and Nigel Trevena publishes the first collection of facts about this almost unknown group.

For, despite their flirtation with *exposé!*, the Velvet Underground are distinctly ill-exposed. During their lifetime, there are a few record reviews and a few articles by hard-core fans, like Sandy Pearlman and Jonathan Richman, in tiny magazines. Explicitly opposed to the gigantism of the San Franciscan scene – which dominates American pop as it turns into rock during 1968–9 – the Velvet Underground are not featured in the dominant media of the time.

Out of the mid-1970s, the story enters our time. Punk brings the Velvet Underground into their own: they are one of the three pre-1976 groups – DollsStoogesVelvets – that will be admitted to the pantheon. The group became a rock journalist's touchstone: the subject of discographies, esoteric research, myth-making in articles by Giovanni Dadamo, Mary Harron. The trickle of Velvet Underground bootlegs becomes a flood: the 'Foggy Notion' and 'Cycle Annie' 45s, the *Skydog* LP, *The Velvet Underground Etc.*

This is only natural, as English punk is a fantasy of the Warhol Factory, proletarianized and transposed to London 10 years on. From 1980 onwards, the whole period becomes part of an established history thanks to books like Andy Warhol and Jean Hackett's *Popism: The Warhol 60s* (1980), Jean Stein and George Plimpton's great *Edie* (1982) and finally Victor Bockris and Gerard Malanga's *Up-tight: The Velvet Underground Story* (1983). Now we have a positive industry: books from Ultraviolet, Nat Finklestein, biographies of Nico and Warhol, Velvet Underground discographies, passionate fan hymns like *What Goes On* magazine.

And the music is picked over until only the bare bones remain. The punks went for the glamorous New York nihilism, with minimalist tendencies, while post-punk groups like Orange Juice and Josef K went for the acoustic quiddities and guitar mantras of the third album. And then, no one had any new ideas for about 10 years: as successive generations of indie boyrock groups diluted the Byrds and the Velvet Underground, it became hard to listen to either by the early 1990s.

Everything in pop is over-*exposé!d* now, isn't it? Each accretion brings a slow shutting down of the Velvet Underground's original strangeness, original promise. Once they shocked; now the 'shocking' is at the media industry's heart. Once they sounded like nothing else; now they sound like everything else. Fixed like flies in amber, whether they re-form and play is almost irrelevant, except in the individual transactions between musicians and audience. But should this be regretted? Isn't this inevitable? Isn't it better to know that the Velvet Underground were major liberators?

The first Velvet Underground album is such a concentrated package that it is not surprising that pop culture took 20 or so years to catch up. *The Velvet Underground and Nico* straddles pop and the avant-garde with the decisive quality of a pre-emptive strike. Encoded within the record are references to authors like Leopold von Sacher-Masoch and Delmore Schwartz, Lou Reed's teacher, whose *In Dreams Begin Possibilities* is the definitive account of the immigrant experience in the first quarter of this century.

Best of all was the way the thing looked: a blurred picture of the group with five separate shots underneath, so lit that you could hardly tell which were the boys and which were the girls. This severe androgyny went further than English attempts at the same game, which had the innocence of childhood. Here the deadpan, blurred look is matched by lyrics about matters that were not hitherto the subject of pop songs. Underneath everything, is John Cale's viola, penetrating enough to bring down the walls of Jericho.

At one stroke, the Velvet Underground expanded pop's lyrical and musical vocabulary. This was deliberate and gleeful: contemporary interviews had Reed blithely discussing a Cale composition 'which involved taking everybody out into the woods and having them follow the wind'. Reed was in love with the doo-wop that we'd never heard in Britain, the Spaniels and the Eldorados, intense, slow ballads with a lot of heart from the early/mid-1950s. Sometimes, these songs would slow down to the point of entropy – an oasis of calm and a refusal of the harshness of ghetto life.

Reed's own sentimental strain would surface later, on 'Pale Blue Eyes', or on his own doo-wop songs, 'I Found a Reason', but as he well knew, by then, both he and pop had the curse of self-knowledge: 'And I've walked down life's lonely highways, hand in hand with myself.' For 'The Black Angel's Death

Song' on the first album, however, he wrote lyrics so phonetic that it would take their appearance in print some 25 years later, to decipher them. ('The idea was to string words together for the sheer fun of their sound, not any particular meaning,' he writes in 1992's *Between Thought and Expression*.)

In 1963, LaMonte Young recorded 'Sunday Morning Blues' with a group called the Theatre of Eternal Music, which comprised Young on soprano sax, Marian Zazeela (voice), John Cale (violin), Tony Conrad (viola) and Angus MacLise (drums). MacLise played beatnik style, while Young improvised over the drones provided by the other three musicians. The result stands at a critical juncture between Near and Far Eastern music, contemporary bohemianism, minimalism and the whole panoply of art rock that the VU would set in motion.

So many great stories here. Angus MacLise became the first drummer for the Velvet Underground in 1965, leaving when he realized that the group would sign and end at a stipulated time. He rejoined briefly when Reed was hospitalized with hepatitis in June 1966, playing in Chicago at Poor Richard's (rough tapes from the show have wonderful Cale vocals on 'Venus in Furs' and 'Heroin'). MacLise decamped for the East, and died in Tibet, reportedly of malnutrition, in 1978. His son is in line to become Dalai Lama.

Conrad produced an album of drones and minimalist percussion in 1972 called *Beyond the Dream Syndicate* – the last three words being the embryo Velvets' first name. Reed made his own drone move with 1974's *Metal Machine Music*, so forbidding that it was rumoured to be a contract-breaker with his record company. Legend apart, there is no reason to believe it was not sincere. In concert, the first VU would routinely play a piece called 'Melody Laughter', 25 minutes of minimal Maureen Tucker percussion, guitar/viola drones and Nico wails. In the mid-1980s, some enterprising soul put a November 1966 performance on LP, and it's fabulous.

When Cale left the VU in 1968, he took these drones to his production of Nico's *tour-de-force*, *The Marble Index*; the first Stooges album; and his collaboration with Young's contemporary Terry Riley, *The Church of Anthrax*. LaMonte Young himself has remained a hermetic figure, only partly by choice: currently performing in New York with a rock group, he finds record company backing elusive. He has just released *The Second Dream of the High Tension Line Stepdown Transformer*, drones for eight trumpets and mutes. (For a good introduction to his work, find his five-hour *The Well Tuned Piano* set with accompanying booklet.)

Hearing this wealth of esoteric materials is one way to recapture the first, taboo thrill of hearing *The Velvet Underground and Nico*, and is one way of disentangling the group's music from its legend. The simple fact is, to hear the record in 1968 was to be let into a secret world, once you'd got past the first frisson of pure evil. (Like the sleeve notes say, 'the flowers of evil are in full bloom with the Exploding Plastic Inevitable'.) Like every other great pop record, it changed your life.

Hooked, you then followed the Velvet's epigrammatic, spiritual path, through the sheer, physical rendering of nihilism ('I Hear Her Call My Name'), through the solipsism of faith ('There are problems in these times/Oooh but none of them are mine') to the transcendence of pure pop ('Despite all the computations/You could just dance to a rock 'n' roll station/And it was all right'). This simple affirmation was all the more powerful for being hard-won.

So what did the Velvets do? They helped to enfranchise women as instrumentalists in pop groups, as opposed to featured singers (although they had one of those as well, and what a strange tale that is); introduced a whole strain of contemporary American classic music to a wider market; wrote some wonderful, enduring songs; laid down a marker which no art-pop group has since been able to ignore. Without the Velvet Underground I would never have visited New York; read Delmore Schwartz; heard LaMonte Young and Terry Riley; perhaps even become a writer.

The Velvet Underground stand at the point where the archaic, immediately post-war culture of repression and *exposé!* meet the full implications of the 1960s: sexual freedom, social mobility, pop as the motor of the culture industries. If tragedy stalks their story, it's because they, and many others at the same time, were exploring uncharted waters. Now that we think we know everything about pop, it is easy to tie up their story into a neat, *Late Show*, style package: this omits any account of the group's first time courage, which is the reckless courage of all those who, both then and now, refuse to be content with the world as it seems.

Frieze, Summer 1993

Sounds Dirty:
The Truth about Nirvana

Situated on 51st Street and Broadway, in the heart of the old entertainment area, Roseland is a New York institution. In the 1920s it was the city's largest dance hall – 'The downtown headquarters for such urban dance steps as the Lindy and the Shag', according to a contemporary guide – but the formerly plush decor has been stripped and painted black for tonight's more brutal conditions of entertainment. The dance floor is a war zone: a simulated war zone

Nirvana, July 1993 (photo: Jesse Frohman/Katz)

to be sure, but still not for the faint-hearted. Hundreds of young men ricochet off each other at high speed in the 'moshpit', creating flows and eddies that take on a life of their own. And then, by a combination of individual effort and group will, one of them will crest on the surf of this human tide, splaying his body out in pure abandonment before disappearing again. It's a communal, physical release.

This is Nirvana's first New York show for almost two years. Expectation is high: this group arouse curiosity and passion like few others. Since their second album, *Nevermind*, went to the top of the US charts in January 1992, they have found themselves in a situation similar to that which the Sex Pistols experienced in 1977: to some, they are rock prophets, standard bearers for a generation; to others they are, as their drummer Dave Grohl drily summarizes, 'Cynical slacker little fuckin' punk jerks.'

Although they are exceptionally successful, Nirvana are not quite as successful as some other artists on the Billboard chart. As *Nevermind* faded out of the top 200 in July this year, 13 acts had sold more records, including Garth Brooks, Kriss Kross, Michael Jackson, Boyz II Men and Michael Bolton – the country, R&B, rap and rock that are the staples of American pop. But *Nevermind* and its breakthrough single, 'Smells Like Teen Spirit', have come out of left field to create a cultural and aesthetic impact that goes far beyond statistics.

Nirvana stir up deep emotions. Most obviously, *Nevermind* has changed American music: music media like MTV are now full of post-Nirvana groups like Pearl Jam, Soul Asylum and Stone Temple Pilots, who play that mixture of rock, metal and punk now known as 'alternative'. (Two years ago it would have been grunge, but following Perry Ellis's autumn 1991 collection, grunge has entered the language of fashion: the anti-fashion of sloppy T-shirts, flannel shirts and old Levi's that poverty-stricken bohemians and students have worn for years suddenly became high style.)

Nirvana have also been seen in sociological terms: as defining a new generation, the 20-something 'slackers' who have retreated from life; as telling unattractive home truths about a country losing its empire and hit by recession; as representing the final, delayed impact of British punk on America.

They have also shocked people by trashing male gender codes: kissing each other on the national network show *Saturday Night Live*, appearing in dresses in the video for their single 'In Bloom', doing pro-gay benefits. We may be more used to this in Britain, but America is a country with much more machismo in its popular culture. A sensational appearance on last year's globally broadcast MTV awards, where they smashed their equipment and mocked rock competitors Guns 'N' Roses, sealed their status as America's bad boys.

Nirvana have become an issue in America: they have attracted all the hostility and scapegoating that goes with this territory. A September 1992 *Vanity Fair* article, which alleged that singer Kurt Cobain and his wife, Courtney Love, had taken heroin during Love's early pregnancy (an allegation hotly supported by the writer of the article, Lynn Hirschberg, and even more hotly denied by Love and Cobain), crystallized the couple's ascent, or rather descent, into celebrity. Suddenly, the stories were about Kurt 'n' Courtney – the 20-something Sid Vicious and Nancy Spungeon – rather than Nirvana the group. As Cobain now admits, 'It affected me to the point where I wanted to break up the band all the time.'

When American groups are successful, they tour constantly to keep their records in the charts. As *Nevermind* went on to sell 4 million copies in America (double platinum in the UK; 9 million world-wide), Nirvana withdrew. In the vacuum caused by their disappearance, the rumours flew: Cobain was dead; they were splitting up; they were recording a new album so unlistenable to that no one would buy it. Kurt 'n' Courtney were hardly out of the news or the gossip columns: threatening unofficial biographers on the telephone and in person; being arrested for domestic unruliness in Seattle; fulfilling the media demands of punk couplehood.

By doing nothing, Nirvana have become mythic figures, a process that will be accelerated by three Nirvana-related books this autumn. This is a lot of baggage for anyone to carry, let alone three scruffs from the hinterlands of America. As they reappear in the public eye with an album called *In Utero* – 'in the womb' – the group are surrounded by an atmosphere of high tension, a magnetic charge which both attracts and repels.

Nirvana take the Roseland stage with everything to prove – straight into a sequence of crunching numbers: 'Serve the Servants', 'Come As You Are', 'Lithium', 'School'. Most alternate quiet, almost whispered verses with wild choruses which crackle like a power surge. The audience goes mad, cheering alike old favourites and new songs – the punning 'Penny Royal Tea' (named after a concoction used to induce abortions) or the pathologically personal 'Heart Shaped Box' ('I wish I could eat your cancer when you turn black …').

Nirvana are famed for leaping around on stage, even smashing their instruments, but tonight they don't do much except play hard and accurately. Bass player Chris Novoselic dominates stage left with all his six feet seven inches but, despite his lack of mobility, or in fact because of it, it is Cobain who holds your attention. He hunches into the microphone and croons, growls and then screams from the pit of his stomach. You might think this was just a teenage tantrum, but then you watch the group's control, you hear the Beatlesesque tunes and the smart, elliptical lyrics, and you realize that Nirvana are serious.

Just when you think that they're relaxing into the home run, something extraordinary occurs. Cobain leaves the stage for a couple of minutes. He reappears with a female cellist, who sits centre-stage, a dramatic contrast to all the boys moshing in front of her. The group start an acoustic song, 'Polly' – the story of a rape victim who outwitted and escaped her captor. It's a harrowing lyric, and Cobain sings it very quietly. Too quietly: the audience ignores him.

When it becomes clear that Nirvana are not going to rock, an abyss opens between the group and the audience: you can hear it as the buzz from the crowd threatens to drown out the acoustic instruments. Suddenly, Nirvana look vulnerable, but each song is more harrowing: 'Dumb', where sarcasm masks deep hurt; or 'Something in the Way', based on Cobain's experience of sleeping rough. Then they begin Leadbelly's 'Where You Gonna Sleep Tonight', which Cobain sings with all the keening notes of a Childe ballad, Appalachian style.

'Where You Gonna Sleep Tonight' has that quality of desolation which haunts the most powerful American rock, from Leadbelly to Bob Dylan to

Neil Young to R.E.M. As Cobain circles round the lyrical repetitions, his voice becomes more and more racked ('When you feel bad in America,' Cobain tells me, 'it's like losing your stomach'), and he pushes the words so hard it's as though he's trying to vomit them out. Then it's suddenly over: the group leave an audience nonplussed into an eerie silence. It takes a while before the calls for an encore become persistent.

Push me, pull you. Nirvana do encore, with 'Smells Like Teen Spirit', but follow it with several minutes of deliberate feedback by Cobain, who remains on stage long after Novoselic and Grohl have left, crouched in his own world. As an industry showcase, it's unprecedented; in America, success demands conformity and repetition of what made you great. It's a total punk rock show: a bitter, dogged stand-off between the group's insistence on doing what they want, and the audience's expectations of what they should do.

The sleeve of *Nevermind* shows a baby swimming under water towards a dollar bill on a fish hook. The intended meaning is clear: the loss of innocence, the Faustian contract that usually comes with money. Take it, but if you do, you're hooked for life. It's a parable of Nirvana's current dilemma: they've taken the bait, but the contradictions of their success are threatening to tear them apart.

How can the members of Nirvana retain their integrity, which is very important to them, in a situation which demands constant compromise? How can they sing from the point of view of an outsider now that they're in a privileged position? How can they suffer relentless world-wide media exposure and still retain, in Grohl's words, 'the spontaneity and the energy of something fresh and new' that has marked their career?

Kurt Cobain materializes in the lobby of a smart midtown hotel. It's a quiet entrance, but an entrance nevertheless: for all his fragility, Cobain is very much the star when he needs to be. He is of medium height and painfully thin. His garb of baggy patched jeans, woman's acrylic cardigan and shredded red and black jumper – exactly like the one worn by the *Beano*'s Dennis the Menace – would cause him to be thrown out in any other circumstances.

It's the day before the Roseland concert. Dave Grohl and Chris Novoselic have already passed through on their way to the appointment for which Cobain is somewhat late. This is in character: Grohl is a straight-ahead 24-year-old, the son of an Irish/American family, who wears layers of ripped casual clothes and has the physique and stamina of the sportsman he was before punk took over his life at 15. 'Let's do it,' he says when we go to talk; he is the youngest of the three, and his precision and fire give Nirvana much of their attack. He would not be habitually late.

Nor would Novoselic, but for a different reason. A tall, rangy man, dressed in black, who handles his size with care, he is the bottom of the group – the bridge between Grohl's energy and Cobain's spaciness. The Los Angeles-born son of Croatian immigrants, he has the air of someone who has fought his demons: his teenage conversion to punk rock has matured into thoughtful application of his political beliefs. It was Novoselic who arranged Nirvana's last major show, a multi-artist benefit for Bosnian rape survivors, after visiting what had been his mother country.

If Grohl and Novoselic are definitely in the world, Cobain tunes in and out: 'You haven't been waiting for me, have you?' he asks the assembled company. 'If you have, shout at me.' But everybody is resigned to the fact that when you do anything with Kurt Cobain, you have to wait. This is partly due to the nocturnal time that many musicians keep, partly due to the fact that, like many parents with a one-year-old child, Cobain is sleep-trashed, and partly due to the fact that, like it or not, much of the pressure that surrounds Nirvana comes to rest on Cobain's frail shoulders.

His response, in public at least, is simultaneously to court and to flee attention. This is a quite understandable human trait and does not denote insincerity. At first, he mumbles and is vague. His straw-coloured hair is shoulder length, centre-parted, and falls over his beard: together with the white 1950s sunglasses – the sort which people wore in the UK during the punk days – it means that you can hardly see his face. When you finally see his eyes, you understand why. They are of a startling blue sensitivity.

When you prise him away from the mania of his situation, Cobain is courteous, intelligent, quiet. Nirvana's success has meant validation for this German/Irish 26-year-old from Aberdeen, 180 miles away from Seattle in America's North-West. Whereas once he might not speak for days, now thousands hang on his lyrics and public pronouncements; where once he was an outcast, attacked on his home streets, he can now flaunt his difference in the eyes of world, and be loved for it. He has also learned that if you're loved, you're hated, and that if you provoke, you get a backlash.

The three members of Nirvana were all born between 1965 and 1969, a fact reflected in their name, a sarcastic comment on hippie pieties: in some ways, they are acting out the freedoms and failures of that time. All three are the children of divorced families. 'A lot of people have this theory that we cling together because of that,' Dave Grohl says. 'We all basically grew up with our mothers, although Chris and Kurt went back and forth.'

Cobain was the youngest of the three at the time his parents divorced, and it hit him hard. 'It was when I was seven,' he says. 'I had a really good childhood and then all of a sudden my whole world changed, I remember feeling ashamed. I became antisocial and I started to understand the reality of my surroundings, which didn't have a lot to offer. It's such a small town that I couldn't find any friends who were compatible. I like to do artistic things. I like to listen to music. I could never find any friends like that.

'I felt so different and so crazy that people just left me alone. They were afraid. I always felt that they would vote me Most Likely to Kill Everyone at a High School Dance. I could definitely see how a person's mental state could deteriorate to the point where they could do that. I've got to the point where I've fantasized about it, but I'd have always opted for killing myself first.'

The three were born into an environment where pop was the way of interpreting the world. As the youngest, Dave Grohl was entranced by the first American response to English punk music: the new wave of groups like the B-52s and Devo.

Novoselic 'listened to hard rock radio like Judas Priest and Black Sabbath, then I wasted good years of my life listening to Ozzy Osbourne and Def Leppard.'

Cobain grew up with the Beatles, graduating to the American rock/pop of the day at high school: Cheap Trick, Led Zeppelin. 'My mother always tried to keep English culture in our family,' he says. 'We drank tea all the time.' Much later, he immersed himself in the English Gothic of Joy Division: 'I've always felt that there's that element of Gothic in Nirvana.'

Sometime in the mid-1980s, the two Aberdeen outcasts met. 'I saw that Kurt was enlightened,' says Novoselic. 'I liked him because he was funny, he was an artist, he was always drawing stuff. He was always a bohemian, for sure, but he always had trouble with rednecks. I think that was just his bad luck. One time this redneck just held him down and tortured him.'

'For a long time I had no male friends that I felt comfortable with,' says Cobain. 'I ended up by hanging out with girls a lot. I just always felt that they weren't treated with respect. Women are totally oppressed in small towns like Aberdeen. The words bitch and cunt were totally common; I mean, you'd hear them all the time. It took me years to realize that these were the things that were bothering me.

'I thought I was gay. I thought that might be the solution to my problems at one time during my school years. Although I never experimented, I had a gay friend, and that was the time I experienced real confrontation with people. I got beaten up. Then my mother wouldn't allow me to be friends with him any more, because she's homophobic. It was devastating, because finally I'd found a male friend who I actually hugged and was affectionate to. I was putting the pieces of the puzzle together, and he played a big role.'

All three were empowered by punk, as it slowly filtered from Britain through the US. Whereas in Britain punk groups were guaranteed major label and indeed national media attention, in America they were shut out after a disastrous Sex Pistols tour and the coincidental rise of disco: just after the Sex Pistols broke up on the West Coast in January 1978, the soundtrack to *Saturday Night Fever* went to the first of its 24 weeks at America's number one.

Punk's commercial twin, new wave, had limited but significant success in the early 1980s, but the pure stuff went underground, burrowing through America, city by city, like a termite. Each major city had its own 'scene': Washington, San Francisco, Los Angeles, Minneapolis. Few groups from this culture had any major label attention until the mid-1980s: instead, a network of independent labels and clubs developed similar to the one that had fuelled the growth of rock 'n' roll in the 1950s. Cobain had read reports of the Sex Pistols' US tour: 'I'd just fantasize about how amazing it would be to hear this music and be part of it. But I was 11; I couldn't. When I finally heard American punk groups like Flipper and Black Flag, I was completely blown away. I found my calling. There were so many things going on at once, because it expressed the way I felt socially, politically, emotionally. I cut my hair, and started trying to play my own style of punk rock and guitar: fast, with a lot of distortion.'

Cobain started Nirvana with Novoselic in 1986. He'd always written – thoughts, scraps of poetry – and they became lyrics. The group played locally in Seattle and Olympia, a college town that was home to a particularly thoughtful punk scene. This was based around the group Beat Happening, who ran their own club, set up their own label and recorded songs like 'Bad Seeds'. 'A new generation from the teenage nation,' they sang; 'this time let's get it right.' In 1987, this generational rhetoric was charming but absurd: nobody thought like that any more. In 1992, it seemed like prophecy.

In the last years of the nineteenth century, Seattle had been an important transit point for the Klondike Gold Rush in Alaska. The city bore for many years the legacy of this boom-or-bust phenomenon. Nearly 100 years later, another gold rush occurred when the major record labels and national media descended on Seattle. What they found, and have since placed in the mainstream of American pop, was a form of music where, as Novoselic says, 'people had outgrown hard-core and were rediscovering rock – Blue Cheer, the Stooges'.

By 1987, two young college graduates, Bruce Pavitt and Jonathan Poneman, had a Seattle label up and running, Sub Pop, which released records by local groups like Mudhoney, Nirvana and Soundgarden. They were good at marketing books, slogans, and wanted a word to describe, half mockingly, half seriously, the noise that these groups made. It wasn't punk exactly, although it was steeped in punk attitude and politics; with the gut-wrenching downer pace of heavy metal and the tuneful tension of rock, it was … grunge.

Sub Pop released Nirvana's first single, 'Love Buzz', in 1988. Their first album, *Bleach*, was recorded a year later for just over $600. Dave Grohl joined in 1990 – the final focusing of the group. By this time, the American music industry was beginning to take up punk groups: the signing of New Yorkers Sonic Youth to DGC, a subsidiary of Geffen, had made waves, and it was DGC A&R man Gary Gersh who finally signed Nirvana in 1991. A bidding war put the advance up to £287,000, and the group went in to record *Nevermind* with producer Butch Vig. By the time *Nevermind* was released, Nirvana were being handled by industry insiders: management by John Silva of Gold Mountain, record company by David Geffen – a highly privileged position. *Nevermind* made everything else the Seattle groups had recorded sound like a demo: although Cobain now disavows the record as 'too slick', Vig's production gave Nirvana a power and a clarity which enabled people to finally hear what was going on. Even with such a strong team and such a strong product, Nirvana were not expected to sell more than about 250,000 records: *Nevermind* sold over a million within six weeks of release.

Success brought Nirvana more problems than it solved. 'I didn't anticipate it all,' says Novoselic. 'I didn't know how to deal with it.' 'We were touring constantly,' says Cobain, 'so I didn't realize what had happened until about three months after we'd become famous in America. It just scared me. I was frightened for about a year and a half: I wanted to quit. It's only after the birth of my child that I decided to crawl out of my shell and accept it.'

Nirvana should have been on top of the world but instead they freaked out. Part of the problem had to do with the culture from which they came, which had celebrated the outsider – 'Loser', read an early Sub Pop T-shirt slogan – and which was fiercely anti-major label, pro-independent. One of Nirvana's first acts on joining Geffen Records was to print a T-shirt which read 'Flower-sniffin' kitty-pettin' baby-kissin' corporate rock whores'.

The group's unnecessary agonizing about 'selling out' and 'corporate rock' cloaks the more serious problem of how they can retain their strong ideals. Exposure to the mass market tends to iron out subtleties and ironies, as Nirvana have found to their cost. Nirvana are pro-gay, pro-feminist. Cobain's irritation at the attitudes of some of his new fans spilled over in the sleeve notes for their post-success compilation, *Incesticide*: 'I have a request for our fans,' he wrote. 'If any of you in any way hate homosexuals, people of different colour or women, please do this one favour for us – leave us the fuck alone! Don't come to our shows and don't buy our records.'

'As a defence, I'm neutered and spayed,' Cobain sings, and, indeed, he presents a view of men that runs against the macho self-image of a country where the ethic of youth, health and personal self-improvement is still strong. He presents himself, like Morrissey once did, as old, ugly, still ill.

Rock requires personal authenticity, and Cobain embodies what he sings. 'My body is damaged from music in two ways,' he says. 'I have a red irritation in my stomach. It's psychosomatic, caused by all the anger and the screaming. I have scoliosis, where the curvature of your spine is bent, and the weight of my guitar has made it worse. I'm always in pain, and that adds to the anger in our music. I'm grateful to it, in a way.

'My stomach was so bad that there were times on our last tour where I just felt like a drug addict because I was starving. I went to all these different doctors but they couldn't find out what was wrong with me. I tried everything I could think of: change of diet, pills, stopped drinking, stopped smoking. Nothing worked, and I just decided that if I'm going to feel like a junkie every morning, vomiting every day, then I might as well take the substance that kills that pain. That's not the main reason why I took heroin, but it has more to do with it than most people think.'

Self-destruction haunts youth culture aesthetics: the myth of 'Live fast, die young and have a good-looking corpse' that runs from Thomas Chatterton through to James Dean and Sid Vicious. The grunge generation was not immune: patterns of local drug supply meant that heroin was easily available and, cloaked in the loser ethic, many Seattle groups succumbed.

'I'd taken heroin for a year and a half,' Cobain says, 'but the addiction didn't get in the way until the band stopped touring about a year and a half ago. But now things have got better. Ever since I've been married and had a child, within the last year, my whole mental and physical state has improved almost 100 per cent. I'm really excited about touring again. I'm totally optimistic: I haven't felt this optimistic since my parents got divorced, you know.'

For the whole of the twentieth century, America has thought of itself as, among other things, a young country. In this, the person of the child – from foetus to late adolescence (which, post-baby-boomer, can last until the mid-forties) – has become of prime importance. Out of this has come the youth culture which has colonized the world.

Now that America is in a crisis of recession, corruption and indeed social cohesion, the child and the teenager have become sites of struggle: the intense abortion battles, star revelations of child abuse, teen suicides, teen violence. Whether consciously or not, Nirvana have slipped into this national obsession, with their album concepts and messages from within the emotional front line.

Nirvana sing as traumatized children who have been empowered by the freedoms within popular culture. Their courage and talent have made them beacons for anyone who has felt the same way but have also placed them in the eye of the storm. Their growing pains are intense and are conducted in public: *In Utero* is a dark record, finely poised between self-destruction and optimism, and the Roseland show makes it quite clear just how much they are struggling.

Before the racked finale, however, Cobain does a wonderful thing: 'Ye-eh-eh-eh-eh,' he shouts over five notes during 'Lithium', and, as the roaring crowd back him up all the way, the hairs on the back of your neck stand on end.

Observer, 15 August 1993

Identity Crisis: The Ramones

May 1977: I came to on the floor. The floor was carpeted, brown. There was a skylight above me. I was home, that was all right. Punk was playing at deafening volume from the room on my right. 'Things don't last for ever, and somehow baby, they never really do.'

My head was in suspension, that pre-pain shock. I sat up, and a thick liquid started running down my face. That's when I realized something was wrong. After some frantic exploration, I found the cause of the trouble: a deep indentation, about two inches long, several inches in from the hairline. My fingers were covered in blood: I had to do something. I got up, and felt very peculiar indeed.

There was a hole in my life: it lasted from the middle of 'Gimme Gimme Shock Treatment' (from *The Ramones Leave Home* side 1, track 2) to the middle of 'I Remember You' (side 1, track 3). 'Gimme Gimme Shock Treatment' is a killer example of what the Ramones do best: moronic riffs, blank lyrics delivered in an affected accent and a rhythm section that thunders like a herd of buffalo – all within one minute 38 seconds. The album's opener, 'Glad to See You Go', is pretty cool, but only as a foretaste: limbering up with a few exploratory leaps, I'd hit overdrive as soon as Joey Ramone kicked off: 'I was feeling sick, losing my mind.' Yeah! Let's go!

How to dance to the Ramones had always been a problem. The previous summer, my friends Julian and Pam Wernick and I had been bound together by an obsession for the first Ramones album that bordered on the religious. We'd sit in Meanwood, Leeds, singing along to 'Loudmouth, 53rd & 3rd' and 'I Don't Wanna Go Down to the Basement'.

Like me, Julian was training to be a solicitor; like me, he hated it. Bonded by distortion, we sneered at all the other students. 'Ey daddio, don't wanna go …' We'd even tried to make it to the Ramones' 4 July 1976 concert at the Roundhouse – a long journey from Chester via Leeds. At Birch Service Station on the M62, I poured petrol into the radiator of Julian's Hillman Imp van. At the exact point that the Ramones went on, we were stuck in a recriminatory mess on the very top of the Pennines. It could have been worse: it was hot, the scenery was very beautiful; but WE DIDN'T WANT IT! We wanted urban mayhem, New York degradation, dominance and submission.

The best we could do was carry the album around with us like a sword. We behaved badly, but were evangelical: we wanted everyone to fall at the feet of our Ramonic God – pure monolithic noise. We'd gatecrash parties, wrench Steely Dan off the turntable, and start to, well, twitch very quickly. Not many people knew about punk at that point: there were no rules. People would object, but we'd keep on twitching. They'd laugh, but we didn't care, lost in a new world.

Soon after, I started going to punk events and noticed that the audience, if it moved at all, didn't twitch but leapt up and down on the spot. It was a claustrophobic time, with long periods of boredom broken up by frantic spasms of movement: most of the places you'd see these groups in were tiny, and were full of people foaming at the mouth on bad speed, whom it was sensible to avoid. So you moved on the spot: and the only way was up. There wasn't much room at home either … so I leapt.

Staggering down the stairs, I pieced together the lost minutes. There was no getting around it: I'd leapt straight out of my room into the top of the door, the lintel, and knocked myself flat out. I was alone. Somehow I got myself into the car, a white Morris Marina (bad cars were another feature of this period) and drove up Ladbroke Grove to St Charles's Hospital, where I was patched up with a bandage around the face. Then the headache started.

Worse than the pain was my mortification: thank God no one was around to see what a fool I'd made of myself. The next day, however, I decided that the whole affair was quite punky. It had to be recorded. I put on my favourite shirt – 1960s, white, doubly overprinted with the famous Op image of Bridget Riley with Continuum – and zipped off to play vacant in a Paddington photo booth. I Xeroxed it and sent it to all my friends: look what punk rock did to me! Knock yourself out and become a new person!

The scar is still there, still tender. Somewhere in my subconscious, 'Gimme Gimme Shock Treatment' is still playing on a loop. The only song I've ever blacked out to, it reminds me of the time when listening to the Ramones was as much an act of commitment as of pleasure. Seventeen years on, it's hard to remember just how powerful their early records were: so startling, so brutal, so *right* that they wiped the slate clean and made new life possible. I'm proud to be marked by them.

Mojo, December 1993

Suede: Dream On

In early June 1992, I go to see Suede at the Boardwalk, a venue converted out of a Victorian Church at Knott Mill, an old warehouse area of Manchester. Tonight is being filmed by Granada TV for a new group slot which will be transmitted at 4 a.m. months later: Suede are top of the bill over two other hopefuls, Wonky Alice and Swirl. It's the post-Madchester moment: nobody knows what's going to happen, but everybody can feel the tide ebbing away from this city. Suede are one possible future: touted as the next big thing on the basis of one *Melody Maker* cover, presented as everything that the Madchester groups are not: pervy rather than laddish, spiky rather than luv dup, defiantly un-dance. Tonight they have to play in front of the very people whose dominance they are usurping.

Suede's first single, 'The Drowners', has just been released, and all the three cuts are terrific: witty, passionate, with seemingly explicit gay references to thrill my heart. Live, they are not terrific: in front of a cold crowd, they fail to project into the hole that has opened up at stagefront. The TV lights ruthlessly expose their failings. They look disgusted and yet, even at their worst, you can see that they are reaching for something; that, given the right audience, they could burst into flame.

Over a year later, Suede play in front of a capacity crowd at the Clapham Grand, a refurbished music hall. The occasion is a benefit for the Red Hot Aids Charitable Trust. It is hosted by Derek Jarman, the man whose courage in openly living with AIDS is a beacon to us all. His projections for the Pet Shop Boys' 1989 tour open the show; the crowd cheer when he delivers an impassioned speech about the age of homosexual consent in this country.

There are plenty of added attractions once Suede begin: Chrissie Hynde and Siouxsie come on to perform 'Brass in Pocket' and Lou Reed's 'Caroline Says'. Jarman's 1970s Super-8 films, snapshots from an age of untroubled experimentation, provide a poignant backdrop to Suede's extravagant tales of twisted sex and casual brutalities. They give Suede's 1970s flourishes an extra resonance: how things have changed in those 20 years.

You look at the audience, and they are all there for Suede, teens and 30-somethings, of all persuasions. The great thing about the group is that they are not afraid of playing slow songs, but the highlight is the flat-out rocker, 'Moving'. As Brett Anderson and Bernard Butler spin, bump and grind, their bangs tossing from side to side in counterpoint, the Grand shimmers in pure white light. Time is frozen in an everlasting present.

Five months later, Suede are frustrated, fidgety. The first album reached number one in March, and has been digested and marketed to the max. Since then, no new material, but a video summing-up: *Love and Poison*. They have just toured America, and have come back with the irritation that any sensible person has when they compare American possibility with English stagnation. Every move of theirs is still picked over in the music press, engendering an unnecessary hypersensitivity on everyone's part.

Suede have become superstars within a limited firmament: they want to take on the world, but London is dragging them. Adolescent at a time when pop was already self-conscious, they've talked the language too well and have become *slotted in*. They have been over-eager, greedy almost, and they've paid the price. 'It pisses me off,' Brett Anderson says. 'When you're being written about as we have, everyone has to have their theories about you. The theories are fine, but they limit you by defining you.'

Suede are a modern phenomenon in that they are suffering from the modern discourse surrounding pop culture, which is to talk solely in terms of marketing, strategies, hype. This ignores one firm pop rule: you can only persuade people to buy something rotten once. It also ignores another firm pop rule: that what appears to be the result of deep calculation is usually the result of an accident, a mistake, a quick decision. As Anderson says, 'When you're young and inexperienced, you have to learn in public.'

The great pop records happen for many reasons, but one reason is that they connect with their audience in a direct and fundamental way. Like the Sex Pistols sang on 'Pretty Vacant', 'we know what we feel'; pop works by bringing to the surface, enacting, transcending, the feelings repressed by our static, conservative, still-puritan society. In this, the over-used word 'love' is simply the most obvious key with which to unlock our over-defended psyches.

'Music works from the heart rather than the head,' says Mat Osman. 'It's become very hard for us not to think about what we're doing. It isn't ideas music at all; it's romantic, heartfelt music, which is why there's a sexual element to it. To be a pop band, you have to be open to people, to housewives and schoolgirls, and to do that you have to sing about how you feel, not what you feel about it. Brett's always tried to write in a way that's completely open.'

Musicians always talk about not being defined but they have a point. In most societies, the musician has an ambiguous role: as Jacques Attali writes in *Noise*, he/she 'is simultaneously reproducer and prophet … an integral part of the sacrifice process, a channeller of violence; simultaneously *excluded* (relegated to a place near the bottom of the social hierarchy) and *superhuman* (the genius, the adored and deified star)'. The musician is always in motion: that very thing that much of the media is there to prevent.

In preparing for this interview, I decide not to look at the cuttings but to play the album I've lived with this year. It's all there. What I love about Suede is their ambition, their tarty rockers, their ballads, their gleeful trashing of sex and gender boundaries, their Piaf/Genet evocations of beauty in ugliness. And they come from a place that I know very well: the stretch of England that goes south from London, through Haywards Heath, to Brighton and then Worthing – the tragic seaside resort above all.

Pop (and rock's) rhetoric is of the inner city, but scratch the surface of most English pop stars and you'll find a suburban boy or girl, noses pressed against the window, dreaming of escape, of transformation. Suede are no exception: drummer Simon Gilbert comes from Stratford-upon-Avon; guitarist Bernard Butler from Leyton, east London; Brett Anderson and bassist Mat Osman grew up in Haywards Heath, a dormitory town 40 miles south of the capital.

'The reason it's there is because of the railway line,' says Osman. 'Everyone works in London. It's a holding pen, nothing more, and because of that there's no community. The generation that grew up there weren't planned for at all. It was meant for people aged 25 who worked in London, and you moved when you got to 35. Me and Brett came from a generation who never got out. You weren't meant to be young there.'

'It's so unappealing,' says Anderson. 'Everything is wrong about it. It doesn't have the charm of small town life, it's too big for that. I never completely withdrew. I met people of similar minds, there were four of us, and we stuck together and took mushrooms and acid, sat in the park off our faces, and people maintained their distance. We did all those things you did when you're young: visit haunted houses and get the Bostik out.'

From their account, it's clear that the pair fell between the cracks of class. Anderson's father is a taxi driver who, contrary to stereotype, is 'a classical musical buff, obsessed with Liszt and Mahler and Berlioz'. Both his children are artists: Brett is a singer and lyricist, Justine a painter. Mat Osman's father is 'a businessman, but my parents divorced when I was about 13. I suppose I'm middle class, but we were very poor for the area.'

Osman and Anderson are one prime unit in Suede. They've known each other the longest: it's 10 years since they met as teenagers in Haywards Heath. 'I knew of him,' Osman says. 'It's a small enough place that anyone who's slightly out of the ordinary is pretty well known. I met him at parties, when he used to look like a young stockbroker, with a striped shirt and tie-pins. You had this room full of people with huge bushes of hair, and Brett had this little blond flick, a yellow suit: he looked like Tommy Steele's son.'

We're in a rambling, Victorian warehouse just off Old Street which doubles up as a rehearsal room and the office of Suede's management, Interceptor Enterprises. Osman is straight-ahead, combative – tall in a brown suede shirt and purple velvet trousers, with a cynicism beyond his years. Anderson is more reserved, slowly opening up as the light fails in the Old Street rehearsal studio. Thinner than his photos, with watery eyes, he wears a tight blue T-shirt and old black Levi cords. He picks at lunch: a cheese salad baguette.

Now in their mid-twenties, Osman and Anderson were teenagers in the early 1980s: time of new romantics, post-punk and, which has now been completely forgotten, near nuclear war (triggered by the Russian invasion of Afghanistan). Both went wholesale for pop: 'You re-create yourself as an outsider,' says Osman. 'Because you know you're not going to have fun as the centre of the gang, you're immediately drawn to people like David Bowie. When you live in an environment where there is nothing elegant, nothing lasting, you're bound to be drawn to someone like that.'

Anderson: 'I've always been obsessed with the two things that make up pop music, which is image and musicianship. I started very early to play around with my image, cutting and dying my own hair, and making myself look absolutely dreadful. I had a punk phase, cut my hair like something that had survived a laboratory experiment. It's such a temptation to fucking stir everyone up a bit. Everyone is so complacent, and the status quo is maintained so forcefully: you just want to stick a spanner in the works. 'I used to go to Brighton and to Worthing too: my sister went to art college there. It's a funny place, just old people and this art college stuck in the middle. It's like a page out of *Sgt Pepper*: the old English world next to the art and drug-taking world.'

'Everyone who lived in Worthing was about 95,' Osman adds. 'They were dropping like flies, and the second-hand shops are the best in the country – if you don't mind wearing dead people's clothes. 'I can't remember the time I didn't think that being in a band was the thing to do. I think that's something very British. It wasn't something I talked about, because I wasn't the kind of person who became a pop star. This was the time of Wham! and Spandau Ballet, who obviously came from the big city, were obviously different from you. The most important thing about the Smiths was that Morrissey was obviously one of us lot. Here was someone who was squarer than you at school.'

'The first Smiths song I heard was "What Difference Does It Make,"' says Brett Anderson. 'The line that got me was "But still I'd leap in front of a flying bullet for you." I love the way it's not obscure. It's inches away from being a cliché, but it's so straight and so beautiful: it wasn't a cultural reference, it's just something that everyone has felt. Just the fact that you were into the Smiths meant that you didn't have to worry about anything else. It was an incredibly calming strength at the time.'

Like any music capital, London has long been the subject of song – from the low-life vignettes of the late-nineteenth-century music hall, right through Noël Coward and Hutch to the present pop era. Many are simple celebrations, – like the rapid transits of punk: 'Let's Submerge' by X-Ray Spex, 'In the City' by the Jam; others, like the Pet Shop Boys' 'West End Girls' and the Smiths' 'London', present the metropolis as the site of an ambiguous glamour and excitement.

We tend to think of the 1960s as a golden age, a period of pop enthusiasm, of Swinging London, but many records of the period give a harsher view: David Bowie's great 'The London Boys', Them's bitter 'You Just Can't Win' ('You're up in Park Lane now/And I'm somewhere round Tottenham Court Road'). Perhaps the bleakest is the Kinks' Dickensian 'Big Black Smoke', the story of a teenage runaway: 'She slept in caffs and coffee bars and bowling alleys/And every penny she had/Was spent on purple hearts and cigarettes.'

Of all contemporary groups, only the Pet Shop Boys and Suede have had the ambition and the talent to sing about London in a convincing way – one that acknowledges that the city is in recession after the febrile boom of the 1980s, and that recession does terrible things to people. Neither are native Londoners. Just think of the PSBs 'Kings Cross' and 'The Theatre' – the latter written after a reported comment by a Tory MP that the homeless were the people you stepped over on your way to the opera.

Suede's attitude is summed up by a line from 'He's Dead': 'I have the look of a son/With all the love and poison of London. 'Part of the enjoyment is the pain, isn't it?' says Brett Anderson. 'Like a drug addiction, you don't enjoy the drug until you're addicted, that's the dichotomy, and it's the same with London. There's so much depression and yet so much warmth as well. On the bad side, there's a strength and a romanticism which I draw from.'

Osman and Anderson both moved to London in the mid-1980s to pursue further education: Mat studying politics at the LSE, Brett architecture at University College, London. 'I lived in a place called Richmond Gardens, just off Tottenham Court Road,' he says. 'Three of us in a miserable room. I'd go to Heaven and places like that, the E scene: gay nights and mixed nights. I've had some horrific times, but when I first moved here it was such a complete change it's like I was on a cloud.'

Neither was satisfied either with higher education or the jobs on offer to them. 'I'm not a very good worker at all,' Osman says. 'I'm not lazy, just not very good at it.' In 1989, according to Anderson, 'We placed an ad in the *NME* saying, young guitarist wanted for, I dunno, important band. We listed our influences as … Pet Shop Boys were in there.' They got Bernard Butler. 'Suede started the day that Bernard walked into the room,' Osman adds. 'Everything before that is a bit pointless to discuss.'

'We didn't start by writing anything good,' Anderson says. 'A lot of the first things we did were total rubbish. It took a while for us to click. A lot of dodgy little songs that will never turn up, or hopefully they won't. Bernard was a great musician when we met, but I didn't concentrate on singing. I used to play the guitar a lot, so I couldn't forge ahead with my voice. It wasn't until he took over with the guitar that I could actually think about it.'

Bernard Butler, infamously, never talks to the press, and he doesn't. It's quite obvious, from their live performances and records, that he forms another prime unit of Suede with Brett Anderson: that of close musical collaborator, the person who can alchemize the singer's fascination with the tawdry, with the extravagant, into gold. If you want him to speak, just listen to the way that 'Animal Nitrate' unfolds, the heavily treated guitars curling with a sickening thrill, into Anderson's yelping vocal.

And then they were five: as well as guitarist Justine Frischman, Suede added drummer Simon Gilbert in January 1990. 'I always dreamed about living in London when I was in Stratford-upon-Avon,' he says. 'Fantasized about being on the Tube, because of the Clash and the Sex Pistols. I thought, that's where it's all happening. I was always in bands: I knew what I wanted to do.'

Gilbert is quiet, friendly, enthusiastic. When the interview finishes, he asks, quick as a flash, 'What were the Sex Pistols like live, then?' For him, Suede was a lifeline. 'I'd been in so many bands, 12 or something, and this was the first one where it didn't matter if there was only one person watching, because it gave me such a thrill every time we played. We really had something that the others had, the mixture of personalities.'

Despite their early promise, the world did not listen to Suede. 'We were incredibly lucky to get away with two years in the wilderness,' says Mat Osman. 'I've had friends who've had record companies at their fourth *rehearsal*. One of the reasons we're quite disliked is that when we did arrive, we knew what we were doing – what we were talking about, our sound, our own look, only because we'd had two years of learning, of terrible howlers.'

Suede arrived, fully fledged, in May 1992. Heralded by that April *Melody Maker* cover, 'The Drowners' was a landmark single: with their stuttering guitar lines, massive drum beats and exaggeratedly cockney vocals, Suede span a whole section of English rock music into a new orbit. On the sleeve was the famous 1970s picture of female model Veruschka body-painted with a man's suit; inside, Brett Anderson sang of a masculinity where surrender, being 'taken over', was the most pleasurable thing possible.

In retrospect, you could say that everything was in place, and this is the accepted view: Justine Frischman had left in autumn 1991, leaving the group as a four-piece; they had a sympathetic record company run by industry veteran Saul Galpern (who oversaw their signing to Sony later in the year); they had a sharp PR team, Phill Savidge and John Best, who wrested coup after coup: 19 magazine covers over the next year. None of this would have meant anything without the records, without the dramas that Suede conjured out of today's Britain.

The first was the most immediate: Suede presented a severe androgyny, all 1970s waisted angles and pouting, with lyrics that told stories of explicit gay sex – 'The Drowners', 'Animal Nitrate' – or at least went in for a good blurring, like on 'Moving': 'So we are a boy we are a girl.' The Madchester groups had been laddish, blokeish, even homophobic: in contrast (and they had to be in contrast) Suede were haughty, girly, quite possibly homosexual. There was that infamous Anderson comment that he was 'a bisexual who'd never had a homosexual experience'.

This is a sore point. Osman moans about how 'frustrating the androgyny thing' is. Anderson sighs: 'I said, I *feel* like a bisexual who'd never had a homosexual experience. I was talking about low-life songs, about quite a spiritual thing. I was trying to express something universal, and it was taken out of context and used to categorize me. It has as much depth as 'shut that door' or something like that. But if it sparks off some debate, then it's worth it.

'My songs tend to come from my experiences, and the experiences of people who are close to me. The songs that specifically revolve around the gay world, like "Animal Nitrate", are written because I'm involved in it through my friends – about 50 per cent of whom are gay: they are love songs for their feelings. When people say that I'm just using gay imagery, it depresses me, because my friends go through emotional turmoil. I've felt that on their behalf, and written songs for them. The idea of gay love songs being tender seems alien to most people. I want to redress that.'

'It was so sad to see that quote flashed around the world,' says Osman. 'It was an attempt not to get boxed in, and he may have put himself in one of the smallest boxes of all time. Brett singing "he" and being seen as a gay love song is just as constricting as the boy–girl thing. I'm surprised that people haven't thought he might be singing from a female point of view, or a gay point of view, or writing about himself. It's quite an egotistical thing to write a song: most end up as being about the person who wrote them.'

But much of Suede's power to date has been to do what pop can do best: bringing to the surface the lives of people who have been ignored by the mainstream media, the dispossessed. It's still hard to talk about homosexuality in rock: dance, disco and Techno are fine, because record companies are quite happy to initially market a tune at the gay market. But rock? No, never. 'In an ideal world, it shouldn't matter whether somebody is gay or not,' says Simon Gilbert. 'Unfortunately, at the moment it's still taboo.'

In June last year, Gilbert came out in an *NME* interview – still one of the very few people in rock music to do so. As it happens, he is the only gay members of Suede. 'Before I did the interview I just knew I was going to do it,' he says. 'My mum didn't know before, and I called her up because she always buys the press, and she was brilliant about it. I got a letter yesterday saying, "Thank you for coming out because it gave me confidence to come out myself." When that happens, it's brilliant.'

On their album, released this March, Suede also laid down their claim to write about a fresh generation: formed by recession, 'with sweet FA to do today'; if you want, the English equivalent of the slackers. With its mocking Shadows twangs, 'So Young' presented a view of youth that you don't often hear in the media, and certainly never from within: trying to connect in an environment turned cold, where obliteration through drugs is a constant temptation, where the smallest mistake can result in casual brutality.

'I see everything around me as quite violent,' Anderson says. 'It comes from being vulnerable. When you sit and watch what's going on, you see it from very early on – when you're a kid and your parents have arguments. England is a very violent country, more mentally than physically, and it manifests itself in extreme cases, like, for example, the East End at the moment. The pressure is intense. I find it difficult getting out of my door sometimes.'

Next month, Suede release their first new record for nearly a year: an epic, eight-minute single called 'Stay Together'. The lyric speaks of a desperate need to make physical, emotional contact in a world approaching Armageddon: at the climax, Anderson gabbles desperately, buried in the mix. The postscript is a desolate wail of feedback. On another level, it sounds curiously innocent: a sweet goodbye to Suede's 'new suburban dream'. It's clear that they're moving on.

'I'm thinking quite apocalyptically at the moment,' Anderson says. 'I've been writing a song about everything going mad at the moment. Do you get that sense? There's so much instability in this country – race riots, everything's falling apart. Apart from people in the band and the music business, every single friend I've got is on the dole. It's ridiculous. So I'm thinking terrible things.

'It's the way I used to feel when I was 13 or 14, actually. In the early 1980s it was very much like this. When the Russians went into Afghanistan, there was real tension going on, and all the newspapers had articles explaining who would be maimed, and who would be immediately burned. It was frightening, being that age. Then there was massive insecurity, this enormous threatening thing, but now I'm less afraid of death, more concerned with the hellish consequences of something like a nuclear war.

'We were flying to Las Vegas on our American tour: the night before I'd had a dream about this character called Little Boy. I didn't know who it was, I thought it was a character from a book. I asked Mat about it, and he told me that Little Boy was the name of the bomb they dropped on Hiroshima. We were flying over the Nevada desert where they did all the nuclear testing. I turned and looked out of the window and, no joke, there was a mushroom cloud there. It was actually caused by a big fire that was going on in the desert, and these things happened in the same day. It was one of the most frightening things in my life.'

Mojo, January 1994

Delicious Surrender: British Psychedelia

A brief drum fanfare leads into a loose shuffle; a foppish singer begins to croon, so languid that he's barely there:

> Everybody's got new clothes
> Makes me feel kind of old
> But in my heart I really know
> No, clothes don't buy my soul …

In the second verse, we get to it: 'Now my head is really spinning' – a simple statement authenticated by heavy echo, whooshing cymbals, and a droning, feedbacked guitar solo so psychotropic it makes you feel dizzy, 26 years after it was recorded.

It's 1967: the group is Tintern Abbey, the song is 'Vacuum Cleaner'. A year before, they'd have been a tough R&B band, all sexual urgency, but now all is delicious surrender: 'Fix me up with your sweet dose/Now I'm feeling like a ghost.' On the flip, they sing about bees, but they're not buzzing round your hive: the insects are chided for their industriousness, while the middle eight introduces a 'virgin of humble origin [who] knew of no sin'. Again, there is no overt sense, just the shimmering, elusive wash of sound. Released in November 1967, 'Vacuum Cleaner' disappeared without trace, as did Tintern Abbey, of whom we know nothing except their names, and the fact that they appeared with a live owl perched on the singer's shoulder. Yet with their 'medieval' name, in their very elusiveness, Tintern Abbey are the perfect example of a lost pop era, now exhumed – in our past/present pop loop – in dozens of compilations, both semi-legal and official. And what a revelation they are.

British psychedelia has always been regarded as a faded Xerox of the 1967 West Coast boom. Compared to the Americans' messianic qualities – serious drugs, confrontational politics, a Dionysian wildness with Wild West trappings – the English groups were inner-directed, childlike, too domestic for these new conditions. Psychedelia here was seen by contemporary commentators like Nik Cohn as a summer 1967 blip, while in the US, it connected with the counterculture and laid the foundation for the rock industry of the 1970s.

The 1960s were a time of vast overproduction: in the UK, 1964 saw the biggest singles sales ever, and a similar pattern occurred in the US. Bewildered by the beat boom, record companies released more records than they could promote: the failure rate was astronomical. The last 20 years have seen a thorough recycling of this wastage: beginning with Lenny Kaye's wonderful *Nuggets* and the process has continued in the UK with series like *The Perfumed Garden* and Phil Smee's mammoth 11-CD *Rubbles* project.

These compilations tell a story that begins in black music and ends in the whitest music possible: progressive rock. The earliest records here push R&B structures to the limit: the Birds crunch a Tamla cover, 'No Good Without You Baby', into a Kinks/Who slab of hard rock, while The Syndicats' Joe Meek-produced 'Crawdaddy Simone' speeds up the Yardbirds' rave-up into total bedlam. These are amphetamined, supremely confident records from the time when English pop was exploding world-wide.

The point here, as Chris Cutler notes in his memoir of the period, *File Under Popular*, was pure noise:

> The disordering effect of sheer volume seemed somehow to express directly, ecstatically, something beyond words or previous experience. Perhaps it was a sense of destruction, of the self being destroyed, of ceasing to be outside and becoming part of something bigger and communal.

And yet even at the fullest freakout, there was an R&B discipline: almost all these songs fall within the three minutes watershed.

Cutler suggests this 'unbroken line of uniquely English development: the Shadows, the Yardbirds, the Who, the Pink Floyd'. Common to all was a concentration on sound and texture: 'the whole of the British psychedelic movement was first and foremost experimental – none of the relaxed California hedonism here.' It was also pop: other major influences were the Kinks (a return to the music hall) and the Beatles (instrumentation: just listen to the 'Penny Lane' horns on the 23rd Turnoff's peerless 'Michaelangelo').

As the drug of choice changed from amphetamines to marijuana and LSD, the pop mood was overtaken by reverie: if you followed the hipster in Smoke's 'My Friend Jack' (who 'eats sugarlumps'), you could get aboard Kaleidoscope's 'Flight from Ashiya': 'cigarettes burning faster and faster/Everyone talking about the everafter.' But each trip was a journey into the unknown, and there was always the chance that you couldn't find your way back home: as Kaleidoscope repeat on 'Ashiya''s fade: 'nobody knows where they are'.

Other examples are the Attack's 'Colours of my Mind', the Human Instinct's 'A Day in My Mind's Mind' and Fairytale's 'Guess I Was Dreaming': all contain a blurring of the real and the fantastic, aurally reproduced by detuned, raga-style guitars and fey voices. Sometimes lyrics would be inspired by Grimms' fairy-tales, Edward Lear and Lewis Carroll. Fleur De Lys and

Boeing Duveen and Beautiful Soup went all the way with their redactions of 'Dong with a Luminous Nose' and 'Jabberwock'.

Acid made you see the world anew, just like a child, but the stripping away of the ego could cause another return to childhood, the battleground of therapy: that was the point of 'Strawberry Fields Forever'. Syd's Pink Floyd explored this simultaneous wonder and terror; their success pulled in groups like Virgin Sleep, who, on 'Secret', boast of having tea with a teddy bear, or Fire, who offer this sterling excuse for delinquent behaviour: 'My father's name was Dad, my mother's name was Mum – how can I take the blame for anything I've done?'

Childhood wonder became one way of transcribing the LSD loss of ego; another was time-tripping into medievalism and beyond. Songs like Poets' 'In Your Tower' and Tintern Abbey's 'Beeside' evoked a mythic, pagan past, while the success of Procol Harum's monkish 'A Whiter Shade of Pale' occasioned the church-organ drenched 'Dream Magazine' by Svensk and 'Meditations' by Felius Andromeda. 'Millennial delusions were rampant in the rarefied atmosphere of London in 1967,' writes David Dalton; psychedelia began in style and ended in mayhem. All were released as singles: most of them are great, and none of them were hits. Why?

1967 was a watershed year: the success of *Sgt Pepper* skewed a section of the pop market into buying albums. The very wildness of these records alienated the pop audience (the biggest-selling 45 of 1967 was Engelbert Humperdinck's 'Release Me'); and the most influential medium for breaking innovative records was removed by the Marine Offences Act which, on 15 August 1967, made pirate radio stations illegal.

Such an intensity couldn't last. 1968 was the year that everything started to go wrong. The volume of money poured into the music industry meant that groups could develop (or spread) ideas that had been contained within three minutes over the 40 or so minutes of an album. Commitment took the place of inner exploration. There was a division within pop, between singles and albums, between 'bubblegum' and 'progressive': for groups like Tintern Abbey, the Attack or Fire, who combined both, there could be no place.

Yet the influences of these oblique, obscure records has passed into British pop. You can trace a line from David Bowie (whose *Pin-Ups* recycled this period), through the gleeful playfulness of mid-1970s Eno, to late 1980s dream-pop groups like My Bloody Valentine and the lush interior landscapes of Ambient Techno. The noises are not the same, yet they reveal a constant in British pop culture: a flight from claustrophobia into an inner and outer-space, a flight so headlong as to transform the world.

Mojo, March 1994

Pop Art: *Backbeat* and Stuart Sutcliffe

There is an old photograph taken at the Hamburg Art Carnival of three teenage artists: Klaus Voormann, Astrid Kirchherr, Stuart Sutcliffe. As it happens they are a romantic triangle, but you'd never know it from their gang-like similarity, their defiantly androgynous collusion against the world.

They look magnificent. Astrid dominates, wearing a low-slung black T-shirt and a black leather suit. Both men have the swept-forward, bouffant *pilzenkopf* (mushroom head) haircuts that the Beatles later adopted: Voormann is in a velvet tunic with lace cuffs and ruff, Sutcliffe sports black leather trousers, pointed winklepicker shoes and a shirt tied high around his midriff. Carnations adorn his shirt and trousers, through which rises a huge hard-on.

It is 1961: a world away from ours. Popular music was pure show business; the Beatles were unknown outside a few people in Liverpool and Hamburg; the elements which fuel today's pop culture – weird clothes, sexual nonconformism, drug delirium, emotional intensity – were so far underground as to be invisible to mainstream society. The photograph freezes the three at that point where our world becomes visible, because they helped to create it.

1961 is also a year elided from most youth culture histories, and it is this forgotten moment which Iain Softley's *Backbeat* seeks to excavate. On one level, the film is a rite-of-passage teen movie – boys bond together, meet girls, go through changes, pass into death/adulthood/stardom – but its meat is the life of Stuart Sutcliffe, the fifth Beatle who died just eight months before the group's first UK hit, the brilliant painter who symbolizes that moment when the British finally put their own spin on American pop culture.

Sutcliffe was born on 23 June 1940 – the first full year of World War II – to a Scottish family living in Liverpool. His younger sister Pauline, an American-trained psychotherapist, speaks of a family that, unlike many others, held together during the war: 'Unlike many boys of his age, he had the full-time presence of his father until he was seven – developmentally through the most critical years.' Entering Liverpool Art College in autumn 1956, Stuart quickly showed a confidence beyond his years. Art school has long been the place where weird kids go to hang out, but Sutcliffe was serious about his painting. His early canvases were figurative, but encouraged by teacher Arthur Ballard he went further – into Abstract Expressionism. Fired by the example of painters such as Modigliani and De Stael, whose brief, intense lives reflected their work, he started to burn up for his art.

Klaus Voorman, Astrid Kircherr and Stuart Sutcliffe at the Top Ten Club, 1961 (Christies)

By 1959, Sutcliffe was a powerful presence in Hope Street – well known for his productiveness as a painter and for his sartorial individuality. According to flatmate Rod Murray, 'Everybody tried to look like Jean-Paul Sartre. If you had a black poloneck, a black beret and a small beard, you were in. Otherwise you wore an art school scarf, khaki denims – KDs – jumpers and a duffle coat. Most of us wore what was left over from school: nobody had the money to be really fashion-conscious then.'

Sutcliffe and Murray shared a first-floor flat in Gambier Terrace, a row of decaying nineteenth-century houses between Liverpool's Anglican cathedral and the city's red-light district. With his friend John Lennon, a less committed art student, Sutcliffe started to dress like a Teddy Boy in tight jeans, pointed winklepickers and with a swept-back, quiffed hairstyle. 'Stuart liked to look moody and deep like James Dean,' says Murray. 'Then he'd turn round and grin at you: "Do I look all right?"'

Lennon and Sutcliffe were pop fans, a rarity at the time. Although Elvis Presley had had a huge impact in the UK in 1956, pop music was seen as a transient, working-class fad. The mainstream student musical taste was for traditional jazz: a hearty English revival of New Orleans Dixieland, historically important but unlistenable to now. Sutcliffe painted to the Everly Brothers and Buddy Holly, while Lennon had his own group that he had formed with fellow Southsiders Paul McCartney and George Harrison. This was something new.

Eduardo Paolozzi and Eddie Cochran

When rock 'n' roll first came to the UK in 1955, it was so strange and powerful it might as well have come from outer space. For British teenagers, who had grown up with rationing until 1954, America was the future: a science fiction land of tailfins, supermarkets, sun-kist bodies and bizarre crimes. Within the greyness of austerity, these Technicolor fantasies were a beacon to artists such as Paolozzi, who with collages like *I Was a Rich Man's Plaything* (1947) laid out the imaginative parameters of pop art.

What Paolozzi did was to rip up American magazines and put them together in montages that juxtaposed sex, death and consumerism: the great equation of the century's second half. Although these ideas gained wider exposure through the Independent Group – with whom Paolozzi contributed to the 1956 exhibition 'This is Tomorrow' – they were not in art school currency; as Rod Murray says, 'London was a long way off from Liverpool.'

A generation younger than the Independent Group, Sutcliffe and Lennon didn't have to come to terms with pop, they bathed in it – and indeed the proletarian elements of the Ted style provided a suitable metaphor for their middle-class rebellion. The problem was how to do it for themselves. The bridge was provided by that now-forgotten fad skiffle, which after Lonnie Donnegan's massive 1956 hit 'Rock Island Line' reduced American folk-blues into a racket that anybody could play as long as you had a guitar, a washboard, a stand-up bass and an attitude. Like punk 20 years later, skiffle's do-it-yourself ethic offered access to production. It is where British pop begins.

1960 is the year when everything comes into focus. In January Stuart sells a painting in a student exhibition for £65. Encouraged by Lennon, he spends the money on a Hofner President bass guitar – a radical move for the time, as Pauline Sutcliffe explains, 'Very few people had then the electronic equipment we have now. A guitar then was an exotic object of a different kind of desire.' Stuart then joins Lennon's group, renamed the Silver Beatles. He takes the *nom de guitare* Stuart De Stael.

In March Stuart and John go to see leather-clad Gene Vincent and Eddie Cochran in one of the first hard-core American rock 'n' roll package tours of the UK – according to Nik Cohn, 'the starting point from which British pop really began to get better'; a few days after the tour Cochran is killed in a car accident. On 24 July, along with his Gambier Terrace flatmates, Sutcliffe is pictured in a sleazy tabloid exposé – 'THIS IS THE BEATNIK HORROR: this bizarre new cult imported from America is a dangerous menace to our young people' – the first public documentation. A few weeks later the Beatles drop the Silver and head for Hamburg.

This is where *Backbeat* kicks in, with the chance meeting between British rockers and German existentialists. As Astrid Kirchherr explains: 'It was so soon after the war. All the struggles we had then: the guilt of war, the guilt of our parents. Then meeting some youngsters from England who felt, the Germans are our enemies. They came to play music, and suddenly they found Germans were creative people that they could be inspired by – and we felt the same. When these two parties came together the feeling was indescribable.

'Before we met the Beatles, we used to go to little bars and hear jazz. There were about three or four gay bars in Hamburg: I used to go to them, they had a nice atmosphere. All our influences were from France: literature, paintings, artists. The Beatles' influence was more from the States: that's what was so exciting about our relationship. We could pass our experiences on to each other.'

Goodbye to all that

Backbeat hones in on two triangles as a way of charting how the Beatles sloughed off their British puritanism. As Astrid now says, 'In Germany, our parents gave us the freedom to show our feelings. Klaus had been my boyfriend. When Stuart became my boyfriend he didn't feel right in Klaus's company but after Klaus told him how happy he was that I was happy, they became very close friends. The guilt was taken away from Stuart.' The other triangle, between John, Stuart and Astrid, spills over into violence when Stuart wants to leave the group: as the two young men pound each other, their punches dissolve into hugs. 'John and Stuart had a hard time,' says Astrid. 'They needed and loved one another, but though there was nothing homosexual about it, Stuart had a hard time expressing his feelings to John and vice versa. I told him it was not a problem: you can tell someone you love them.'

The most obvious symbol of this emotional breakthrough came in 1961, when the Beatles swapped their macho, stiff quiffs for the androgynous, brushed-forward cut inspired by Astrid and Klaus. 'Think of a Greek statue, think of Michelangelo's *David*: picture this hairstyle,' she says. 'Jean Cocteau put it in *La Belle et la bête*; Jean Marais had this hairstyle but shorter. All the art students had this style too, but wanting to be different, we made it longer. Without grease, it was very poetic.'

As the film climaxes, the Beatles are on the brink of a success, the magnitude of which no one could grasp. Their screaming fans are cut against Stuart's own frenzy: having left the group, he is studying under Paolozzi at Hamburg's state school of art. His paintings are becoming more sculptural, more violent – twisted heads, fuzzy with pain, or grid-like abstracts in primary red and black. Looking at them 30 years later, you can see infinity.

Stuart Sutcliffe died on 10 April 1962 from a brain haemorrhage, cause unknown. At the film's end, Stuart's spirit enters John as he stands with Astrid in her studio. 'John became something else after Stuart's death,' she says. 'To most people he was still the hard Lennon, but to those of us who knew him, he allowed his heart to become gentle. He became a man.'

Hamburg gave the Beatles the emotional intensity to match the black music they loved. By April 1963 they were an English sensation: two smash hits written by the group themselves was an unheard-of feat in an industry that still copied America. A year later they occupied the top five places in the US singles charts as the spearhead of what is now called 'the English invasion' of America. A bewildered media focused on their hairstyles as a symbol of this phenomenon.

As the Beatles relaxed into their success from 1965 on they refused the example of rock stars such as Elvis Presley who had become all-round entertainers. And the more successful the Beatles were, the weirder they became. They took drugs and sang lyrics of an unprecedented emotional intensity: you can hear echoes of Stuart in Lennon's 'You've Got to Hide Your Love Away', 'Norwegian Wood' and 'Girl'. In 1967 Lennon acknowledged his debt by immortalizing Sutcliffe on the Peter Blake cover for *Sgt Pepper's Lonely Hearts Club Band*: high pop art for the mass market.

Backbeat roots this extraordinary story not just in the emotions you would expect, but in a time and place: Liverpool and Hamburg, 1960 and 1961. This closer look at the Beatles' story places them not as an accident or a *deus ex machina*, but as harbingers of a major tectonic shift in British society: increased access to higher education, better health thanks to the Welfare State, the opportunity for foreign travel – a real democratization, predicated partly on consumerism, that has since been reneged upon. It also illuminates the 1960s shift that was both surprising and all-encompassing: during that decade what had formerly been the preserve of outcasts – tiny minorities on the edge of the known world – became the motor of the Western consumer machine.

The Beatles and their German friends took a journey into the unknown that had effects that still resonate today: it is *Backbeat*'s achievement, within the disciplines of a commercial film, to capture what this felt like. As Astrid Kirchherr now says, 'It was such a wonderful time: we put our knowledge into a big pot and just drank from it.'

Joy Division: Someone Take These Dreams Away

Here are the young men, a weight on their shoulders
Here are the young men, well where have they been?
We knocked on the doors of hell's darker chambers
Pushed to the limits, we dragged ourselves in
Watched from the wings as the scenes were replaying
We saw ourselves now as we never had seen
Portrayal of the trauma and degeneration
The sorrows we suffered and never were freed
Where have they been? Where have they been?

August 27, 1979: Joy Division are headlining a ridiculous festival in a field outside Leigh, half-way between Liverpool and Manchester. The leading independent labels of both cities – Zoo and Factory – are meeting to showcase their talent: A Certain Ratio, Orchestral Manoeuvres in the Dark, Echo and the Bunnymen, the Teardrop Explodes. To the local police, this is tantamount to an alien invasion: they've closed down the town and are searching everyone on entry for drugs. One of my carload is already in custody.

In the twilight, Joy Division start their journey. What you get is this: at the back, a lanky drummer who pounds out rhythms at once intricate yet simple. At climactic moments, Stephen Morris attacks a syndrum for those 'pou pou' noses that you're starting to hear on disco records like 'Ring My Bell'. On the left, a slight person with the face of a debauched choirboy and the clothes of a polite young man – Bernard Dicken as he is then called – hunches over a guitar which is issuing rhythmic, often distorted blocks of noise. The sound scythes through the air.

On the right is the bearded bass player with his dyed blond thatch, engineer boots and double-breasted jacket: bent at the knees, he swings his instrument round like an offensive weapon. Peter Hook's basslines are prominent in the mix: Joy Division use them to carry the melody as so much else is texture. In the centre stands the singer: very pale, sometimes sweaty, tall, dressed in different shades of grey. He has the severe haircut of a Roman emperor.

At the beginning, Ian Curtis is still, singing as if with infinite patience. Then, as the group hit the instrumental break, it's as though a switch has been flipped: the stillness suddenly cracks into violent movement. The running joke is that he does the 'dead fly' dance – the leg and arm spasms of a dying insect – but he is more controlled than that. As the limbs start flying in that semicircular, hypnotic curve you can't take your eyes off him for a moment.

Then you realize: he's trying to get out of his skin, out of all this, for ever, and he's trying harder than anyone you've ever seen. This is extraordinary. Most performers keep a reserve while they're onstage; only giving a part of themselves away. Ian Curtis is holding nothing back: with the musicians behind him every inch of the way, he's jumping off the cliff.

Near the end of the set comes a new song, 'Dead Souls', which begins as a roller-coaster of soaring guitar and lurching basslines. After a couple of minutes Curtis starts to sing: 'Someone take these dreams away.' He's seeing visions, of figures from the past, of mocking voices – a terrible beauty. By the time that the song reaches its coda, he's shrieking, 'They keep calling me, keep on calling me, they keep calling me', and the hairs on our necks stand up. This is it, no way round it: Ian Curtis is raising the dead.

'I was into, I suppose nowadays you'd call it slacking, but in those days I called it being a lazy twat,' says Bernard Sumner today. 'I couldn't believe that I was now a professional musician: my whole ambition was to do something that I enjoyed, but not actually work hard at it. Just let the ether flow through me – ha! – and I'd be this medium for this music from the spirits that came through me. I'd just lie there and the music would come through my fingers, because I'd imagined that's what art was.

'It's difficult to speak for everyone, but one of the funny things was that we never talked about the music: we had an understanding which we never felt the need to vocalize. I felt that there was an otherworldliness to the music, that we were plucking out of the air. We felt that talking about the music would stop that inspiration. In the same way, we never talked about Ian's lyrics or Ian's performance. I felt that if I thought about what he did, then it would stop. I thought, If something great is happening, don't look at the sun, don't look at the sun.'

> Shadow that stood by the side of the road
> Always reminds me of you.

Just over 14 years ago, in the early hours of 18 May 1980, Ian Curtis died by his own hand. It came as a total shock. The group were due to go to America a day later. With a single 'Love Will Tear Us Apart', and the album *Closer* ready for release, Joy Division were poised for a breakthrough. As Chris Bohn wrote later, 'The suicide didn't so much bring [their] journey to the heart of darkness, to an abrupt halt as … freeze it for all eternity at the brink of discovery.'

Manchester is a closed city, Cancerian like Ian Curtis. The main participants didn't openly mourn, but carried on under a different name, New Order, into the group we have known and loved during the 1980s. The label that Joy Division had helped to build, Factory Records, became the model of non-metropolitan success. Everything culminated in the summer of 1990, the last summer of love, when Happy Mondays broke through and New Order finally went to number one with the World Cup theme, 'World in Motion'. Grey and black had turned into Day-glo, darkness into light.

Yet Joy Division have remained a powerful presence, or indeed, absence. They have been recently cited by writers as diverse as Kurt Cobain, Courtney Love, Donna Tartt and Dennis Cooper, who entitled his second novel *Closer*. They also inspired the comic artist James O'Barr, who saturated the three parts of his novel *The Crow* with Joy Division lyrics, character names and an open dedication to Ian Curtis, 'who showed me the indescribable beauty in absolute ugliness'. It was during the filming of this dark story that Bruce Lee's son, Brandon, was killed by an accidental shot.

I began regularly visiting Manchester again after 1990, and experienced Curtis's absence as a powerful event that I hadn't yet come to terms with. As things turned sour for both Factory and New Order, it was hard not to feel that his death remained unresolved for the main participants. It seemed like a good time to tell his story. I contacted Curtis's group, manager, label owner and wife – Bernard Sumner, Peter Hook, Stephen Morris, Rob Gretton, Tony Wilson, Deborah Curtis – and they all, except Gretton who hardly ever does, agreed to speak.

In her forthcoming biography, *Touching from a Distance*, Deborah Curtis writes about the reality behind the performance, the fact that Ian's mesmeric style mirrored the ever more frequent epileptic spasms that she had to cope with at home. As she says now, 'People admired him for the things that were destroying him.' Ian Curtis's death was a personal tragedy with wider implications. Couldn't it have been prevented? Was what we thought to be artistic exorcism sheer, unrelenting autobiography? Where did such darkness come from and why did we so willingly enter it?

A change of scene, with no regrets
A chance to watch, admire the distance
Still occupying – though you forget
Different colours, different shades
Over each mistakes were made
I took the blame
Directionless, so plain to see
A loaded gun won't set you free
So you say.

Joy Division began, as did so much else, on 4 June 1976. Invited by the fledgling Buzzcocks, the Sex Pistols played their first northern date, in a tiny hall above Manchester's Free Trade Hall. In a super-8 film shot that day, Johnny Rotten twists around the small stage in an already stylized ritual of aggression and withdrawal. 'It was dead exciting and dead heavy, real laddish,' says Peter Hook. 'Something was happening and the music was secondary.'

'I went with Hooky and Terry Mason, our roadie,' says Bernard Sumner. 'He'd read somewhere about the Sex Pistols having a fight onstage and he dragged us down to see them. I didn't think they were good: I thought they were bad, that's why I liked it. I thought they destroyed the myth of being a pop star, of a musician being some kind of god that you had to worship.

'I first met Ian at the Electric Circus. It might have been the Anarchy tour, or the Clash. Ian was with another lad called Ian, and they both had donkey jackets: Ian had "HATE" written on the back of his but I remember liking him. He seemed pretty nice, but we didn't talk to him that much. About a month later when we decided to try to find a singer – because Hooky and I had formed a group – we put an ad in Virgin Records. Ian rang up and I said, "Right, OK". We didn't even audition him.

'Ian brought a direction. He was into the extremities of life. He wanted to make extreme music: he wanted to be totally extreme on stage, no half measures. Ian's influence seemed to be madness and insanity. He said that a member of his family had worked in a mental home and she used to tell him things about the people there: people with 20 nipples or two heads, and it made a big impression on him. Part of the time when Joy Division were forming, he worked in a rehabilitation centre for people with physical and mental difficulties, trying to find work. He was very affected by them.'

Ian Curtis was born on 15 July 1956, the elder son: his father worked in the Transport Police. During his teens, his parents moved from Hurdsfield on the outskirts of Macclesfield to the huge 1960s blocks of Victoria Park, near the station. Although only just beyond the furthest Manchester suburb, Macclesfield is an older, small town, where the looming Pennines offer both an escape and a witchy emptiness: 'It's actually quite nice, the hills around,' says Sumner. 'But if you drive round there on a winter night, and I've done it, you won't see a soul on the street.'

According to Deborah Curtis, who met him when he was 16, Ian had a normal bohemian adolescence. Like many teens growing up in the early 1970s, he was fired by David Bowie, who placed in pop culture a whole set of self-destructive references both musical and literary: Jacques Brel, Lou Reed and the Velvet Underground, William Burroughs. At the time, this seemed like little more than the standard teenage dramatization of misery: after leaving the King's School, Curtis went to work every day and, in August 1975, got married. It seemed as though he was settling down.

With hindsight, it's now clear that things went deeper. When he was 14, Ian would, like many teens do today, raid the medicine cabinets of anyone they visited, and try out the combination of drugs as a leisure option. In the summer of 1972, there was an ambiguous overdose with his friend Oliver Cheaver, where both boys had their stomachs pumped: overdose or suicide attempt? 'I think he wanted to be like Jim Morrison,' says Deborah Curtis. 'Someone who'd got famous and died. Being in a band was very important: he was very single-minded about it. He'd always said that he didn't want to live into his twenties, after 25.'

'Everyone says Joy Division's music is gloomy and heavy,' says Bernard Sumner. 'I often get asked why this is so. The only answer I can give is my answer, why it was heavy for me. I can only guess why it was heavy for Ian, but for me it was because the whole neighbourhood that I'd grown up in was completely decimated in the mid-1960s. I was born and raised in Lower Broughton in Salford: the River Irwell was about 100 yards away and it stank. At the end of our street was a huge chemical factory: where I used to live is just oil drums filled with chemicals.

'There was a huge sense of community where we lived. I remember the summer holidays when I was a kid: we could stay up late and play in the street, and 12 o'clock at night there would be old ladies outside the houses, talking to each other. I guess what happened in the 1960s was that someone at the council decided that it wasn't very healthy, and something had to go, and unfortunately it was my neighbourhood that went. We were moved over the river into a tower block. At the time I thought it was fantastic; now of course I realize it was an absolute disaster.

'I'd had a number of other breaks in my life. So when people say about the darkness in Joy Division's music, by the age of 22, I'd had quite a lot of loss in my life. The place where I used to live, where I had my happiest memories, all that had gone. All that was left was a chemical factory. I realized then that I could never go back to that happiness. So there's this void. For me Joy Division was about the death of my community and my childhood. It was absolutely irretrievable.

'When I left school and got a job, real life came as a terrible shock. My first job was at Salford Town Hall sticking down envelopes, sending rates out. I was chained in this horrible office: every day, every week, every year, with maybe three weeks' holiday a year. The horror enveloped me. So the music of Joy Division was about the death of optimism, of youth. Just before Joy Division was a time of total upheaval for me: it came very early.'

The group took shape. Sumner claims they were always known as Joy Division. Peter Hook disagrees, and for the first few months they were more generally known as Warsaw – after Bowie's 'Warszawa'. 'We had so much aggro then,' says Peter Hook. 'Most of the musicians in Manchester then were very middle-class, very educated: like Howard Devoto. Barney and I were essentially working-class oiks. Ian came somewhere in the middle, but primarily we had a different attitude. We felt like outsiders: it was very vicious and backbiting.'

Warsaw dithered with drummers until another Macclesfield native, Stephen Morris, joined in summer 1977. 'Ian was a year or two above me in the King's School,' he says. 'He remembered me because I got kicked out with a couple of friends for drinking cough medicine, and the older boys were advised to go round checking the pupils' pupils.' The group played the Electric Circus and recorded a four-song EP, *Ideal for Living*, which showed them moving away from thrash to a more measured, heavier sound. 'We were just having fun,' says Sumner; 'learning where to put your fingers on the guitar and what sort of amplifiers to use.'

By the time the record was finally released, they were known as Joy Division – a name taken from the book that inspired the EP's final cut, 'No Love Lost': Ka-Tzetnik 135633's *House of Dolls*, a pulp nightmare diary of Nazi terror. The sleeve featured drawings taken from World War II: a drummer boy, a Jewish boy in the Warsaw ghetto. 'Ian had always been interested in Germany,' says Deborah Curtis. 'At our wedding we sang a hymn to the tune of the German national anthem. We went to see *Cabaret* a dozen times.'

'For me it was about World War II,' says Bernard Sumner, 'because I was brought up by my grandparents. They told me about the war, about all the sacrifices people had made so that we could be free: we had a room upstairs with gas masks and sand bags and English flags, tin helmets. The war left a big impression on me, and the sleeve was that impression. It wasn't pro-Nazi; quite the contrary, I thought, fashionable or unfashionable, what went on in the war shouldn't be forgotten, so that it didn't happen again.'

It would help to put this period into some kind of perspective. Punk was primarily libertarian, anarchist even, but there was a persistent right-wing trace that came from its opposition to the power politics of the day – the end of consensus socialism. In both English and American avant-garde rock – whether it was the Ramones or Throbbing Gristle – it had become important to say the unsayable, to examine the right-wing, to try to come to terms with the darker side of the human psyche. This is not a wise thing to do in pop culture, which is notorious for flattening out complexities.

Ian Curtis was a bundle of paradoxes: he was a Tory, yet he liked the writing of bohemian authors like J.G. Ballard and William Burroughs. At the same time as he wrote haunted lyrics and gave mesmeric performances, he was a great practical joker. He could be both a charismatic leader and highly suggestible: he hated confrontation and could be all things to all men. Even the people closest to him will disagree: according to Peter Hook, 'Ian was interested in the occult.' Sumner says he wasn't.

During 1978, Joy Division left their naïvety behind; they started to get good. In January, they played the infamous Bowie/Roxy disco, Pips. 'That was the first time I saw Ian being Ian onstage,' says Stephen Morris. 'I couldn't believe it: the transformation to this frantic windmill.' Their appearance at the chaotic Stiff Test/Chiswick Challenge battle of the bands in April brought them to the attention of their future manager, Wythenshawe native Rob Gretton, and their most persistent propagandist, Tony Wilson.

'Every band in Manchester played that night,' he remembers. 'I sit down and then this kid in a raincoat comes and sits next to me and goes, "You're a fucking cunt: why don't you put us on television?" That was Ian Curtis. At the very end of the night, Joy Division went on and after about 20 seconds, I thought, this is it. Most bands are onstage because they want to be rock stars. Some bands are onstage because they have to be, there's something trying to get out of them: that was blatantly obvious with Joy Division.'

During the spring of 1978, the group recorded an 11-track album for RCA under the auspices of Northern Soul DJ Richard Searling, but they were moving so quickly that it was obsolete almost as soon as it was recorded. 'There was suddenly a marked difference in the songs,' says Peter Hook. 'We were doing a soundcheck at the Mayflower, in May, and we played "Transmission": people had been moving around, and they all stopped and listened. I was thinking, What's the matter with that lot? That's when I realized that was our first great song.'

Everything was coming together. Rob Gretton took over the group's management: his first act was to commission a sequence of designs from Better Badges – this was the era of the badge as underground communication. Tony Wilson put them on *So It Goes*, a Granada TV programme (their performance of 'Shadowplay' was overlaid with negative footage from a *World in Action* documentary about the CIA), and had them as headliners when the new Factory club opened in Hulme. After the group had sweated out their contract with RCA, they went into the studios with Martin Hannett to record what would become *The Factory Sample*.

'I'd seen them in Salford Tech,' Martin Hannett told me in 1989. 'They were really good. It was a very big room, they were badly equipped and they were still working this space, making sure they got into the corners. When I did the arrangements for recording, they were just reinforcing the basic ideas. They were a gift to a producer, because they didn't have a clue. They didn't argue. *The Factory Sample* was the first thing I did with them: I think I'd had the new AMS delay line for about two weeks. It was called digital; it was heaven sent.'

'Joy Division had a formula, but it was never premeditated,' says Bernard Sumner. 'It came out naturally: I'm more rhythm and chords, and Hooky was melody. He used to play high lead bass because I liked my guitar to sound dis-

Bernard Sumner, Ian Curtis, Peter Hook and Stephen Morris, central Manchester, July 1979 (Pennie Smith)

torted, and the amplifier I had would only work when it was at full volume. When Hooky played low, he couldn't hear himself. Steve has his own style which is different to other drummers. To me, a drummer in the band is the clock, but Steve wouldn't be the clock, because he's passive: he would follow the rhythm of the band, which gave us our own edge. Live, we were driven by watching Ian dance; we were playing to him visually.'

'Ian used to spot the riffs,' says Peter Hook. 'We'd jam; he'd stop us and say, "That was good, play it again." We didn't have a tape recorder then: imagine! He spotted "Twenty Four Hours", "Insight", "She's Lost Control" – all of them. If it hadn't been for his ear, we might have played it that once and then never played it again. You didn't even know you'd played it, half the time. It's unconscious, but he was conscious.'

'Ian was a writer,' says Bernard Sumner. 'He would always have a file box with him, full of lyrics. He'd sit at home and just write all the time, instead of watching telly. He'd stay up: I don't know this, I'm just surmising, because he'd come in with reams and reams of lyrics. He never wrote any music but he was a great orchestrator. I'd arrange the songs and we all wrote the music, but Ian would give us the direction. He was very passionate at those moments: if we were writing a song, he'd say, "Let's make it more manic!"'

While *The Factory Sample* slowly sold out its 5,000 copies, Joy Division proceeded apace – in traditional industry terms. In late December 1978, they played their first London date, at the Hope and Anchor. The next month they recorded their first, four-song session for John Peel. In March they did four demos for Martin Rushent, preparatory to their signing to Rushent's company, Genetic, a subsidiary of the WEA-owned Radar Records. It never happened.

'The more we went into it, the more we realized that it was going to be very difficult to work with these people,' says Peter Hook. 'Genetic were offering us a lot of money, like £40,000, which was flattering, but so far out of our comprehension that it didn't matter. Rob just decided that the toing and froing with Tony was (a) more interesting and (b) more frustrating, but (c) ultimately more rewarding. He decided it was better to work with someone you could just walk down and get hold of. Factory, for all its failings, if you had a beef, you could just walk in there and yell.'

The group were busy recording with Martin Hannett at Strawberry Studios in Stockport. When they'd finished *Unknown Pleasures*, they took it to Factory. There was no contract, but, as Peter Hook says, 'We had a sheet of paper saying that the masters would revert to us after six months if either of us decided not to work with each other. That was it. It was amazing that the agreement lasted so well.'

This was Joy Division's first breakthrough: '*Unknown Pleasures* was our first outing into the real world,' says Bernard Sumner. 'I've been waiting for a guide to come and take me by the hand,' Ian Curtis sings on the opening 'Disorder', and the following nine tracks are a definitive Northern Gothic statement: guilt-ridden, romantic, claustrophobic. On 'Interzone' the group take a Northern Soul riff, N. F. Porter's 'Keep On Keepin' On', but blast off to another place entirely: 'Trying to find a way/Trying to find a way/To get out.'

The standout song was 'She's Lost Control', a live favourite with its Stooges guitar and swooping bassline, quickly covered by gay disco diva Grace Jones. 'It was about a girl who used to come into the centre where Ian worked to try to find work,' says Bernard Sumner. 'She had epilepsy and lost more and more time through it, and then one day she just didn't come in any more. He assumed that she'd found a job, but found out later that she'd had a fit and died.'

I'd just moved to Manchester that spring, and *Unknown Pleasures* helped me orient around the city. I reviewed it for *Melody Maker* in typically overheated style:

Joy Division's spatial circular themes and Martin Hannett's shiny, waking dream production gloss are one perfect reflection of Manchester's dark spaces and empty places: endless sodium lights and semis seen from a speeding car, vacant industrial sites – the endless detritus of the nineteenth century – seen gaping like teeth from an orange bus …

Unknown Pleasures is one of the strongest debuts ever, defining not only a city but a time. Martin Hannett: 'Ian Curtis was one of those channels for the Gestalt: the only one I bumped into during that period. A lightning conductor.' As Biba Kopf wrote last year, 'No other writer so accurately recorded the corrosive effect on the individual of a time squeezed between the collapse into impotence of trad Labour humanism and the impending cynical victory of Conservatism.'

The group hated the record. 'We'd played the album live,' says Bernard Sumner. 'The music was loud and heavy, and we felt that Martin had toned it down, especially with the guitars. The production inflicted this dark, doomy mood over the album: we'd drawn this picture in black and white, and Martin had coloured it in for us. We resented it, but Rob loved it, Wilson loved it, and the press loved it, and the public loved it: we were just the poor stupid musicians who wrote it! We swallowed our pride and went with it.'

Existence, well what does it matter
I exist on the best terms I can
The past is now part of my future
The present is well out of hand.

There were problems. 'Ian was primarily a fun guy, a good laugh,' says Bernard Sumner. 'But in a weird way. He wasn't a straight person. Let me start with his moments of intensity, which was when he got frustrated. I remember him having this argument with Rob Gretton at our rehearsal room, T. J. Davidson's. He got so frustrated that he picked up the garbage bucket, stuck it over his head and started running up and down the room, screaming at Rob, and he was just completely mad. He had an explosive personality, but most of the time he was cool. He really was.

'His performance was a manifestation of this frenzy. He was Ian, Mister Polite, Mister Nice, and then suddenly, onstage, about the third song, you'd notice he'd gone a bit weird, started pulling the stage apart, ripping up the floorboards and throwing them at the audience. Then by the end of the set he'd be completely and utterly manic. Then you'd come offstage and he'd be covered in blood. But no one would talk about it, because that was our way: we didn't think he knew why he got himself worked up that way.

'One day we were doing a gig at the Hope and Anchor in London. I was really ill with flu, and they had to come and drag me out of bed. Every time Steve hit the cymbals the whole room turned upside-down: literally, in my head, my eyes turned upside-down. It was horrible. There were only about 20 people there. We were driving back home in Steve's car: I was really ill, shivering, covered in a sleeping bag. Ian just grabbed the sleeping bag and pulled

SHIMMY....

Collage by Jon Savage, July 1979

JOY DIVISION~FACTORY~JULY 13

it off. He'd been moaning about the gig, the audience, the sound: he was in a really negative mood.

'So I grab the sleeping bag back, and he grabbed it back again and covered himself with it, and started growling like a dog. It was scary. He suddenly starts lashing out, punching the windscreen, and then he just went into a full overblown red-state fit, in the car. We pulled over on to the hard shoulder, dragged him out of the car, held him down. Then we did about a hundred miles an hour to the nearest hospital, somewhere near Luton. We were in this horrible casualty ward and the doctor said, "You've had a fit; you'd better go and see a doctor when you get back."'

This attack, which occurred in the early hours of 29 December 1978, marked the full onset of Ian Curtis's epilepsy. Throughout the whole of a demanding period for the group, Curtis was receiving medical treatment for what was becoming a serious condition. 'With Ian it was the full-blown grand mal,' says Stephen Morris. 'They put him on heavy tranquillizers; the doctor told him the only way he could minimize the risk was by leading a normal regular life, which by that time wasn't something he wanted to do. He liked to jump around onstage, and to get pissed: it was one of the reasons he got into the band in the first place.'

The pressures were building up at home, as Deborah Curtis explains: 'With Joy Division it all came together for him. I told myself at first that it was all part of the act, but it was all wrong. There wasn't an Ian at home and an Ian in the world, it became like that all the time. The trouble started when my pregnancy began to show: he had that first fit. It sounds awful, but he liked to have the attention. One of the things he liked about me was that I did stand behind him, 100 per cent, whatever he did. When I got pregnant, everybody made a fuss of me, and I think he was a bit jealous.'

Natalie Curtis was born on 16 April 1979. Just over a month later, Ian had the most serious in a sequence of grand mal attacks, which involved hospitalization. His solution to the pressures at home was, according to his wife, withdrawal, but there was no escape from the momentum of Joy Division's success. 'With being so young, you think of yourself as being invulnerable,' says Peter Hook. 'We were being driven by this thing called Joy Division, and basically you just did your damnedest to keep it going.'

Unknown Pleasures broke new ground in several ways. In staying with Factory, the group showed that a non-metropolitan, independent label sector was viable. There was Peter Savile's brilliant, baffling sleeve design. Despite releasing a powerful record full of raging emotions, Joy Division refused to open themselves up any further in print: after a couple of mistakes, they did no interviews. 'Rob thought the music was such a beautiful notion that he didn't want us daft bastards fucking it up for anyone,' says Peter Hook.

Joy Division were on a roll, constantly writing new songs, some of which are collected on *Substance* and *Still*: 'Something Must Break', 'Sound of Music' and a trio of classics – 'These Days', 'Dead Souls', the Spectorian 'Atmospheres'. 'That was the best track that Martin ever mixed,' says Sumner. 'I thought that was beautiful.' In October, they began a 24-date UK tour supporting Buzzcocks, which enabled them to give up their day jobs. In a break from the tour, Joy Division played their first concert abroad, at the opening of a new arts centre, Plan K in Brussels. It was there that Ian Curtis met Annick Honoré and fell in love. 'Ian wasn't having a very good time with Deborah,' says Peter Hook. 'They were married before the group came in, and they had had a reasonably normal life. The sad thing about your girlfriends is that you leave them behind. You move on and you're subject to temptations.'

'Annick loved him and understood him,' says Tony Wilson. This triangle dominated the last months of Ian Curtis's life. 'I knew something was desperately wrong,' says Deborah; 'but I didn't think it could be that. He was so possessive with me, that it didn't occur to me that he might go the other way.' The affair resumed during Joy Division's short January 1980 European tour. On his return to the house that he shared with Deborah in Macclesfield, Ian Curtis collapsed after drinking a bottle of Pernod and cutting his wrists.

> This is the crisis I knew had to come
> Destroying the balance I'd kept
> Doubting, unsettling and turning around
> Wondering what will come next.

At only 23, Curtis was facing one of the most difficult life situations of all: falling in love with another woman while he had a child. 'I've been through it as well,' says Peter Hook. 'You do get very confused, and it's easy to lose your head, especially where kids are concerned.' In March, the group spent two weeks in London's Britannia Row Studios, recording what would become their second LP, *Closer*. Ian stayed with Annick in London, while Deborah had finally found out what was going on.

'Factory was like a family,' says Deborah. 'They'd exclude anyone who wasn't what they were looking for. I remember when I was expecting Natalie, standing at the door of the Factory, Tony looked me up and down. It was obvious what he was thinking: how can we have a rock star with a six months pregnant wife standing by the stage? It wasn't quite the thing. Then this glamorous Belgian turned up: she was attractive and free. I don't blame Ian: most people need a partner and if you exclude that partner you have to find somebody else. It's only natural. He needed someone to look after him.'

It's easy to see now that Ian Curtis's torment went into the songs: those that didn't refer to his emotional dilemma were taken from the darker sources of literature – 'Colony' from *The Heart of Darkness*, 'Atrocity Exhibition' from J. G. Ballard's novel – or his own experience. '"The Eternal" was about a little mongol kid who grew up near Ian,' says Sumner. 'He could never come out of the house: his whole universe was the house to the garden wall. Many years later Ian moved back to Macclesfield and by chance he saw this kid. Ian had grown up from five to 22, but the kid looked the same. His universe was still the house and the garden.'

Ian might have been, as Tony Wilson says, 'trancelike' during the sessions but the group remember them with pleasure. 'Hooky and I always felt that Martin Hannett did his best stuff when he did it quick,' says Sumner. 'We recorded a lot of it by direct injection, straight into the board, but we wanted some real life ambience on it, so Martin put some big speakers in the Britannia Row games room. We pumped most of the album out through the speakers and recorded them, to make it sound live.'

'When we heard the lyrics, we knew they were very very good,' says Peter Hook. 'They were very open, weren't they? He was telling us a lot about himself, his fears and his doubts, but you were too young and caught up with the excitement: it was like a snowball going downhill. It's a great shame because you should have been able to just hear it and say, "Ian, can we have a chat with you? What's the matter?" But when you're young, you don't notice things.'

'The mood he was in when he wrote that stuff is a very big question,' says Tony Wilson. 'It's almost as if writing that album contributed to his state: he immersed himself in it, rather than just expressing it.' In many ways, *Closer* stands as the definitive Joy Division album, not the least because the sheer pleasure of the music – which looks forward to New Order's electro – buoys up the often bleak lyrics and vocals. It was also the group's most successful record – reaching number six, UK in summer 1980 – by which time it had been overtaken by events.

> Through childlike ways, rebellion and crime
> To reach this point and retreat back again
> The broken hearts, all the wheels that have turned
> The memory's scarred and the vision is blurred
> Just passing through till we reach the next stage
> But just to where, well it's all been arranged
> Just passing through but the break must be made
> Should we move on or stay safely away?

'It ended up with Ian having fits onstage,' says Bernard Sumner. 'In early April we did two gigs in one night: supporting the Stranglers at the Rainbow, then the Moonlight Club. At the first gig he started dancing, but he didn't stop at the end of the song. We were trying to stop the song, and he was dancing faster and faster, went into a spin, span into the drums and knocked the kit over. We realized he was having a fit and we had to carry him offstage. By the time we got him to the dressing room he'd come out of it, and he just broke down in tears. He was so ashamed. We didn't know what to say, or what to do.'

Peter Hook: 'For him to get up there, suffering from epilepsy, perform like that, be exposed, must have been absolutely awful. I think we were to blame for railroading him into doing it. He was in a no-win situation: he didn't want to let us down, he didn't want to let himself down, and yet it was making him ill. It's our own weakness, we make ourselves ill. But to have the brains to realize that if you carry on doing it, one day you're not going to wake up. That takes a lot of guts.'

Three days after the Rainbow concert, on 7 April, Ian attempted suicide with an overdose of phenobarbitone. The next night, he was expected onstage, at Derby Hall, Bury. 'It was a complete disaster,' says Bernard Sumner. 'We had to pull Ian out of psychiatric hospital. He came to the gig, couldn't go on, and Simon Topping of A Certain Ratio went on instead. The crowd freaked, and a full-scale riot went on. A lot of people got bottled. Ian saw this and of course thought it was all his fault. He just broke down again.

'He was in hospital for another four days. His wife already knew what was going on. He needed to get out, so he stayed at my house for two weeks. During that time I tried to drum into him what a stupid thing it was to take an overdose. We came to an agreement: He wanted to leave the band, he wanted to buy a corner shop in Portsmouth or somewhere, he wanted to go off and write a book. We didn't want him to, but we understood his predicament. The agreement was that he wouldn't do any gigs for a year, that we'd just write.

'But around this time, he'd agree to anything you told him. His reaction to a problem had been rage: he was like a human blowtorch and he'd burn you out of his way. Now his other solution was someone would come along and play God, tell him what to do. You can't do that with a person's life. We were loath to advise Ian, because whatever we'd have said, he'd have done it. I remain convinced to this day that if someone is going to commit suicide, they're going to do it, no matter what anybody says to them. Ian was going to do it.'

During April, Deborah Curtis instituted divorce proceedings. Ian stayed with Bernard and Tony Wilson, finally ending up with his parents. He continued with his hospital and therapeutic visits. It was business as usual for Joy Division – some concerts were cancelled, but the group were busy shooting a video for the forthcoming single, 'Love Will Tear Us Apart', and preparing for their first visit to the US on 19 May.

'The way they described Ian dying was so far from the way I perceived it that it's not worth getting annoyed about,' Rob Gretton says in Johnny Rogan's *Starmakers and Svengalis*. 'There was no great depression, no hint at all. The week before, we went and bought all these new clothes; he was really happy. A lot of his problems were personal: we could advise him, but we couldn't do anything about it. I wasn't worried as a manager, I was worried as a friend.'

'I don't think Ian was worried about the American tour,' says Bernard Sumner. 'I would have been extremely worried. If we'd agreed that we were going to keep the band together, but we weren't going to do gigs any more, how come a month later we were going on an American tour? It wasn't right. People start getting all the wrong priorities once you start becoming successful. They don't know when to leave you alone and give you a rest. You need more than one kind of sleep in this profession.'

To the other members of the group there was no indication of what was to come. 'If he was depressed, he kept it from us,' says Peter Hook. 'On the Friday I drove him home to his parents and we were in the car, laughing away: "Yessss! We're going to America on Monday!" Screaming with excitement, so happy. I think he was mood-swinging because of the drugs. When he got out of the car and I went home, I could barely contain myself. I was so excited.'

'The Friday night we went out with this lad I used to work with called Paul Dawson,' says Bernard Sumner. 'He called himself the Amazing Noswad. He was a psycho: we took him out to observe him. I know it sounds horrible, but we were fascinated by this lad. I was supposed to see Ian the next night, but he rang up and told me he was going to see Debbie. He said he'd meet me the next day, as we were going over to Blackpool to water-ski. He never turned up.'

On the Saturday, Ian Curtis returned to the Barton Street house he shared with his wife. When Deborah returned from work late in the evening, they had a discussion about the divorce. Deborah returned to her parents; Ian insisted she should. 'I'd had enough,' she says now. 'I was working so hard and my mum was looking after Natalie. I could have stayed with him that night, but he made it clear he didn't want me there. I was dead on my feet. I could have woken up the next morning and he'd have done it while I was asleep. I think he'd decided, and he was just trying to pick his moment.'

Ian had been watching Werner Herzog's *Stroszeck*, the plot of which concerns a German musician who travels to America, is swamped by the alien culture and commits suicide. After Deborah left, it was the early morning of Sunday the eighteenth. Curtis played Iggy Pop's *The Idiot* incessantly. After writing a note to Deborah, he went into the kitchen, put the rope from an overhead clothes rack around his neck, and jumped. Deborah found him the next midday, by which time any attempt at resuscitation was too late.

> Hangman looks around as he waits
> Cord stretches tight then it breaks
> Someday we will die in your dreams
> How I wish we were here with you now.

'I was the first of the group to be told,' says Peter Hook. 'I was just about to sit down and have my dinner and the phone rang: "I'm Sgt so and so, and I'm sorry to inform you that Ian Curtis committed suicide last night." I went back in, sat down and had my dinner. I didn't say anything for about an hour. Shock. It was such a huge thing to cope with. I don't think you ever really come to terms with it.'

'I went water-skiing anyway,' says Bernard Sumner. 'I came back to my friend's house and the phone rang. It was Rob. He said, "I've got a bit of bad news for you. Ian's committed suicide." "You mean he's tried to kill himself?" "No, he's done it." And it was like the cymbals at the Hope and Anchor: the whole room just turned upside-down. I put the phone down, went and washed my face with cold water. Then I got back on the phone and took it like a man.

'It was the breakdown of his relationship, accentuated by the amount of barbiturates he was taking to subdue his epilepsy. Barbiturates make you so you're laughing one minute, crying the next. He'd had a physical breakdown, a relationship breakdown, which caused an emotional breakdown. I came to terms with it straight away, because I could put my reason on why I thought he'd done it. Now I accept these things: if it's going to happen, it's going to happen. Also I don't really believe it ends there.'

'I went to great lengths to push everything to the back of my mind at first,' says Deborah Curtis. 'I threw things away, mementoes I wish I'd kept now: I thought it would help. How can you be angry with someone who's dead? They aren't there, you can't shake them. You're totally impotent: it's horrible. I felt angry with him because he had the last word. Seeing articles that dismissed his death as "Oh, he had marital problems" really annoyed me. He didn't commit suicide because he had marital problems. He had marital problems because he wanted to commit suicide.

'I think Ian invented scenarios that would come true. Annick could have been anybody: he needed to find a justification for doing what he was doing. It was something he talked about from when we met, but as we got older, and it got nearer the time, the more I had the feeling that he hadn't forgotten about it. But he wouldn't talk about it: when I tried to once, he actually walked out of the house. I think he enjoyed being unhappy, that he wallowed in it. When we were kids, lots of people were miserable; they grew out of it. I thought Ian would.'

'We all knew quite early that we wanted to carry on,' says Peter Hook. 'The first meeting we all had, which was the Sunday night, we agreed that. We didn't sit there crying. We didn't cry at his funeral. It came out as anger at the start. We were absolutely devastated: not only had we lost someone we considered our friend, we'd lost the group. Our life basically. He isn't someone I will ever forget. In my studio at home, I sit writing between two massive pictures of Ian. He's always there, always will be.'

'Our first album as New Order, *Movement*, was really horrible to make,' says Stephen Morris. 'We said we had to carry on, but it was a real struggle. I couldn't listen to *Movement* for ages: making it was hard because Martin took Ian's death harder than we did. He took it really badly. I don't think you notice the day you get over a death like that: I had a dream about Ian just before we made *Republic*: telling us not to be cruel, which I thought was really odd.'

'Ian made it all more serious,' says Tony Wilson. 'It made it something that wasn't just a business, a game that was played. Bizarrely enough, several deaths followed: their US agent Ruth Polski, Dave Rowbotham of the first Durutti Column, Bernard Pierre Wolff, who shot the *Closer* sleeve. Outside of Ian's personal family, the worst affected was Martin Hannett: he was an inspirational producer and a remarkable man. When Martin died, I was terribly upset.'

'Suddenly we didn't have any eyes,' says Bernard Sumner. 'We had everything else, but we couldn't see where we were going. I was really depressed after Ian died, very unhappy and disillusioned. I felt that I didn't have any future. I was listening to Lou Reed, Street Hassle, really down music. I started smoking draw, and found that electronic music sounded great. Mark Reeder, a friend from Berlin, sent me over records like "$E=MC^2$" by Giorgio Moroder, Donna Summer, early Italian disco. I discovered a new quality in music, which was to pep you up. Suddenly, this was the new direction.

'With Joy Division, I felt that even though we were expecting this music to come out of thin air, we were never, any of us, interested in the money it might make. We just wanted to make something that was beautiful to listen to, and stirred our emotions. We weren't interested in a career or any of that. We never planned one single day. I don't think we were messing with things we shouldn't have done, because our reasons were honourable.'

Mojo, July 1994

Boomers and Busters: US Teen Movies

Today's 15–24-year-olds were born between 1965 and 1974, when birth rates were dropping. Because of this drop, those 37 million people are referred to as the baby bust. In contrast, because of high rates between the end of World War II and 1964, the 75 million people born during that period are called the baby boom. Birth rates alone account for the formal identification of these two groups as distinct generations. So the baby-bust generation has grown up in the shadow of the baby boom. We had the numbers and enjoyed a more affluent economy too. By the 1980s, the American economy had suffered severe economic setbacks and high unemployment. Divorce rates rose, and the patriarchal structure of the American nuclear family eroded. The 1980s kids feel cheated – they feel that our parents loved us more, that schools were better, that life was easier.

Donna Gaines, *Teenage Wasteland* (1990)

In a very short time, I went from thinking (as I had been told over and over again) that my generation had nothing to say to thinking that it not only had everything to say but was saying it in a completely new way.

Richard Linklater, introduction to *Slacker* (1992)

It's not enough just to be young – nor has it been since World War II, when the spread of American-style consumerism, the rise of sociology as an academic discipline (and market research as a self-fulfilling prophecy) and sheer demographics turned adolescents into teenagers. No longer in between, youth was at the centre, caught in a web spun out of others' projections.

Encoded into the idea of youth-as-consumer were and are all manner of hopes and fears. Teenagers may well have been the key to this brave new postwar world, but they harboured potentially uncontrollable energies which had to be policed. In a classic puritan dichotomy, youth was simultaneously celebrated and derided, exploited and punished: the language of criminology and sociology – key phrase: 'juvenile delinquency' – is there right at the beginning of the teenage dream, the shadow of all that hopeful marketing speak.

This shadow continues to haunt mainstream accounts of youth. Youth cannot exist for itself; it has to be emblematic – keyword: 'generation'. In Ben Stiller's *Reality Bites*, Vickie talks about the AIDS test as being the rite of passage of her 'generation'; in Richard Linklater's *Dazed and Confused*, a bunch of stoned kids start talking decadism: 'It's like the every other decade theory y'know. The 1950s were boring, the 1960s rocked. The 1970s, ohmigod, they obviously suck. I mean, c'mon. Maybe the 1980s will be radical, y'know.'

Both *Reality Bites* and *Dazed and Confused* exemplify a self-consciousness which has been present in teen movies since the 1950s. A 'sophisticated' comedy set in 1994, *Reality Bites* presents Winona Ryder as Lelaina, a college valedictorian (i.e. top student) trying to make her way in the media. Will she choose conformism or ethical individuality, as epitomized respectively by the two men with whom she forms a love triangle? *Dazed and Confused* is a straight period re-creation: a teen rite-of-passage movie set on the last day of high school, May 1976; a multiple-storyline mood piece.

Looking to either film for some kind of authentic youth expression may seem a waste of time: if you want the real teenage news, listen to pop. For many years, film has lagged behind pop music – always the motor of youth culture – for obvious reasons: higher budgets, tighter controls, longer production time. Yet with the vertical integration of the media industries, film and music have become intertwined as never before: both *Reality Bites* and *Dazed and Confused* are suffused by carefully selected soundtracks to be released on disc and cassette. The connection between film and music goes deeper. For the last 15 years, pop has subsumed the filmic techniques of cut-up, flashback and time-tripping into its very fabric. One listen to 'Grandmaster Flash on the Wheels of Steel' takes you into the last 10 years of black and white music, where the introduction of sampling technology has rendered both past and present into a huge data bank – instantly accessible to make something new out of the old.

There's another way of looking at youth movies. As in other industries, most notably TV, 'youth' is shorthand for research and development: it's the place where people play and, if they come up with something, they're let out into the real world. Film, TV, music: all these are industries which deal with perception. So if we expect youth movies to contain new information, let's take *Reality Bites* and *Dazed and Confused* on their own terms and ask: what are they telling us? Is it anything new?

The immediate key to both is 1970s retro – just as peculiar to baby-boomers today as their own obsession with 1950s retro was to their parents. In *Reality Bites*, as you'd expect from a determinedly light romantic triangle, it's set dressing. The camera pans around the room of style maven Gap manager-ess Vickie (Janeane Garofalo) and picks out Kiss posters, *vinyl* records on the wall. Lelaina makes out with nerd MTV-style executive Michael (Ben Stiller) to the sound of Peter Frampton's 1976 US hit 'Baby, I Love Your Way'.

There is a code here. That Michael is shown only to pick at the surface of 1970s retro is part of his lack of commitment; in contrast, when Lelaina makes out with archetype Troy (Ethan Hawke) – 'He'll turn this place into a den of slack' – it is to the sound of roots acoustic guitar music. Seventies retro is one appurtenance of the powerless baby-busters, for whom 'the 1990s dream' is to find a job and your own place, which 'could take years'. Both musically and sartorially, the style informs the development of the new American rock – grunge, flannel shirts, Neil Young guitars. But there's no Nirvana here, of course. *Reality Bites* has cameo appearances from second-rank grungistes such as Dave Pirner (Soul Asylum) and Evan Dando (the Lemonheads). Troy has a band: his centre-parted haircut, sound (Woody Guthrie meets hip-hop and hard-core punk) and hit tune 'I'm Nuthin'' are taken from this year's big post-grunge star, Beck, whose hit 'I'm a Loser' ('baby, so why don't you kill me?') inverts a generation's perceived powerlessness with a wit lacking in this over-telegraphed film.

With its contemporary pop-cult references covered, *Reality Bites* essays today's perception. It is determinedly media-postmodern: intertextual to the max. In-joke 1: cameo-player Dave Pirner is Winona Ryder's real-time boyfriend. In-joke 2: creepy Michael works for In Your Face TV, an exact replica of youth channel MTV. We see Lelaina's serious documentary about her generation trashed into an entertainment but obvious parody of MTV's quick blip sensationalism. The parody is full onscreen, taking up minutes of film time, as does Lelaina's original camcorder documentary.

The camcorder sequences ably telegraph baby-buster concerns: divorced parents, McJobs, pop-cult time-trips – so much so that they read like the definitions contained in Douglas Coupland's novel *Generation X*. Less successful is the film's attempt at radicalism: Lelaina gets back at her TV presenter boss ('Good Morning Grant') by substituting his cue cards. She knows he won't look at them, and he doesn't: 'Personally I've had a preference for young girls' he announces to a stunned studio audience – yet this is left hanging. Cut to Lelaina moaning about losing her job.

Such cowardice leaves you irritated at the film's obvious manipulations: the pop-cult intertextualities fade to leave an essentially conservative story-line. As Lelaina says at her graduation: 'What are we going to do now, how are we going to repair the damage they [the previous generation] have left? (*Rustles papers*) The answer is, the answer is: I don't know.' Oh, come on! Indeed, the film's form – pastiche, quick cutting, the variety of sources – often recalls nothing so much as that quintessentially 1980s product: the pop video.

Dazed and Confused is a more thorough exploration of the 1970s, named after an infamous Led Zeppelin song and soundtracked by the arcana of mainstream US boy taste from the period – Ted Nugent, Aerosmith, Nazareth, Black Sabbath. The kids are loose, long-haired, flared, stoned; like the music's tepid funk inflections, they ape the jive of black style without any questioning. Like the just-graduates of *Reality Bites*, the stoners of *Dazed and Confused* are haunted by the baby-boomer privilege epitomized by the 'political' 1960s.

Take this dialogue between two of the film's three intellectuals: 'You know what I said last year, I was talking about going to law school so that I can be an ACLU lawyer and be in a position to help people who are getting fucked over and all that stuff. I was standing in line at the post office yesterday and I'm looking around and everyone's looking really pathetic and they got drool and this guy was bending over and you could see the crack, it was like eeurgh! and I realized it sounds good but I've got to confront the fact that I just don't like the people I've been talking about helping out. I don't think I like people period.' 'So you're not gonna go to law school, whaddya wanna do?' 'I wanna *dance*.' Right, and the next big fad, disco, is just around the corner.

Unlike the baby-busters of *Reality Bites*, all the high-school teenagers in *Dazed and Confused* have cars. Mobility is easy, and is translated into form. Director Linklater decompresses the rigorous psycho-geography of his *Slacker* (1991), where the camera follows each character for a few seconds, even minutes, before following another, each linked by a chance passing – not a road movie, but a walk movie – into the more gentle cruising motion of that groundbreaking 1973 time-tripper, *American Graffiti*. There's even an explicit homage to that film's older character, John Milner, in the sleazy Wooderson ('That's what I like about high-school girls: they get younger, I stay the same.') Both films use the past to tell us about the present. In the case of *American Graffiti*, the innocent, uncomplicated pleasures of on-the-cusp 1962 America (Beach Boys supplanting rock 'n' roll, the 1960s about to happen) are undercut by Vietnam. As the DC-6 drone intensifies at the film's end, there are biographical captions: Terry Field (Toad) is 'missing in action near An Loc December 1965', while Curt is 'a writer living in Canada' – someone who crossed the border to avoid the draft, this edge is informed by baby-boomer confidence: as the film's shaman, DJ Wolfman Jack declaims, 'It's a great big beautiful world out there.' Although there is darkness at the edge of town, there is also possibility (symbolized by Curt's visit to Wolfman Jack's radio station). In looking back at a conservative time, Francis Ford Coppola and George Lucas capture not just the dark side or the repressiveness but an almost luminous immanence: America is still rich and still hopeful.

We know from *Generation X* movies like *Slacker* and *Reality Bites* that this hope has long disappeared. Draw a teen-movie time curve from *American Graffiti* (1973/1962) through to these early 1990s films and, somewhere between *Over the Edge* (1979) and *River's Edge* (1986), the parents have gone. We're talking divorce, trailer parks, teen homelessness, maybe child abuse – the psycho-damage of Roseanne Barr, Axl Rose, Eddie Vedder, Kurt Cobain. As they keep saying in *Reality Bites*: 'psychotic'.

But nobody in *Dazed and Confused* (1993/1976) is psychotic. They still have parents who live together. There's some violence, and a deep anger running beneath the surface, but this is American pop at its moment of greatest self-indulgence: the chemical/sexual freedoms of the 1960s redacted into thoughtless hedonism. Like the stoner guy says, in his marijuana-leaf T-shirt and pimp hat, about George Washington's wife, 'She was a hip, hip lady. She'd have a big fat bowl waiting for him when he got home.' Linklater is too sharp to let this boy stuff pass without comment, and there are some neat pro-fem touches – a feminist reading of *Gilligan's Island* in the ladies' loo being one highlight. Yet *Dazed and Confused* so convinces you that these kids are apolitical – the brutal freshman-hazing rituals are barely commented on – that when Pink (Jason London) hits his own moral crossroads (should he sign an anti-drug testament and lie?) you shrug your shoulders and walk away: *is that all he's got to worry about?* Uh, cool. Let's, like, have another joint.

If *Dazed and Confused* veers into *Beavis and Butt-head* territory, that's because it's more attuned to the *zeitgeist* than the shamedly aspirational *Reality Bites*. I know Lelaina walks out on Michael, but did you notice that Troy is wearing a suit at the film's end? Do you think he's going to inherit a trust fund from his dead daddy? The 1976 teens don't care, and they get away with it; the adult figures in the film, just as in *American Graffiti*, are benign, firm or outwitted. The drug of choice is marijuana, the petty hooliganism just a lark. *Carrie* this is not. At the end, as in *Slacker*, Pink and his friends break out of the film's location, Austin, and hit the open road – still a powerful shorthand symbol for freedom, in Linklater's words, 'a spiritual breakthrough'. This sense, however ambiguous – you know that Pink could well be in Narcotics Anonymous or dead of AIDS by now – offers a fragile hope which, together with the seductiveness of the teen paradise depicted (the music almost sounds good) shows that form can dictate content, that the past can bleed into the present.

These films aim to deal with one of our central dilemmas: the lack of rituals that bridge adolescence and adulthood. Beneath all their pop culture and teenage encumbrances, they represent two approaches to this perennial rite of passage. *Reality Bites* internalizes 1980s forms into a cynicism which says: conform. *Dazed and Confused*, on the other hand, feels like a gentle victory; experimentation may mean self-indulgence, but it doesn't mean death. A refusal, however small, can make a difference.

Both films deal with the perception of today's teens – where the past is not linear but serial, constantly changing and informing the present. *Dazed and Confused* carries this through the internalized powerlessness of the late 1980s – that hangs around *Reality Bites* like a miasma – into the sense of liberation that still existed in the mid-1970s. Just as the excitement of *Poison* was undercut by its sourcing of a repressive period, so the dazed and confused kids of 1976 are given grace by their brief moment in the sun.

Sight and Sound, July 1994

The Death of Kurt Cobain

Depending on your view of the afterlife, suicide may or may not be the solution to a life that has become unbearable. For those left behind, however, suicide can be seen as a selfish act which creates problems. Those problems are magnified exponentially when the suicide is a parent, a husband, a group member, a world-famous performer. Death may be a full stop: it is all the more devastating when it is a question mark.

Kurt Cobain's suicide two weeks ago has become a mythological event – a drama acted out in the media. Two weeks ago, 10,000 fans held a vigil in Seattle Center, where Cobain's wife, Courtney Love, reduced the audience to tears with her impassioned reading of the suicide note. Sales of Nirvana albums have tripled; shops in England and America have created their own consumer shrines – dump bins showcasing the group's four albums with the legend 'Kurt Cobain: 1967–94'. At least one fan has already died in a copycat suicide.

Media interest has been compounded by the mystery surrounding Cobain's last days. There is little firm record of what he did between Thursday 31 March – when he walked out of the Los Angeles clinic where he had gone to beat his heroin addiction – and the morning of Friday 8 April, when his body was found in his Seattle home. This is an awful image: of a young man so determined to die that he shuns friends, family, colleagues, or any one of the thousands of people who might feel connected to him through his music. What had brought him to this point?

Nirvana are an extremely successful group – over 20 million records sold in 30 months. Cobain was part of a great team – with bassist Krist Novoselic and drummer Dave Grohl – but he was the front man, the conditions of which are that his voice, presence and face be reproduced millions of times. Because he was a rock star, people looked to him for answers. Cobain hadn't expected this attention, and grew to hate it at the same time as he believed in it. He hid himself behind a wall of hair and sunglasses; already committed to drugs for pleasure, he took them to obliterate himself.

In this light, his last few days can be seen as a terrible refusal – by walking off the face of the earth, Cobain attained the invisibility that he craved. The method of death reinforced this refusal. Cobain left his i-d open on the ground, held the gun to his chin, and pulled the trigger, destroying the face known to millions. (Identification was made through fingerprints.) His respite was brief; Cobain's face has been reconstituted in media reports the world over, but now he has become an object, an abstract.

Nirvana were outcasts, small-town American stoners who believed in punk because it articulated their resentments and offered them freedom. Fame and success are thought to bring freedom, but they also bring bondage. Your life is no longer your own; every move is watched. If you've become successful young, you're encouraged to stay that way. You've got to the top of the mountain, but the view isn't quite what you expected. You're faced with a whole new set of problems, not the least of which are the expectations of others and your own expectations of yourself. As Cobain wrote to his fans in his suicide note: 'The fact is I can't fool any of you: it simply isn't fair to you or to me.'

Nirvana hit the media as punks. When I visited America in February 1992 to promote a book about the Sex Pistols, I was asked dozens of times about the group whose breakthrough record, *Nevermind*, directly echoed the title of the Sex Pistols' *Never Mind the Bollocks*. As punks, Nirvana were bound into several scripts. One had to do with the failure that was encoded into the original punk – through the very word itself, American slang for the worthless, the victimized.

On another level, punk was a hit-and-run engagement with the mass media that was doomed to failure: becoming successful with lyrics and attitudes which express an impassioned criticism of society has been a perennial youth culture contradiction – from the Rolling Stones on. Making money out of protest and alienation is enough to confound the angriest young man or woman. Most rebels finally make the transition to adulthood by coming to some accommodation with the power structure – Mick Jagger's involvement with high society and high finance being a good example.

But for punks in particular, conventional success meant failure: the founding myth was laid down by the Sex Pistols, who fell apart as soon as they became successful. Cobain found that his desperation had been turned into a mass consumption which made him rich and reinforced that very image of desperation. Cobain wrote a sarcastic song about it, 'I Hate Myself and I Want to Die': 'It was making fun of ourselves,' he told *Rolling Stone* earlier this year. 'I'm thought of as this pissy, complaining, freaked-out schizophrenic who wants to kill himself all the time.'

Nirvana were quickly slotted into the punk storyline. At the height of *Nevermind*'s success, it was rumoured that they had broken up. When, at around the same time, Kurt Cobain met his future wife, Courtney Love, the couple were quickly seen as a new Sid Vicious and Nancy Spungeon. Cobain and Love playfully toyed with this, checking into hotels as Mr and Mrs Simon Ritchie (Sid Vicious's birth-name), at the same time as they rejected it: 'You'd think that people would be evolved enough not to be entertained by a carbon copy of what happened 15 years ago,' Cobain said last year. 'Haven't we progressed?'

Cobain's attitude to the media became paranoid and confrontational: as he wrote in 'On a Plain', 'the black sheep got blackmailed again'. After a hostile *Vanity Fair* profile in September 1992, he regarded himself as 'cursed'. 'I almost feel like people don't believe me,' he said. 'Like I'm a pathological liar at times, because I'm constantly defending myself. People haven't evolved enough to question anything that is printed. Even I'm really bad at that too: I believe a lot of the things that I read.'

Just like the Sex Pistols, Cobain became immersed in a struggle, trying to free himself from the self-destruction involved in the whole punk script. And now he's caught, like a fly in amber. The most popular media trope came from his mother, who referred to the 'stupid club' – Jim Morrison, Jimi Hendrix, Janis Joplin – the drear baby-boomer litany that serves to reduce Cobain's individuality. It's hard to take, because Nirvana were so full of life. Their success, at the end of 1991, offered hope – not the least because their music joyously cut free of circumstances. 'I'm the product of a spoiled America,' Cobain said; his story is an American tragedy.

It's July 1993. Nirvana haven't played New York in a year and they're doing what is a music industry showcase at Roseland in mid-town to preview their new record *In Utero*. The pressure is on them, to an extent that I haven't encountered since the punk days, 14 years before. It's as though Nirvana have become negative, psychic lightning conductors: rumours are rife, nobody knows what they are going to do, all eyes are upon them.

The night before, I've interviewed Cobain in a hotel room in the clouds – with midnight Manhatten outside, we achieved that state that Paul Morley describes in his book *Ask*: 'Did you reach that uncanny, disorientating point where you *float*, right over the edge of a revealing all-round contemplation? Was it just a cover-up? What appeared to be the trouble?' Time stands still. Cobain doesn't look well – his skin is sallow and rough – but he is articulate, polite, witty, sensitive, occasionally sarcastic. He is also surprisingly frank about his heroin use, so much so that I accept his reasons (it was for constant stomach pain) and his statement: 'Now things have got better. My whole mental and physical state has improved almost 100 per cent. I'm totally optimistic.' I wanted to believe him. The next morning Cobain overdoses.

The concert is extraordinary, one of the most powerful I have ever seen. Cobain has the magic, the ability to keep you concentrated on him at all times, the ability to make it seem as though he is singing exactly what you're feeling – that emotional empathy that pop does so well. The young crowd go wild. But then a very odd thing happens. The group stop for a moment, then bring on a cellist Lori Goldstein to accompany them on several acoustic numbers.

Nirvana aren't carrying the crowd with them. Cobain doesn't shrink from the situation; his performance of the Leadbelly standard, 'Where Did You Sleep Last Night', is both exorcism and challenge. The song is chilling –

> My girl, my girl, don't lie to me
> Where did you sleep last night? In the pines, in the pines
> Where the sun don't shine
> I shivered all night long

and Cobain pushes and pushes this lyric to his limit of endurance. After his last scream, there is a brief silence.

This is not the end. Nirvana regroup for another song, 'Something in the Way', sung by Cobain in an intimate whisper. The lyric refers to the period when, in late 1985, he was estranged from both his divorced parents, and homeless in his home of Aberdeen, a decayed logging town about 100 miles south of Seattle in the Pacific North-West. According to official biographer Michael Azerrad, Cobain was unemployed and penniless that winter. He passed the time in the town library reading and writing poetry, sleeping in friends' cars or, at one point, under the North Aberdeen bridge.

Here is one aspect of Nirvana's triumph: 1985 was the high point of the Reagan years in America – a time when many Americans felt like exiles in their own country. Through talent, luck and guts, Nirvana epitomized the rehabilitation of these exiles when they broke through in 1991. As Gina Arnold wrote in *Route 666: The Road to Nirvana*: 'Nirvana's music reflects a time, my time. And my time has its own history, its own leaders, its own rules.'

Nirvana put an underground bohemianism into the mainstream American music industry. Their breakthrough single, 'Smells Like Teen Spirit', took its last two words from a deodorant marketed at young women, containing the magic word of mediated youth, 'teen', it opened up a gap between the 25–45-year-old baby-boomers and the 15–24 baby-busters. Less in number than their 1960s counterparts, early victims of economic outbacks and late 1980s recession, baby-bust teenagers have almost no societal value in the US. The result is a frustration and anger which, lacking an adequate outlet, turns into drug- and alcohol-induced oblivion, if not self-destruction.

That's where Nirvana – all children of divorced parents – came from; that's what they transcended on *Nevermind*. Songs like 'On a Plain' explore feelings of powerlessness – 'As a defence I'm neutered and spayed' – which is undercut by the music's sheer glee. The group sound welded together as they negotiate the classic verse/chorus/verse structure at high speed. The guitars are recorded so high that all you hear is distortion – as the sound breaks up, you feel it could go anywhere – but this wildness serves the discipline of a strong melody: sweetness in chaos.

In January 1992, *Nevermind* toppled Michael Jackson's *Bad* at the top of the American charts. Then it was grunge – high-fashion and low-concept music industry, the media hurricane. Nirvana came from a small, inward-looking culture in the Pacific North-West. Kim Warnick, of the label Sub Pop, which released Nirvana's early recordings now says, 'I don't think they knew what they were doing. How do you survive something that successful?'

An early sign of trouble could be found in Cobain's combative sleevenotes for the *Incesticide* compilation. Nirvana played pro-abortion, pro-gay benefits, wore dresses and presented an androgyny rare in the machismo of American rock. At high school, Cobain had been victimized for having a gay friend; he was shattered to find that the people who had once bullied him were now his fans.

Just by being themselves, Nirvana made enemies. There were spats with the most important American rock media: small beer but indicative of the compromises which Nirvana were called on to make. The American music industry is a conformist medium, and Nirvana were in the centre, signed to David Geffen's record label, managed by industry insiders Gold Mountain. When Nirvana locked horns with MTV – they wanted to play a new song, 'Rape Me'; MTV wanted the hit – it was explained to them that refusal could have repercussions for other Gold Mountain artists. The group backed down.

Things began to unravel quickly. In spring 1992, the group nearly split up in a dispute over music writing royalties – Cobain got an increased share. Both he and Courtney Love were taking their heroin use to the point of scandal. A September 1992 *Vanity Fair* article broke the story, after Geffen had signed Love's group, Hole, for a much-hyped $1 million advance. One result was that the couple's daughter, Frances Bean, was taken from them two weeks after birth. Cobain and Love weren't allowed to be alone with her for a month, until they showed that they were off drugs.

It now seems as though this was the turning point: 'I wanted to break up the group all the time,' Cobain said. Just after the *Vanity Fair* article, Love and Cobain threatened two unauthorized biographers – the first public sign of a violent streak. 'There was a lot about guns on *Nevermind*,' Gina Arnold says. 'People took it as irony, but he wrote about it effectively because he knew it.' In July 1993, Cobain was arrested after police seized guns and ammunition at his Seattle home. Nirvana struggled with their new record, *In Utero*, which on release in late summer 1993 rose to number one in both the US and the UK. Nirvana had worked constantly for the few years preceding a success the magnitude of which had taken them by surprise. The pressure had got to them immediately – instead of touring to support their success, they withdrew for six months. Their front man was in an obvious state. But Nirvana were locked on the roller-coaster, a hot property which had to work in a media environment of unparalleled relentlessness.

During the past 10 years, the music industry has integrated as never before with the other industries that are usually part of the same conglomerate: publishing, TV, advertising, films. The demands of these industries are voracious, the arena is not just national, but global. There is so much more music media than there was 10 years ago. As a result, time has accelerated. Their UK representative Anton Brookes says, 'What happened to Led Zeppelin in 10 years, happened to Nirvana in three.'

The campaign around *In Utero* sought to present Nirvana as a group which had hit a troubled patch but was now through it. Just after it was released, *Rolling Stone*'s Michael Azerrad published a warts-and-all official biography – which, while detailing Cobain's heroin use, did so as if to state that it was all over. As we know now, it wasn't and the Roseland show caught a group finely poised between optimism and self-destruction.

The group rallied for an autumn American tour which showed that they still had the magic: 'Nirvana played the Seattle Arena in January,' says Kim Warnick. 'It's a 6,000-seater venue and it's hard to come across in somewhere like that. As soon as they hit the stage, it felt like nothing else existed. You felt: "This is the only place I want to be."'

It's now clear that, beneath the surface of business as usual, things were going very wrong: 'I haven't felt the excitement … for too many years now,' Cobain wrote in his suicide note. 'I feel guilty beyond words about many of these things, for example when we're backstage and the lights go out and the manic roar of the crowd begins, it doesn't affect me in the way in which it did for Freddie Mercury – who seemed to love and relish the adoration from the crowd, which is something I totally admire and envy.'

On 4 March, Cobain overdosed in Rome, on a combination of champagne and 60 pills of Rohypnol, a Valium-like drug. This was Cobain's fourth overdose in a year. He came out of coma the next morning and flew back to Seattle on 8 March. The rest of the European dates – including an English tour – were cancelled. Damage limitation went into effect: the incident was presented as an accident, excessive high spirits, that kind of thing. Recent reports in the *Los Angeles Times*, denied by Gold Mountain, suggest that this wasn't an accident but a suicide attempt.

The Rome incident was the last straw for the band and Courtney Love. The couple had an intense, sometimes abusive relationship, but no one doubts that they were in love, and that they cared for each other. 'After Rome,' Love said, 'I couldn't take any more.' Ten days after his return to Seattle Cobain locked himself in a room after a row with his wife, who called the police, fearing a suicide. The police confiscated four guns and 25 boxes of ammunition.

The next weekend, Cobain's wife and group performed what is called, in drug-counselling speak, an 'intervention'. As Gina Arnold explains: 'Friends and family get together to confront the addict. It's the idea of "Tough Love": if you don't go into rehab, we won't support you. We will withdraw our love. We're not a band, a family unless you come clean.' Cobain denied that he had a problem but on Monday 28 March, he checked into a rehabilitation programme Exodus at the Daniel Freeman hospital in Los Angeles with Courtney Love. On 31 March, he walked out into thin air.

Many people have wondered why there was nobody he could turn to, but if everyone around Cobain made it clear that they didn't want to know him if he was taking heroin, then his state of mind when he relapsed must have been despairing. On returning to Seattle, it seems that he made for the city's drug district, Capitol Hill. Some claim to have seen him in town that Friday and Saturday. The rest is supposition.

Cobain spent three days in the wilderness. A missing persons report was filed by his mother, Wendy O'Connor, on 4 April. It stated that he was a suicide risk. Even so, nobody found him – he may have made a trip to his country house in Carnation, several hours away from Seattle. Sometime on Tuesday 5 April, Cobain went into the large room above his garage, pulled up a chair to get the full view of Puget Sound, and pulled the trigger.

His body was found by electrician Gary Smith at 8.45 on the morning of Friday 8 April. The news hit England that evening. The rest was mayhem, as the international media descended on Seattle, offering money for exclusives about Cobain's heroin use, publishing photos of the dead body. On Sunday 10 April, a small memorial service was held at Seattle's Unity Church, quickly organized, and attended by about 200 of Cobain's family and friends. At the same time, there was a public vigil in Seattle's principal public space, under the space needle, which, built in 1962, had promised a better world.

The tone was set by Courtney Love's rendition of her husband's suicide note, taped and broadcast to the crowd. 'He left a note for you,' she said. 'It's more like a letter to the editor. I mean, it was going to happen. It could have happened when he was 40. He always said he was going to outlive everybody and be 120.'

Love then called Cobain an 'asshole' for what he did and asked the crowd to call him an asshole too. Gina Arnold remembers: 'After a couple of hours, I went up the needle, I wanted some time to myself. The sun was setting down over Puget Sound; the crowd was yelling over and over: "Asshole! Asshole!" It was the perfect mix of anger and sympathy.

'They had written messages and were building shrines. The notes were heartbreaking. "Dear Kurt, you remind me of my dog: his name is Max. I love you both. Mike." About 2,000 kids were in the fountain, screaming Nirvana songs and burning their flannel shirts. The radio was playing Nirvana songs, and they were singing along. When "On a Plain" came on, the roar came up in unison – "Love myself better than you/I know it's wrong but what should I do.'"

Later that night Love came to the vigil in person, distributing Cobain's clothes to the fans who remained. As Ann Powers wrote in the *Village Voice*,

> Love did something that finally made Cobain's death seem real. What Courtney did was argue with him, dispute the terms of his refusal; in doing so, she opened up a view of what he must have really felt, the disorder that consumed him. She would read a little from the note, then curse the words, then express her sorrow … like some heroine from Euripedes, furious at the gods.

Indeed, it is Courtney Love who must now bear the brunt of the attention and blame that follows a shock: one can only wonder what will happen to her and her baby, Frances Bean. A complex, powerful character, she has long been cast as the villain of the piece. Part of this is due to sexism, part a self-willed victim mythology – her songs abound with references to witches and celebrity life – and part her own behaviour: it was reported last week that she overdosed in Los Angeles on Wednesday 6 April, two days before Cobain's suicide was discovered. Love recovered, but was charged with possession of narcotics. She has since denied that she overdosed.

This story already has its own momentum: 'I don't see it going away for a very long time,' says Kim Warnick. 'It will get into stranger and stranger issues.' Many questions will be asked: was this inevitable? Was it in the genes (three uncles died this way), or was it caused by drugs and fame? Where *was* everyone? But they pale before the finality of the act. Cobain meant to die. He freely told interviewers about his propensity towards suicide: it's all over his lyrics. One awful consequence of his act is that what seemed like an internal dialogue is now fixed as autobiography.

Cobain brilliantly embodied the anger and desperation of white Western youth, struggling to make the transition between adolescence and adulthood in societies without adequate rituals. It is an irony that many young people look to rock stars – often invested with religious qualities – to help them with this transition, the very people who often find it difficult to shake off the adolescent attitudes that make them famous. Neither party can ever be satisfied in what is a blocked, and consequently dangerous, ritual.

Cobain's tragedy was to live out that American archetype delineated by Scott Fitzgerald in *Early Success*: 'It is a short and precious time. When the mist rises in a few weeks, or a few months, one finds that the very best is over.' When the storms struck his craft, at the crucial age of 27, Cobain couldn't find his way across.

Observer, 24 July 1994

Lads and Lasses: Fantasy Football

1994 has been the year in which the media fascination with football has reached its height: for the past two years, football writing has been escaping from the sports pages and taking refuge in all kinds of unlikely places, like the *London Review of Books'* Gazza cover, and a number of anthologies; in May, BBC2's Goal TV spewed out six hours of football; and the recent publication of Julian Germain's *In Soccer Wonderland* raises the stakes even higher. This is football's first-ever coffee table book, one which is so highly designed that it might easily exist as an art book. This is soccer as culture industry.

This phenomenon even has its own brand name, the soccerati, a coinage which marks the consummation of football's love affair with the media. One of the more populist offshoots of this alliance is also one of its more irritating, *Fantasy Football*, the new season of which begins tonight on BBC2. The sight of two charmless lads – David Baddiel and Frank Skinner – on a scruffy sofa appears to confirm all the stereotypes held by non-believers like myself.

Indeed, 1994 has also been the year that the Lad has emerged as a force in pop culture; Baddiel and Skinner may be the most obvious ambassadors for this new subculture, but there is also the unexpected success of IPC's new magazine 'for men who should know better', *Loaded*, and the rise to prominence of Essex pop boys Blur. In their different ways, both espouse a male machismo that is not authentic – *Loaded* is staffed by formerly New Man music journalists, Blur are fronted by a former drama student – and which, with a thin veil of irony, seeks a return to more traditional masculine values.

But, as ever, things aren't that simple. The soccerati might well have unleashed the Lad but beyond these fripperies, and the knee-jerk responses that they delight in eliciting, there is something important going on. This concentration on football – and in consequence, the Lad – has highlighted two other profound shifts: an attempt to redefine masculinity after feminism, and the restructuring of the game itself.

According to Rogan Taylor, Football Supporters' Association chair (1985–9) and author of the ground-breaking book on pop and shamanism, *The Death and Resurrection Show*, this embourgeoisification of football (my phrase, not his) has happened for many reasons. The financial restructuring necessary after Hillsborough, the breakdown of the TV duopoly, and breakaway of the Premier League. 'Instead of the old butchers, bakers and candlestick-makers who used to run the game, you had the arrival of post-Thatcherite new men like Irving Scholar at Spurs, David Dein at Arsenal and

Alan Sugar: entrepreneurs with interests across the media.'

The same thing has happened to football as happened to pop from the early 1980s (the birth of the pop video, the emergence of glossy magazines like *Smash Hits*, *The Face* and *i-D*, and increased tabloid coverage of pop). Football has become part of the cultural *zeitgeist* (although for some it always was); it has leapt out of working-class pubs and landed on the dinner-party table. Footballers now do what pop stars do: Ryan Giggs dates *Word* presenter Dani Behr, footballers take E, stalk the catwalk and pose in fashion magazines.

To pop fans, the first signs of this crossover occurred in the 1989–90 E-culture indie/dance marriage described in Steve Redhead's *Footballers with Attitude*: records like Adrian Sherwood's 'The Barmy Army and the English Disease', Gary Clail's 'End of the Century Party' and, of course, the New Order theme tune for Italia 1990, 'World in Motion'. 1990 was the year that football's crossover was confirmed: the World Cup could be reduced, in the media shorthand, to Paul Gascoigne's tears (at England's defeat in the semifinal) and Pavarotti's '*Nessun dorma*', which, as Paul Smith notes in *Playing for England*, was better suited to 'the running patter of television anchormen' than New Order's tune, and thus became the official theme that summer.

Much more interesting than Pavarotti was the *ur*-text of the soccerati, Nick Hornby's *Fever Pitch*. There had been other strong football books – Pete Davies's *All Played Out* and Bill Buford's *Among the Thugs* – but *Fever Pitch* struck the nerve awakened by Gascoigne's vulnerability. It is, above all, a book about heterosexual manhood.

Hornby's own masculinity is more equivocal and troubled than most socerati celebrations – for instance, he admits to being in therapy. No real Lad would do this and I applaud his honesty. He has a sharp eye for the suburban desperation of south-east England, and the masochism of the male (football) obsessive. He's made me aware of something I didn't understand – that many men use football as a method of communication, as a way of expressing their emotions. *Fever Pitch* pulls its punches, however: a long discussion about male faults – 'they die lonely and miserable' – is brought to a halt by a dismissive 'But, you know, what the hell?' When I point this out, Hornby replies, 'Many people who don't usually read books read *Fever Pitch*. It would have been hard to take them with me. I didn't want to veer too far from the straight and narrow. I've never trusted people at the cutting edge; I'm more interested in the people following behind.'

Media trends simplify: the received impact of *Fever Pitch* has also been to legitimize a middle-class populism and a wave of 1970s nostalgia. It has initiated a fad for 30-something confessional pop anthologies – Chris Roberts's *Idle Worship* and John Aizlewood's *Love is the Drug*. Despite some fine pieces – including one from Sheryl Garratt, one of the only two women out of the 17 contributors – *Love is the Drug* is a deadening read; the stories of A-levels, trains missed, lads hung out with, all too often blur into a depressing examination of male bonding without the rigour or detail that could offer a wider interest. Like Hornby's relentlessly nostalgic essay on Rod Stewart in *Idle Worship*, this kind of writing is not social history but experience journalism: the self-deprecating, self-obsessed voice of the south-eastern populist, flattening out oddity and ignoring diversity.

'People in the media have more of a problem with lads than anyone else,' says Nick Hornby. 'People who feel quite comfortable with lads are forced to take on absurd positions.' This is a fair point but, despite Hornby's ambivalences about Lad culture – he likes *Fantasy Football*, doesn't like *Loaded* – these equivocations, and his populism, make it harder to distinguish *Fever Pitch* from the more traditional male products that have followed in its wake.

Since the late 1980s there has been a gradual media process through which the New Man (sensitive, pro-feminist and chore-sharing) slowly became the New Lad (he liked a night out with the boys) to the unreconstructed Lad, which is where we are today. This Lad, as many articles with Baddiel and Skinner point out, is a wilfully untidy, politically uninterested B-L-O-K-E.

Current marketing requires fashionability and a hint of irony to make the old new – that's all much of this is. Frank Skinner, who dismisses political correctness in GQ this month, was emblazoned on the front cover of *Loaded* ('Frank Skinner's World of Smut') and declared inside: 'When it comes to women, my tastes are as straightforward as they come. Basically, when all's said and done, I'm just drawn to sluts with big tits.'

And would David Baddiel find it so easy to persistently insult *Good Sex Guide* presenter Margi Clarke if she wasn't: a woman; working class; and from that city that has been for so long the butt of south-eastern jibes, and policies, Liverpool? There is no recognition here of the fact that heterosexual masculinity is the mainstay of an often oppressive power structure. This realization is often called PC or political correctness, but I call it good manners – indeed, being socialized: thinking about people other than yourself. Let's not forget either that PC began as a new-right coinage to attack the culture of pluralism that is necessary for a bearable future.

Masculinity is in crisis, a fact almost hysterically denied by Lads. The traditional macho lad had been rejected. The New Man (an unreal concoction of creatures who denied an interest in sex, beer and football) has also, rightfully, been rejected. But rather than attempt a coherent response to the very real gains of feminism we have witnessed a desperate return to a crude 'get-them-out-for-the-lads culture' – a shift unwittingly legitimized by the elegant scribblings of the soccerati. I can't help feeling that the re-emergence of the Lads reflects our power politics: pretending novelty, aspiring, as they might say, to hipness, they ruthlessly reinforce the status quo.

Guardian, 17 December 1994

Tricky: *Maxinquaye*

This is a brilliant record: densely layered, full of barely controlled nervous energy, paranoid under surveillance, at times more intimate than you'd wish – like an eavesdropped conversation between estranged lovers. Extrapolating from the intensely personal to the universal, *Maxinquaye* functions as an emotional litmus paper – blue for deep, dysfunctional depression ('brain wants me to'), red for transforming anger.

There is great ambition here. *Maxinquaye* is a definitive statement about what it feels like to be an outsider in Britain: like the best pop, it operates as a parallel communications system. Here is a different language to the tainted speech of news and politics: 'Distortion is the English disaster.' Here you can find the emotions that lie under the statistics: 'I can't breathe and I can't see. TV moves too fast; I refuse to understand.'

The pace is slow, drugged, spacey; the dominant voice is not Tricky but the 19-year-old Martina – who half sings, half raps in a whisper against which Tricky insinuates random phrases – 'switch on, switch off' – and plentiful aphorisms: 'self-preservation keeps the crowd alive'; 'the illusion of confusion'; 'all around the world, brainwashed with the cheapest'. Throughout, there are fragments of conversation: ellipses of thought.

Sometimes they work off each other, most notably on 'Suffocated Love', when Martina sings against Tricky's insistent rap: 'Will you spend your life with me/And stifle me?' When Martina lets rip on 'Pumpkin', she reveals a surprisingly soulful voice. *Maxinquaye* blurs identity, perception and gender: so much so that when Martina flatly states, 'I'll fuck you in the ass/Just for a laugh', the shock is entirely within the record's own logic.

Tricky is more liable to emerge from his confusion with a truth attack: 'How do you like yourself? You don't know yourself!' But a sugar pill sweetens bitter confrontation: a sudden giggle, or a sample that takes your breath away with its audacity – the needle skips that rip through 'Strugglin'' for instance. The overall sound is sparse, based on percussive, almost gamelan textures, over which are slipped perceptual tricks, like the alarm noise on 'You Don't', the high register vocal stutters on 'Feed Me'.

Of the 12 tracks here, five have already been released: single one, 'Aftermath'; single two, 'Ponderosa', with the lines of 1994 –

> The place where I stand
> Gives way to liquid lino
> Underneath the weeping willow
> Lies a weeping wino;

Then there are three versions of 'Overcome' – itself a reworking of Tricky's Massive collaboration, 'Karmacoma'. In addition, 'Hell is Around the Corner' takes the same Isaac Hayes sample as 'Glory Box', but, where Portishead concentrate on the internal dialogue of emotion and love, Tricky is out there, in a hostile world: 'the constant struggle ensures my insanity.'

Despite, or maybe because of this repetition, 'Maxinquaye' is a complete statement. Weaving through the gaps in the social structure, Tricky and Martina have come up with a masterpiece, a survival guide no less. They pose the question – 'How can I be sure in a world that keeps constantly changing? – and, electronically shapeshifting, give the answer.

Mojo, March 1995

Oasis: All I Want to Do is Live by the Sea

By default, pop music has many functions within our society: one of the things that it does is to cast light upon the shadows, those places that the mainstream cannot or will not go. Here is the way to escape your allotted class role, to ram your revenge in the face of the world.

Although they shy away from significance, Oasis put it down for all to hear: as they sing on 'Bring It on Down',

> You're the outcast
> You're the underclass
> But you don't care
> Because you're living fast.

For all the success and acclaim – first album straight in at number one, Best New Band at the 1995 Brits – Oasis still carry themselves like outsiders, or, in the case of singer Liam Gallagher, a particularly feisty middleweight. Scratch the hedonistic surface and there's a sharp social observation; not for nothing is leader Noel Gallagher the first lyricist to mention the *Big Issue* in a hit song – 'Supersonic'.

Where does this powerful drive to transcend their circumstances come from? Now that they're on the point of leaving, where Oasis come from is important and, for all the column inches, little understood. Filtered through a metropolitan media, factoids about the group are accepted as fact: phrases like 'the band look and dress like builders', 'Manc hooligans', 'the battling brothers' may fit a rock PR agenda, but beneath them often lies something less acceptable – the attitude that all northerners are thick, especially if they're Anglo-Irish.

Spend any time in the North-West and you realize that Ireland is a centrifugal force as powerful, if not more powerful, than London: the newly upgraded A55/A5 to Holyhead inexorably pulls you westwards. Another revelation is just how strong an Irish community there is in Manchester. Most southerners are familiar with the idea that Liverpool is an Irish city – with the readily attributed characteristics of humour, fecklessness, extravagant displays of emotion – but you'll find as strong a pattern of immigrants 35 miles to the east, deeper into a potentially hostile country.

There are many booklets about Manchester's history but none easily available about the city's Irish. What information there is is piecemeal. Alan Kidd's *Manchester* traces the major surges of immigration from the early nineteenth century on through the Great Famine of 1846–51 (by the end of which 15 per cent of the city was Irish) to this century. Kidd cites Engels's famous description of Little Ireland (on the site of Oxford Road Station): 'the squalid ghetto … occupied by the newest and poorest of Irish immigrants'.

Emigration from Ireland remains constant, but the last major surge was in the 1950s, when major public projects like Wythenshawe provided a ready market for Irish labour. It was during this surge that the parents of today's rock stars moved to find work: the parents of Morrissey and Johnny Marr, of Liam and Noel Gallagher. Where they settled was a step up from inner-city ghettos like Moss Side or Hulme: the 1930s semidetached suburbs of Burnage and Stretford, the garden city of Wythenshawe.

These are ambiguous zones, far from the city centre; superficially pleasant and peaceful, yet also prone to familiar inner-city problems: broken homes, poverty, unemployment. If you visit Burnage or Wythenshawe, you're struck by the sense of space: these are carefully laid-out areas with a strong civic pride. There are plentiful signs of stability and prosperity, yet scratch the surface and the tension is there all right, much of it centred around the economy of bad drugs.

They are also places of transition. As Noel Gallagher says, 'They were moving a lot of families into Burnage when I was growing up; a lot of old people were dying there. They turned the mini-estate we were on into an Asda. I went back a few years later and where our front room was was the bread section. I get a bit nostalgic every time I go back round there – for a muffin. I ended up by moving back into the city centre, a flat in Whitworth Street, when I was 21, lived there until I was 26, then moved to London.'

You can hear this ambiguity in both the Smiths and Oasis: an aspirant will to succeed – to move on up and out, to go further than your parents have been allowed – allied to a fierce pride and anger about your background. It's a powerful mixture that has wider implications in a country where roots are routinely torn up, where – despite Conservative rhetoric – the class map is ever more tightly drawn. Anti-Irish sentiment is still routine in England: I know, I've experienced it within my own family.

The Irish influence on pop culture is justly documented. Despite the attention paid to the Beatles, the Anglo-Irish contribution is only now being understood – through John Lydon's *No Irish, No Blacks, No Dogs*, the opening chapters of Johnny Rogan's *The Severed Alliance*. There is nothing more thrilling in British pop than the rapier wit and gleeful, lacerating revenge of the Sex Pistols' 'God Save the Queen', Dexys Midnight Runners' 'Dance Stance', the Smiths' 'The Queen is Dead' or John Lennon's nightmare vision of England 'I am the Walrus'.

Surfing on sound, Oasis explicitly put themselves in this lineage, combining the sneer of the Lydon Sex Pistols with the psychedelic drones of 1966–7 Beatles' songs like 'Rain' and 'I am the Walrus' – which they cover with a fine insolent panache. For all their press as horrible lads, their concerts can be warm, communal, agreeably mixed-gender gatherings. If singer Liam can assume a well 'ard persona, then that is, irritatingly, what is fashionable within a Mancunian generation fascinated by the gangsters that have gained ever more visibility in a drug-dominated subculture.

This latest twist on Mancunian dourness is belied by (although they'd hate to admit it) a distinct sensitivity. Noel Gallagher is capable of both defining a generation's hedonism and offering a positive outlook – in joyous exclamations like 'These could be the best days of our lives!' ('Digsy's Dinner'). Then he'll turn round and write something that really strikes home, like the great line from '(It's Good) To be Free': 'All I want to do is live by the sea.' The song ends with an Irish jig, played by guitarist Paul Arthurs.

'It makes a difference to yourself that you have an Irish background,' Noel Gallagher says. 'I think it makes you more passionate about music. Obviously you're always brought up a Catholic and you always end up denouncing it. There's another thing: we'd always get in a van and go anywhere to play a gig, whereas your middle-class groups will say, "I'm not doing that, I've got college in the morning." We just say, "Fuck it, we wanna play." I like working hard, and I'm in this for the rest of my life now.'

Guardian, 17 March 1995

Blank Regeneration: Beavis and Butt-Head

When Elastica lifted the organ melody from the Stranglers' 'No More Heroes' for the central guitar riff of 'Waking Up', they reopened the debate that has been going on in black music – above all rap – for years: in sourcing the past, what is the difference between legitimate inspiration and sheer plagiarism? Can the present escape the shadow of the past? Isn't all art plagiarism anyway?

Never mind that 'Waking Up' might be just that, a typically sarcastic call to arms; any power that it might have is immediately called into question by the undeniable fact that it sounds like an 18-year-old song. This view was endorsed last February, when, facing a lawsuit, Elastica undertook to pay 40 per cent of songwriting royalties to the Stranglers' publishing company. That's what you might call legitimate.

A whole world of dismissal is enshrined in the sound-alike argument: implicit is the idea that pop no longer has any vigour, that it has all been done before, that it and we are adrift in a postmodernist squall of emotion-free irony, referencing and pastiche. In this mindset, pop is strangled by the weight of its 40-year tradition: all that remains is a series of parlour games for a disempowered generation, doomed to repeat history as farce.

Well, no. Like all the best records, 'Waking Up' contains its own everlasting present. Put it on repeat and you're lost in the moment; then put it next to 'No More Heroes' and you get many more questions than answers. How is it Elastica have made something so hopeful from something so cynical? How is it they've made a triumphal female statement out of such machismo? (The Stranglers were notorious for their misogyny, even in the late 1970s.) Have they not, in fact, used a record, popular when they were pre-school, to make something new and contemporary?

It's like Tricky's curse on *Maxinquaye*: 'brand new, you're retro.' This is not nostalgia, more a perception that the past cannot be beaten: rather, why not take all this stuff and engage with it physically and emotionally, moulding and pummelling it into something that works for you?

This is a problem for the baby-boomers and 30-somethings, brought up on a linear cultural model, who populate the mainstream discourse around pop. Faced with their constant dismissals, isn't it more interesting to tease out the complexities, ambiguities, lacunae inherent in this past-in-present? And, if this is not your perception, can't you abandon the weight of your amazing experience, and, like, use your imagination?

As it happens, Elastica slot into a multimedia matrix that encompasses phenomena as diverse as *Beavis and Butt-Head*; the American resurgence of punk with Green Day, Offspring and Rancid; the fashion world's canonization of Vivienne Westwood and the serial revivals of the look that she helped to originate – punk; a return to the typographic brutalities of the black and white late 1970s; the sheer retro consciousness of Brit rock groups like Blur, Oasis and Suede; the continued marketing of punk itself – Virgin's *Best Punk Album in the World … Ever*.

Beneath the cyclical demands of marketing and fashion lies another common thread, which has to do with a willed stupidity. Like the original punks, many of these cultural workers choose to present a blank, sarcastic façade: the impenetrable eyebrows of the Oasis siblings; the shock of vegetable dye that is Green Day's hair; the vomiting sounds of disgust that punctuate Elastica's breakthrough single, 'Line Up'. And then there's Beavis and Butt-Head; one media cycle on from Bart Simpson, the purest distillation of punk.

> You know me, I'm acting dumb dumb
> You know the scene: very humdrum
> Boredom
> Boredom
>
> The Buzzcocks, 'Boredom' (1976)

> D-U-M-B: everyone's accusing me.
>
> The Ramones, 'Pinhead' (1977)

> I think I'm dumb
> Or maybe just happy
> Think I'm just happy.
> Nirvana, 'Dumb' (1993)

> Maybe it's just jealousy
> Mixing up with a violent mind
> A circumstance that doesn't make much sense
> Or maybe I'm just dumb.
>
> Green Day, 'Chump' (1994)

So why is it that the best and brightest should present a moronic surface to the world? Why is it that obvious over-achievers like Elastica should toy with ideas of failure and laziness: not for nothing is 'Waking Up' a slacker manifesto on a par with Beck's 'Loser' – 'I'd work very hard but I'm lazy/I can't take the pressure and it's starting to show.' Beck himself is not the LA streetkid of popular provenance but the grandson of Fluxus artist Al Hansen, and, as

Beavis and Butt-Head point out in one of their withering commentaries, a kid straight out of gifted class.

Naturally, most media commentary sticks on this surface: 'I turned on *Beavis and Butt-Head* the other night, and it was so much worse, so much more stupid than anything that I had imagined, that I just sat staring in astonishment,' wrote Bob Herbert in the *New York Times*, February 1995. Wow. There's a real inversion here. In fact, as anyone who dares to enter their world knows, Beavis and Butt-Head are first an entertainment, then a theatre of simultaneous identification and revulsion, and further if you will, a satire on male adolescence and its culture.

The intention of creator Mike Judge, a sophisticated 30-something, comes out in a sequence of devastating one-liners. Who can forget the perfectly timed dismissal of arrogant video band Wang Chung: 'Why do these guys look so ... tired?' Stupidity is a constant theme:

> TEACHER: I want you to work with Daria Morgendorffer, our
> straight-A student who won the science award last year.
> DARIA: But Mrs Dickie, Beavis and Butt-Head are complete
> imbeciles!
> BEAVIS: Yeah! She's right!

Place this next to Bob Herbert and the question is begged: so, like who is really being stupid here?

The ascribing of intelligence and stupidity is an explicit tool of social and political control: you only have to look at the relationship of Charles Murray's theories (*The Underclass*, *The Bell Curve*) with the 'whitewash' Republican sweep of American Congress and Senate and the subsequent tightening of the social in the US. If you're not part of the élite, then you're stupid or might as well be, because you're not going to get any resources, or arenas – jobs, communications etc. – within which you might legitimately exercise your innate intelligence.

The history of punk (and pop, and much twentieth-century art) is the return of the repressed, not the least of whom are the people whose intelligence is surplus to requirements: as John Lydon put it, 'the flowers in your dustbin'. To break through this social engineering, it is first necessary to define how you are disadvantaged, embrace it, then throw it in the face of those who would oppress you: whether you're hippies, punks, niggas, queers, whoever. As Jefferson Airplane chanted on their rabble-rousing 'We Can Be Together': 'Everything they say we are, we are, and we are very proud of ourselves.'

One of these magical inversions is the assumption of stupidity by the very intelligent: 'Right, you treat us as stupid – well, we're going to be really stupid.' Like most twentieth-century art stuff, this goes back at least as far as Dada: 'To carry simple-mindedness and childishness to excess is, after all, still the best defence' (Hugo Ball). This impulse goes beyond simple reflexivity into other areas: stupidity as a put-on, a baffle, a process of unlearning, a willingness to traffic in the most degraded cultural products. Anything to get away from the dead hand of the bourgeois canon.

We see the inauthenticity and smart/dumb loop-de-loop in the first foundations of punk, with Richard Hell's justly famous 1975 manifesto 'Blank Generation'. Like a blank canvas – as was the author's intention – the song has come to have many meanings, the most famous in its direct refraction by the Sex Pistols into 'Pretty Vacant': 'We're so pretty, oh so pretty vacant/And we don't care.' Hell later complained that his intention had been hijacked, that his 'blank' meant unlimited possibility rather than vacancy, but then stupidity was encoded in the song itself, directly stolen from Rod McKuen's 1959 beatnik exploitation single 'The Beat Generation' by Bob McFadden and Dor.

For all his literary aspirations, Hell's influence on punk was the look: the baby-bird spiked-up hairdo, the chopped up T-shirts, and the electric shock expression – bug eyes, open mouth. In the bad drug ambience of the mid-1970s New York underground, dumbness and numbness were celebrated in a reaction against hippie pretension, in a new kind of cool. Not for nothing were the Ramones, following a certain natural talent in this area, packaged as the most minimal, cartoon-like human beings possible. The four identikit surnames were only capable of forming a whole together: if you could prise them apart, all you'd get was a quarter of a person.

As the single most influential record of 1976, the Ramones' first album injected this Tom 'n' Jerry dumbdumb straight into the bloodstream of British punk: once you got past its monolithic production and casual brutalities, however, it was possible to detect formal manipulations of great wit – the Beavis and Butt-Head syndrome, in fact. Take 'Judy is a Punk''s numb loser lyrics, squeaked out by Joey Ramone, then suddenly illuminated by this: 'Second verse, same as the first.' You listened, and it was. Then it came again: 'Third verse, different from the first.' Well, so it was too. What were these guys doing – were they being deliberately stupid, or what?

This element of put-on went directly into British punk: the most celebrated instance was the Sex Pistols' appearance on Thames Television's *Today* programme – a classic of the 'you think we're stupid, right, we will be' dynamic. By the spring of 1977, the time when the international media were beginning to broadcast images of this peculiar new mutation, there was a favourite punk look: we all used to do it, in photo booths. Tilt the head back on the neck, open the eyes into a glassy, amphetamined stare, and relax the mouth wide in a slack drool. Duh.

Penelope Houston, the Avengers, early 1978 (photo: Kamera Zie)

It was fun to do because it threw people off. It was a gestural translation of the punk aesthetic; the celebration, say, by an avant-garde, art school group like Wire, of the simplest pop trash – the Troggs' 'Wild Thing' and the Mysterians' '96 Tears'. Check out their splenetic 'Mr Suit' for the funniest punk protest ever – the '1-2-3-4,' countdown into a dumb riff, sarcastic hooligan vocals, an accelerated time scale:

> You can take your fucking money, and shove it up your arse
> cos you think you understand, well it's a fucking farce,
> I'm tired of fucking phonies,
> that's right I'm tired of you.
> No, no, no, no, no, no, Mr Suit.

When these clustered negatives hit what was then Fleet Street, the result was predictable, if entertaining. Punk was tailor-made for the tabloids; it talked the same urchin language (of social realism, the urban underclass) and was stimulated by the same shock principle, even physically encoding tabloid headlines into handbill and fanzine graphics. It's hardly surprising that punk's arrival on the national – and international – stage was announced by a series of screaming headlines: 'The Filth and the Fury', 'The Foul Mouthed Yobs', 'The Punks – Rotten and Proud of it!'

For the mass media the surface was (and is) all; never mind the aesthetics of minimalism, the surface was stupid. Take the packed front page of the *Sunday Mirror*, 12 June 1977:

> Punk Rock Jubilee Shocker: They call themselves names designed to alienate society; Rat Scabies, Dee Generation, Johnny Rotten, Sid Vicious. They swear and spit on stage. They wear stinking sneakers. Their songs cause violence. Fans are injured in riots. They don't give a damn what anyone thinks of them.

The standfirst is fabulous: 'What's burning up the kids? A disturbing report on the amazing new cult.' Cut out, this made a Siouxsie and the Banshees handbill, while the photo of a 'girl fan watching the Stranglers in Manchester', all slackjawed, wide-eyed rapture – not so dissimilar to Man Ray's famous *'Larmes'* – was clipped for the cover of a single by the Snivelling Shits, a deliberately dumb concept formed by a group of music press writers. One side charted the disastrous effects of amphetamines on male sexual performance – 'I Can't Come' – while the other laid out the tabloid transaction for all to see: 'Terminal Stupid'.

It was this mass media interaction which defined punk as it hit the West Coast of America. Lacking the New York tradition, would-be punks, particularly in Los Angeles, were stimulated first by the April 1977 appearances of that most cartoony of all Brit punk groups, the Damned, and secondly, by the media reports broadcast from the summer onwards by the networks. This West Coast punk subculture remains a lost pop moment – charted at the time with wit, smart/dumb paradoxes and perfect visuals by the magazines *Search & Destroy* (San Francisco) and *Slash* (Los Angeles).

Learning their moves from the media – whether firsthand or refracted back into punk style – early West Coast punk show all the signs of dumbdumb in full effect. Take Kamera Zie's picture of the Avengers' singer Penelope Houston: her beautifully posed, model-girl looks offset by a slack jaw, hooded eyes, and carefully ripped stripy T (that beatnik perennial). Below the picture, her art: 'Car Crash', a perfect restatement of the 'too fast to live, too young to die' syndrome:

> Oh no! Car crash!
> Your leg's over here and your head's over there
> But fuck it, baby, what do I care …

Tomata Du Plenty, KK and Tommy Gear, the Screamers, autumn 1977 (Kamera Zie)

Or there's the photo of synth-punks the Screamers: intelligent boys all – performance artists, computer experts, trained musicians, at least two of them were gay – acting braindead in a graffiti-strewn landscape. Here they display, with perfect elegance, the phantasmagoria of punk posture: spiky hair and psychotic stare (singer Tomata Du Plenty), the Sid Vicious uniform of padlock, leather jacket and sneer (synthist Tommy Gear) and, eerily prescient of Beavis, the dumbdumb gesture *in excelsis* (drummer KK).

The archetypal punk group from this period were the Weirdos; a perfect mixture of goofy absurdity, alienation and sheer rock/pop glee. Singer John Denney matched severe thriftstore skills (the 1950s and early 1960s were coming onstream) with a sharp line on his native Hollywood – which, was and is, far more vacant than anyone in Europe (or even the East Coast) could ever realize – in fabulous songs like 'Happy People', 'Idle Life', 'If It Means Nothing'.

Nobody heard them. The West Coast punks were good; they were ambitious – 'Think of it Ellen … a World Full of Weirdos', ran a cartoon calendar made by the band – but what Brit punks took for granted, access to record companies, press, prime-time TV, was denied to the Weirdos, the Avengers, the Screamers, all the others. The only group from this time to make waves, X, wrote about it on their LP *Los Angeles*:

We're locked out of the public eye
Some smooth chords
On the car radio
No hard chords
On the radio
We set the trash on fire.

In America, punk failed. In Britain, it succeeded, to the point that it has dominated music and a particular section of the (youth/style/music) media ever since. Much of this is malign, punk negation long curdling into the programmatic cynicism which is now a major media mode. It's noticeable that the Brit musicians who hark back to this period are either positive (Oasis) or gleeful (Elastica): anything but cynical, because that's what the previous generation did and, sadly, still do.

A forgotten side of 1970s punk was its visionary quality. In a *Search & Destroy* spread on Bay Area group UXA, singer Dee Dee Semrau has the look of now – a near ringer for Donna Elastica – while the caption explicitly projects into the future, in a song called '1995': 'The day is right for invention/Rays of sunlight, break the prisons.' At the time, this might have been the routine Burroughs dystopia; today, it is a record of ideas waiting to be activated at the right moment.

Dee Dee Semrau, UXA, early 1978 (Ruby Ray)

This activation, during the last five years, has either provided a fantastic, creative extrapolation from the primal punk impulse (*The Simpsons*, *Beavis and Butt-Head*) or has been a surprisingly straight replay of 1977–8 complexities: as it happens, there is very little difference between Green Day and the Weirdos. There was no more fitting metaphor for punk's autumn 1991 return in the US than the simultaneous, narcoleptic arrival of Kurt Cobain (*Nevermind*) and River Phoenix (*My Own Private Idaho*): dumbdumb boy passivity taken to its logical, comatose, helpless extreme.

In *The Aesthetics of Disappearance*, Paul Virilio describes the everyday condition which he calls 'picnolepsy'; a frequent but often infinitesimal loss of consciousness – a kind of unconscious tuning out which he relates to the physiological totality constructed by the twentieth-century mass media. The importance of these gaps is shown by the very efforts made by media workers to create a seamless flow, to obliterate any lacunae – but they remain, if only in the popularity of 'blooper' shows like *It'll be Alright on the Night*. Here is the hint of a strategy for survival within the mass media; a withdrawal from their panoptic glare.

In this covert struggle, stupidity is an important site. In the UK, *Beavis and Butt-Head* follows *The Word*, a magazine commissioned by the Channel 4 'youth programmes' department. Overseen by Oxbridge graduates, *The Word* delivers youth up to the market like veal calves to France, catering to its audience with a couple of good bands, then boring them witless with a farrago of organized humiliation, page three sexism and charmless ineptitude. Gaps appear, but will the producers ever relax their control? Never! They're always full on, in this nightmare of overlit, desperate mediation.

Here is another, more pernicious form of stupidity: populism. There are few things less attractive than the sight of over-educated media producers talking down to the public because they fantasize that this is 'what it wants'. The underlying assumption is that the public is stupid and that, if it is not, it is to be rendered so. This is the drive behind the tyranny of market forces. In this light, large sections of the media are the agents of social control, which would be an obvious enough point if it wasn't deliberately masked by falsehoods like market research, freedom of speech and the reaction against political correctness.

Scene: Beavis and Butt-Head, on the couch, watching an unnamed mid-1980s video: post Duran Duran synth pomp style.

BUTT-HEAD: This guy thinks he's like, smart …
BEAVIS: College music sucks.
BUTT-HEAD: I think it's only cool if you, like, go to college … this video's, like, complicated …
BEAVIS: Yeah. It's stupid …

Within this context, it's tempting to see Beavis and Butt-Head as a Janus-faced Candide, wandering unscathed through this polluted landscape. As transmitted on Channel 4, the link between their surface stupidity and the picnoleptic approach to media is reinforced by an accident of editing: black fades where the US ad breaks fall which, like jump-cuts, propel you out of TV's seamless flow. The result, together with the channel hops and video commentary contained within the cartoon, is a kind of interactive, meta-television, where you no longer have to accept the dominance and submission encouraged by *The Word* and its ilk.

Like the original punks, Beavis and Butt-Head are ambiguous phenomenon – as shown by the for and against debate raging on the Internet. In the glare of the mass media, assumed stupidity can easily become real stupidity; irony is flattened out and you become the very thing you are trying to satirize or transcend – viz, the 1993 US moral panic about a pre-teen who, allegedly inspired by B&B, torched his family home. This resulted in the censorshp of Beavis and Butt-Head – a cartoon, remember – which now arrives here in a decidedly peculiar state.

This may well be an index of their success. Like the punks, Beavis and Butt-Head have penetrated the collective subconscious. Holding a mirror to a society in denial is never a popular activity. Despite their sniggering, pre-sexual misogyny, despite their love for retarded American rock like Primus and Pantera, they remain a complex, contradictory phenomenon, if the fantasies and histories projected on to them is any index. Within an increasingly simplistic, monolithic right-wing media, which imposes stupidity from above, they show us that we don't have to take this shit.

Frieze, Summer 1995

The Always Uncjorked Björk

Posing for photographs in a studio high above Covent Garden, Björk is exactly like her image – a little taller, perhaps, but everything else is in place: pigtails, high-contrast clothing, fresh-faced innocence, exuberant sexuality. Bouncing around in yellow sneakers, pink paper trousers and a red T-shirt with Coke-style lettering (ENJOY COCK!), she is every inch the modern pop star – in control of her image, comfortable with the many forms of promotion required by today's music industry, at ease with her celebrity status. As we walk through Covent Garden, she is recognized by two homeless people. Their reaction is disbelieving, then warm. Björk is unfazed but does not linger.

Björk has been performing since age 11, when she made a record in her native Iceland. Raised by what she calls hippie parents, she rebelled in her teens and formed a punk band called Kukl; they recorded for the label run by the hard-core British anarchists Crass. It was a key moment: 'I'm still definely obsessed with the spontaneity of punk. I'm a sucker for energy. Just put all the energy in the world into my ears.' Subsequently, Björk's voice shone through the guitar rock of the Sugarcubes, a group that, like many punk bands, was formed as a joke and ended up an unhappy career. In Iceland, in the early 1990s, with the Sugarcubes disbanded, Björk wrote and recorded much of her *Debut* album – 'songs I had written in the evening when my kid was asleep, almost like a domestic housewife album'. It was her two guest appearances with dance maestros 808 State that had opened up a whole new world for the former punk and paved the way for her collaboration with Nellee Hooper, the producer who, with his connections to Bristol trip-hoppers Massive Attack and Tricky, was in at the ground level of this year's dance-floor boom. He provided the state-of-the-art sheen that made *Debut* so attractive. The CD was a winning mixture of club savvy and more reflective songs that explored nature's mysticism.

Björk's new record, *Post*, develops this fresh mixture. There are the up-to-the-minute dance beats, fused with sharp lyrics in songs like 'Army of Me' and 'Hyper-Ballad'. There is the cover of a vintage show tune, 'Blow a Fuse'. And there are spooky tunes that play with perception: the odd scratchings at the end of 'The Modern Things', with Björk whispering 'no one sees me' in Icelandic, and the psycho-active assault of her collaboration with Tricky in 'Headphones'. Her audacity is one of the most powerful things about Björk. She embodies the sense that anything is possible – in lyrics, in appearance, in gender, and in the very sound of her voice.

JON SAVAGE: When were you born?
BJÖRK: 21 November 1965.
JS: On the cusp of Scorpio and Sagittarius.
B: My mum is heavily into these things, and apparently I'm as much Scorpio as one can be. To me, whether it means something or not – fuck that, I just love the symbolism of it. It's pretty, like Greek and Nordic mythologies. I'm supposed to be run by Pluto. It's like a fairy-tale, it simplifies things.
JS: Is Nordic mythology similar to Greek?
B: It isn't a copy, but it's got the same characters. In mythology wherever you go, you've got the strong guy, the wise woman, the winners and the losers, the travellers and the domestic people. I always like the animals in mythology, like the ravens on Odin's shoulders.
JS: Scorpio is all about life, death and sex.
B: That doesn't surprise me. My three fucking obsessions.

JS: Have you ever had your chart done?

B: My mum did it. I think she took me to all the occult creatures of Iceland, from the age of zero until I was 18, when I became a rebel anti-hippie. I got my fortune told and everything. I think I probably believe most of it, actually. I've got Pluto in a very important place, and that's what I'm about. I have to re-create the universe every morning when I wake up. And kill it in the evening, which is a bit outrageous, but there you go.

JS: Hard work.

B: Heee! Well, maybe not every morning, but maybe twice a year I have to destroy everything. I've also got my moon in the twelfth house, in Scorpio, and my sun in Scorpio in the first house, and also Neptune. Then on my other half, my generational picture, I've got Pluto and Uranus in Virgo, and my midheaven is in conjunct with those two. Virgo is the sign of the nurse, so this means I was born to nurse my generation. I'm still 50-50 about whether this is true, but I was breast-fed on it.

JS: In your lyrics, you seem obsessed with the sea.

B: I am, very much. It's a combination of things – being born on a small island and always having the ocean. It makes your head function completely differently. If I travel, as long as I'm by the ocean, I'm fine. If I'm not, I get claustrophobic.

JS: What do you exactly get from the ocean?

B: First of all, a sense of well-being, like I'm home. I had a really wild upbringing, which I think is the best upbringing anyone could have. My home was by the sea. If I walked down to the sea and sat down by the shore, I was home. That's my mother, the ocean. Nothing can go wrong. I love swimming, another hippie thing. My mum says it's because I'm a water sign. And the sense of space and boats. I'm obsessed with boats. It's freedom.

JS: Do you feel the lack of sea in London?

B: Yeah, it really does my head in. I tried to stay by Little Venice, but it's a canal, so the water doesn't move. I'm only here for work. It's just two hours on an airplane; my kid [eight-year-old Sindri] can go back home when he wants to. I'm only here for a period, to get my little mission done, and once it's finished, it's finished. But after this little job is over, I'm living by the ocean. It doesn't matter where it is.

JS: What do you think your mission is?

B: It took me ages to reason it to myself. I find it very hard to be selfish. I just decided, I'm going to move to London, I'm going to be really selfish, I'm going to get all the instruments I want, all the noises and lyrics I like, and make all the music I can, because everybody's got to express their vision, and no two people are the same. I could happily go and die if I could say, 'I did my best, I made my sacrifices.' It's as basic as that. If I hadn't done this, I would sit in my rocking chair at 85 with my grandchildren on my lap, and say, 'Sorry, I didn't have the guts.' I've become selfish now, believe me. I'll go out to the flower shop and buy flowers just for myself. It's outrageous, isn't it?

JS: What do you feel about moving to London from Iceland?

B: It's a cosmopolitan city. That's the reason I'm here. If I want a dulcimer player, I can get one. If there's a certain photographer I want to work with, more than likely he's going to come through London. I can appreciate London from above, all the rooftops, maybe because I'm a kid and I like Peter Pan. I'm starting to appreciate aimlessness and eccentricity. I've realized that Englishness is about people who have to behave politely all day, and the clothes have to be all proper, but that doesn't mean they're not mad. You have to focus on it, but once you find it and focus on that energy, then you can stay sane. Compared to the English, Icelanders are like people from Sicily or somewhere: *'I'm upset!!!'* Like a volcano, they break things, and two hours later, they're happy. There's a volcanic eruption in Iceland once a year, on average.

JS: Do you think that environment influences behaviour?

B: Very much so. What happens in Iceland is that you get the blizzard in your face, you have to fight the weather all the time, and you stay very alert, you never fall asleep. Your head is always working. People who go there think the Icelanders are really stressed out. They're not, but their energy is on 10. We've got this awkward thing, which is 24-hour darkness in the winter, and 24-hour daylight in the summer. There is snow from October or November until mid-March. It means that in the winter you're just inside and you write all the books you were going to write and get everything done on your own, and then in the summer you go absolutely mad. Like bears after hibernating.

JS: Do you write things down when they occur to you?

B: Yes. I've written diaries for a long time now, and sometimes a whole lyric comes, and I have to pick a sentence here, a sentence there.

js: There's a great lyric on 'Big Time Sensuality': 'It takes courage to enjoy it.' Do you have that courage?

B: I've got a lot of courage, but I've also got a lot of fear. You should allow yourself to be scared. It's one of the prime emotions. You might almost enjoy it, funny as it sounds, and find that you can get over it and deal with it. If you ignore these things, you miss so much. But when you want to enjoy something, especially when it's something you've just been introduced to, you've got to have a lot of courage to do it. I don't think I'm more courageous than most people. I'm an even mixture of all those prime emotions.

JS: Sex does take courage sometimes.

B: I think so, because if it lacks that sensation of jumping off a cliff it would just miss so much. Then again, it has to be pleasurable and enjoyable and lush and all of that. But 'Big Time Sensuality' was actually about when I first met Nellee Hooper. I think it's quite rare, when you're obsessed with your job, as I am, when you meet someone who's your other half jobwise and enables you to do what you completely want … so it's not a sexual romance.

JS: Are you currently in a stable partnership?

B: No. I split with my boyfriend at the beginning of last November, and at that point I'd been with a stable boyfriend since the age of 16, though in different relationships. When we broke up, I thought I might as well enjoy this, which I do and I don't. It's scary at times. The best bit is that you're kind of skinless, you're more vulnerable and emotional and on the edge. There's also that silly thing that I had when I was 15 and 16 – looking around and wondering who it will be! So I'm sitting there on the subway thinking, will you have a long nose or a short nose? Will you enjoy this or that film? It's like a little party game.

There's something really stupid and romantic, thinking that it's just going to be one person. Even though both of us might have five partners before we die, we always think of that one. Then there are all these things saying how brilliant it is to be self-sufficient and not needing anything or anybody and getting all these tools so that you can do everything yourself. It's like you're a little warrior armed with your Walkman and your video and all this technology. Everything's geared toward self-sufficiency. Fuck that. For me, the target is to learn how to communicate with other people, which is the hardest thing, after all. What you should be doing is learning how to live with other human beings.

JS: Do you have visual ideas in your mind when you're writing your songs?

B: Definitely. It's natural for me to express things first musically, then visually, and third, with words. So the words are like a translation of noises and pictures.

JS: 'Army of Me' is a heavy song. Did you have a picture in your mind when you wrote it?

B: I'm a polar bear and I'm with 500 polar bears, just tramping over a city. The lyric is about people who feel sorry for themselves all the time and don't get their shit together. You come to a point with people like that where you've done everything you can do for them, and the only thing that's going to sort them out is themselves. It's time to get things done. I identify with polar bears. They're very cuddly and cute and quite calm, but if they meet you they can be very strong. They come to Iceland very rarely, once every 10 years, floating on icebergs.

JS: Can you tell me about 'Hyper-Ballad'?

B: That's a lyric about being in a relationship, and after a while, say three or four years, you repress a lot of energy because you're being sweet all the time. So I wanted to set it up like a fable, something that happens over and over again. It's about this couple who live on a cliff in the middle of the ocean, and they live in this house, just the two of them, and she wakes up really early, about five in the morning, before anyone else wakes up, and sneaks to the edge and throws a lot of things off: old rubbish, car parts, bottles and cutlery. And she imagines what it would look like if she herself were to jump off. Then she sneaks back into the house, back into bed, then her lover wakes up and it's 'Hello! Good morning, honey!' And she's got rid of all the aggressive bollocks. The chorus goes, 'I go through all this, before you wake up, so I can feel happier to be safe up here with you.'

JS: There are some great subliminal noises on 'Headphones'.

B: That's a track I did with Tricky. He was getting a lot of pressure from his record company, because there was a real buzz about his album, so he was a bit naughty and escaped to Iceland. We drove around in a four-wheel drive and saw the glaciers and swam in the hot spring and wrote this tune. I went into my diary and found a complete lyric about receiving a compilation tape in the post from a friend. It's a very personal thing. You're pissed off with things generally. You save it until the evening, and after you've had your bath and brushed your teeth, you go to bed and take your Walkman and put your headphones on and you fall asleep. The lyric is a letter to that person. I had this idea to do a song that is like a worship of headphones. The chorus is 'My headphones saved my life, your tape lulled me to sleep.' All the noises in the song are just-for-headphones stereo tricks. It didn't need a lot of instruments. Tricky feels really strongly about noises and beats, and that is exactly my weakest point.

JS: Are you in character in a lot of your songs?

B: Most of my songs are written in the first person, from the point of view of my best friends. I find it 10 times easier to express my friends' feelings than my own. If I write about myself, I usually write in the third person. It just feels natural.

JS: Do you sing from your stomach or your chest?

B: My stomach. Most engineers find it quite difficult to deal with me, because most of the singing I did as a kid was when I was walking outside, completely on my own. This is absolutely impossible in London. There is no privacy here. I started singing with the whole of my body, which is both good and bad. The engineers usually end up using the same kind of microphones as they put on a stand-up bass, because it's got a big body.

JS: You've said that you recorded a lot of your vocals on the beach.

B: It was a very sentimental thing. I wanted to sing outside, because I knew everything would fall into place. Nellee made it happen. Compass Point Studio [in Nassau, the Bahamas] was right by the beach. I'd have a very long lead on the microphone and a long lead on the headphones and I'd just sit there at midnight. All the stars would be out, and I'd be sitting there under a little bush. I'd go running into the water and nobody could see where I went.

In the quiet bits, I'd sit and cuddle, and for the outrageous bits, I'd run around. It was the first time I'd done a song like that in about 20 years. I was crying my eyes out with joy, because it was something I so deeply wanted all those years. Almost like you had sex lots of times, and it's gorgeous, and then you couldn't have it for 20 years, and then suddenly you have it. It was completely outrageous.

JS: Do you think that musicians feel and act out emotions on behalf of their audience as a way of helping people deal with emotions?

B: Definitely. It's something I didn't think about until recently. I probably wouldn't have thought about it at all except I had to get my ass over to another country and force myself to think about why I was doing this. It was almost like I wasn't doing it for myself. But if I have any vision of my life, I think I'll be singing until I die, about 90 years old. It's funny, all the attention I'm getting, but I don't think I'm hooked on it. I could just as well move to a little island and live by the ocean and just be the village singer or whatever. Singing on Friday and Saturday nights, writing tunes for the rest of the week. That's my role.

Interview, June 1995

All Pumped Up and Nowhere to Go: Beefcake

The word itself summarizes the perplex of male iconography: have you ever actually eaten beefcake? How would you cook it? What could the mixture of meat and icing taste like? Cheesecake – that peculiarly unpleasant name given to women as objects of desire – is an all too familiar diet item; by contrast, beefcake has an air of impossibility but, like peanut butter and jam, could be so disgusting that it's great.

Here is the familiar contradiction: if the normative masculine role in society is active, what does it do to a man to be put on display, to take on the passivity of the adored object? Hence the persistent stain of femininity and, thus, in our archaic gender equations, homosexuality that attaches to most iconic males, which has to be rigidly policed: in all sports, in men's magazines like *Arena*, in Jason Donovan's libel case. Or, to quote the text, be prepared to enter a world where a motorcycle is regarded as 'the ultimate in male costume jewellery'.

A long-overdue account of 1950s and 1960s gay physique shots, *Beefcake*,

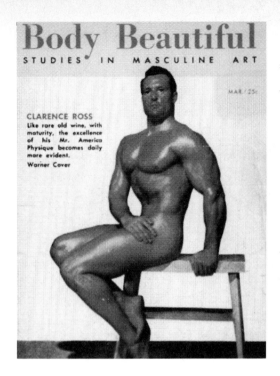

Body Beautiful, March 1954

reveals a hidden history: the homosexual origin of a language with which we are now saturated – a language of Levi's ads (the favour now returned, with thanks, by 1994's 'Taxi'), of boy-zone teen groups, of Hollywood stars, even – whisper it – of sports stars. This is the teenage discourse of youth, sex and death which begins in America's post-war emergence as a world power; it also highlights the polar oscillations of commodity capitalism between exclusion and inclusion.

F. Valentine Hooven III's text is an unashamed fan letter to physique pioneers like Quaintance, Bob Mizer (American Model Guild [AMG]) and Chuck Renslow (Kris Studio). Rightly so. These men – and others – were harbingers of the new perception of which we are the beneficiaries. The fact that, in many ways, we can now live easier in our polymorphous psyches is due to the steps that they took into the unknown. As Hooven writes: 'No serious attempt was made to gloss over the fact that these attractive young men were naked to be looked at and enjoyed. Men, naked for the pleasure of others? That, in the 1950s, was dangerously radical.'

AMG began in 1948 as a document of a particular West Coast phenomenon: the Muscle Beach at Venice, Los Angeles. Selling queer pin-ups mail-order, Mizer expanded into the magazine format. Framed by Quaintance's shamanistic paintings, *Physique Pictorial* became the record of a new aesthetic: covert, masked, incredibly stylized, highly sexual, pagan. The local film industry took note, as here was a language to sell male sex: Tony Curtis posed in AMG model-boy style, Brando as an archaic/mythological pin-up (Julius Caesar), Apollo model Ed Fury (real name: Edmund Holovchik) crossing over into real films. Pictured on set with Richard Widmark, Fury looks just like James Dean.

This aesthetic – the first appearance of camp in the mass media – came from two main sources: Greek *kouroi* and physical-culture magazines dating from the turn of the century. Apart from the work of eccentrics like Wilhelm von Gloeden, these were the only available precedents for the male pin-up. A certain mythological tendency persisted in early gay magazines like *Jason*, *Goliath*, *Hermes*, *Mars*, *Samson*, *American Apollo* – the old Olympus ushering a whole new star system into the Judaeo-Christian West. Now you know why there are so many Doric columns in pictures from this period: they're not just phallic symbolism (although the penis is all-important) but instant signifiers of the new pagan pleasure world.

This met opposition, of course. As the parallel text to this dictionary of bodies, there is a litany of court cases, imprisonments and bureaucratic spite. Renslow went to the US Supreme Court in 1958; Muscle Beach was closed down in 1959; Mizer was sent down in the late 1960s. The codings necessary to avoid prosecution may begin in kitsch antiquity, but end – in Mizer's 1960s layouts – with a bizarre section of runic symbols, used as a psycho-sexual rebus for each model: the precursors of today's personal ad puzzles. It was at this moment of ultimate stylization that the form became obsolete: 1965 saw the first full male nude (i.e. penis instead of posing pouch) in a magazine called *Butch*. After this hesitant, anonymous start, the floodgates opened – not that we'd experience much of it in the UK, with our stringent censorship laws.

There is much here that is archaic: there is now no need for gay marketing to be so covert. These images may very well be sexy because of the repression encoded within them, but then repression is there to be transcended. 'It was more fun when it was illegal' is a familiar trope, but it wasn't. In my experience, the criminalization of homosexuality resulted in lives shattered by alcohol, guilt and self-hatred. The disappearance of this kind of publishing, half bemoaned in the text, was the inevitable consequence of the greater sexual openness – and partial decriminalization – that occurred in the late 1960s. The physique magazines were the victim of their own success.

Yet there is a curious appeal in these images, much of which rests in the persistent idea of the 1950s as a golden age. These young men are frozen in a world that is somehow more leisurely and more confident, which has the fresh bloom of a new vision. Hooven makes a coherent case for these pin-ups as embodying the underbelly of popluxe America, but you can take it further: what they do is give a shadow commentary on the birth of youth culture. It's all here: the re-creation of self (Ed became Fury several years before Billy); the willing feminization of young men; the self-destruction of life in the fast lane (in a spread of models killed by drugs, car crashes, random violence).

It's now clear, despite the weight of exclusion on this subject, that in the creation of a commercialized youth culture, the teenagers and the entrepreneurs who catered to them took much from the psychic and physical spaces occupied by gay men, or, as they'd have had it in the 1950s, queers. The 1990s return of queer as a magical inversion (like punk, nigga) is an index of how much this simple fact – still barely mentioned in most mainstream accounts of pop – needs to be restated again and again.

Frieze, June 1995

Letter from London: Britpop

London, pop metropolis, is periodically celebrated in song and lyric, which in turn redraw the capital's psychic map. You only have to hear Kinks vignettes like 'Dedicated Follower of Fashion' and 'Big Black Smoke', or Clash communiqués like 'London's Burning' and 'White Riot' to freeze the city in 1966 (the spiritual emptiness underneath the 'pretty coloured clothes') and 1977 (inner urban speed surfing; social/political polarization). Here are the twinned totems of white British pop: mod and punk.

The inner city has not been a popular pop location during the last few years: demanding physical and mental space, those touched by the post-House dance explosion sought the very outer suburbs or deep, deep country. Despite the battery of legislation employed against them, ravers are still to be found, but on mountains and wide beaches, places where inaccessibility is the key. So, the cycle turns, back to inner London – this time to Camden Town.

The new syndrome is summarized by Stephen Duffy's new 45, 'London Girls', a heavily stylized pop/rock performance that harks directly back to the early David Bowie of 'London Boys' and 'London Bye Ta Ta'. The lyrics delineate the treadmill of 'the latest Britpop poet laureate':

London town on Tuesday lunchtime: get the papers.
Maybe this time – indiecators [*sic*] indicate a trend: your way.
You're front page news

Welcome to the curiously hermetic, over-mediated world of Britpop.

Initiated by Suede (who, during 1992, sang about 'the love and poison of London'), industrialized by BPI award-winners Blur on their 1994 album

Parklife and fine-tuned by Elastica, this is a febrile, highly specific genre. Within a multicultural metropolis, where the dominant sounds are swingbeat, ragga or jungle, Britpop is a synthesis of white styles with any black influence bled out: the guitar-centricity of the late 1980s indie rock, the social commentary of the Kinks, the laddish dandyism of the Small Faces, the smart-dumb spikiness of Wire, whose presence looms large over Elastica and latest music-press sensations Menswear.

Scanned closer, Britpop reveals itself as an outer-suburban, middle-class fantasy of central London streetlife, with exclusively metropolitan models. Oasis, with their roots in Ireland and the UK's North-West, have a much wider appeal. A petty nationalism hangs over the style, like the pollution that cloaks Camden: it's as though Techno and grunge – both of which originated outside the UK, both of which offered a strong formal/emotional challenge – never happened. In response to the full-on emotion of Nirvana, we get … parlour games: uncomfortably reflective of the flyblown conservatism of John Major's government.

When in doubt, the English retreat into formalized poses, sardonic irony, childlike surrealism: the video for Menswear's new 45, 'Daydreamer', shows this teenage group spooking out amid train sets, bunk beds, ventriloquist's dummies. Laughable you might think, yet, with its queasy, claustrophobic undertow, 'Daydreamer' holds a mirror up to London, a city where people 'flounder drowning', where social life is reduced to the stark choice: 'pull you out or drag you under?' Tinkering around with archetypes, Menswear have hit upon a British pop absolute: the primal, privet hedge perversity of SE suburbia. Out of this seemingly arid culture, exotic blooms will grow.

Artforum, October 1995

The Beatles Outtakes: It'll be All Right on the Night

As befits a myth, nothing with the Beatles is simple. It is in the nature of myth to be paradoxical – it must be easy to understand yet elusive, must begin in the specific and become universal – because these unresolved, internal contradictions keep us coming back for more. Such tensions – between mind and body, art and commerce, Lennon and McCartney – created the group then tore it apart: they remain in the Beatles' vigorous afterlife.

Long after the hysteria has faded into television and newspaper archive, the Beatles' magic is still accessible on disc: 11 original albums and three compilations during their lifetime – not counting a slew of singles and EPs. This is fitting: for the Beatles, the record was always the thing. Like most other Brits in the 1950s, they consumed American R&B and rock 'n' roll on vinyl or shellac rather than in concert (although John Lennon and Stuart Sutcliffe did manage to see Eddie Cochran/Gene Vincent tour in March 1960); when they finally entered the studio, they wanted to make something as complete, as powerful as those extraterrestrial noises that had changed their lives.

The result was an eight-year sequence of releases which remain at once familiar yet enigmatic, near yet far. The Beatles, as Nik Cohn shrewdly noted in *Pop from the Beginning*, were great adepts of that trumping showbiz maxim: 'Leave 'em wanting more.' Already aloof by 1964, they disappeared in the second half of 1966, fuelling world-wide media interest to the point that, by 1968, it had become a morbid cult in North America: nothing excites the media more than withdrawal. By then, the group were seen only in brief TV and film appearances. If you wanted the Beatles, you had to have the records, and the records themselves leave you wanting more.

The Beatles wrote and recorded for maximum impact: there were to be no gaps in the sound, no boring moments – something had to be going on the whole time. In this, they were aided by George Martin, who suggested the speeding up of 'Please Please Me', who helped hone the relentless hooks of 'From Me to You'. Overseeing the mixing process, Martin also curbed their later tendencies to self-indulgence: whether it's the edited feedback note at the start of 'I Feel Fine' (6 to 2 seconds), the 40 seconds cut out of 'Dr Robert', even the early fade on 'It's All Too Much' (which in an early mix pulsates for another two minutes – 8.30 in total). This collective firm hand was only relaxed during their last two years.

Until 1967, everything came in pretty much under three minutes, which, bearing in mind the pressure of new ideas and experiences contained in *Rubber Soul* and *Revolver*, made for some extraordinarily compressed records – so taut as to be addictive. A song like 'She Said She Said' extrapolates from an actual event (Lennon dropping acid with the Byrds in August 1965, getting spooked by Peter Fonda's tale of a near-death experience) into a report from the emotional front line: the struggles of growing up, from the security of childhood to the traumas of adolescence, with LSD-aided personality disintegration as part of the journey – all within 2.31. The unresolved Rickenbacker riff at the fade of 'Hard Day's Night' hints at a whole world of pleasure just out of reach, if only you could hear another half-minute.

With such a perfect career trajectory (i.e. they stopped just as they were becoming boring), interest in the Beatles' recordings has only grown over the last 25 years – an interest slowly but firmly catered to by EMI and the Beatles' holding company, Apple (unlike the Rolling Stones, they seem to have come out of their Allen Klein experience with some degree of control). First, there were the Blue and Red double album compilations of singles and selected album favourites: massively successful in 1973, very poor VFM when reissued on CD last year. Following the bootlegging of 1965/6 live shows (Shea Stadium, the Budokan), the Beatles' Hollywood Bowl tapes were researched and tidied up by George Martin in 1977, resulting in a number one album.

During the early eighties, every Parlophone single was rereleased in sequence, on its twentieth anniverary: hard to imagine, but 'Love Me Do' made the Top 5 in 1982. There were many other repackages and compilations, most notably the *Rarities* LP of alternative mixes and oddities (like the German-language recordings of 'She Loves You' and 'I Want To Hold Your Hand', an early mix of 'Penny Lane', which was slipped out as part of a boxed collection of albums, sold separately thereafter). With the full CD release programme, all these loose ends were tied up by 1988's great 'Past Masters' compilations: everything issued by the Beatles was now available, but that was not enough.

An indication of just how insatiable the market for the Beatles has become was given after the 1988 publication of Mark Lewisohn's definitive *The Beatles Recording Sessions*, which, in detailing every Beatles EMI session, made official the existence of many outtakes – unreleased songs like 'If You Got Troubles', radically altered versions of 'I'm Looking Through You' and 'Strawberry Fields Forever'. An album comprising 15 of these rarities, called *Sessions*, was planned for 1985 but shelved at the last minute. Lewisohn's book was the trigger for the bootleggers, who from 1988 on flooded the market with outtakes from every Beatles period, from their first EMI session ('Besame Mucho', 6 June 1962) to one of their last (the un-Spectored 'I Me Mine', 3 January 1970).

What the market wants, the bootleggers will provide. With the Beatles, the market always wants more: whether 'butcher' sleeves or original memorabilia – the Beatles are auction-house brand-leaders. Until recently, there were up to one hundred unofficial Beatles' titles available. Most cannot be recommended – even if you could find them: in the IFPI crackdown on bootlegging following the recent standardization of international copyright law, UK record fairs and shops have been raided and stripped of illegal recordings, as they contain endless reduplication and are often poor-quality; what the best do highlight, however, is the fact that there is a wealth of unreleased Beatles material, which does nothing less than make you hear a seemingly inviolable canon in an entirely new way.

Bootlegging is morally hard to defend: there is no way round the fact that a performer should get paid for his/her work. Having said that, the history of illegal recordings since Bob Dylan's 1969 *The Great White Wonder* shows that they can create a market for a particular concert or unreleased song. This eventually stimulates the artist and record company to try to get some of this action for themselves, and the official release follows some years later – in Dylan's case, box sets like the *Biograph* and *Bootleg Series* collections. Indeed, it's hard, although everyone involved denies it, to see the recent, highly successful *Beatles at the BBC* double-CD being released without the many exploratory unofficial recordings which preceded it – some of which provided source material for the finished release.

Seven years after the CD floodgates were opened, Apple and EMI are scheduling a comprehensive release programme of Beatles outtakes, unreleased and rare songs. The track-listing is as yet undecided by the three ex-Beatles, Yoko and George Martin, but what seems clear is that there will be up to six CDs' worth of material released before Christmas 1996, with the first two, covering 1958 and 1964, in the shops before the end of the year. This is a bold and risky move. This may be simply too much, too much – rather like *The Beatles at the BBC*, which could have been whittled to one great single album; certainly, what you hear from the outtakes that exist in the public domain is a very different Beatles from the myth – not concise but sloppy, not pristine but raw, not perfect but all too fallible.

All outtakes are like this: that's why they're outtakes. Listening to a selection from 1963 to 1970, you're reminded of the utter boredom that making records induces in all who are not directly involved. Hearing the Beatles crash through small edits on 'Thank You Girl' – in the opening and climax proving particularly difficult – you can hear the tension in the studio: three tunes to do on a day off from a British tour, in one of the coldest winters this century. 'Go on!' hisses Lennon at regular intervals, before the session ends in what can only be described as the Beatles' version of *The Troggs' Tapes*: four March 1963 stabs at a recalcitrant 'One After 909', featuring appalling solos from George Harrison and reaching a full-blown row when McCartney tries some syncopation in take three. He breaks down, provoking an exasperated Lennon: 'What are you DOIN'?' The group then bicker about plectrums and McCartney's luggage.

This is entertaining but hardly serious. What many of the outtakes do make clear, however, is that essential Beatles trait noted by Ian MacDonald: 'their agility in making adjustments from take to take'. On the multiple takes, say, of 'Misery' or 'From Me to You', you can hear the group's incredible drive: beginning tentative and sloppy, winding themselves to a pitch where they're all flying as one. When you get to the patchwork of 'Strawberry Fields' outtakes, one of the few multiple-take sequences worthy of release, you can hear the song shift shape, as it expands from an acoustic sketch into a complex, multiphrenic epic.

Working tapes exist of most Beatles recordings – one of their most famous, 'She Loves You', being a rare exception. The overwhelming bulk of what has come out so far consists of run-throughs (basically, studio demos: an index of the Beatles' privilege, to do this in Abbey Road); early or different mixes (like the Peter Sellers tape, a mono mix of several *White Album* songs collaged with Indian music); actual overdubbing (the famous 'Strawberry Fields Forever' sequence); or, more rarely, completed songs which were either remade or remixed radically ('I'm Looking Through You', 'Across the Universe') or never released.

Trawling across these choppy waters, you begin to understand how the Beatles worked in the studio and just how much fairy dust (EQ, reverb, etc.) George Martin and his team sprinkled in the final mix-down. In the early days, the Beatles would record a take straight through vocal and instrumental, with occasional edit pieces for the tricky bits: getting the intros and outros right, mending an often broken Harrison guitar solo. Lennon often dominates here, although it is McCartney, with his superior sense of timing, who does the count-ins. In most cases, like 'A Hard Day's Night' or 'Can't Buy Me Love', the basic structure is in place from take one, with minor embellishments – like the Lennon/Harrison backing vocals thankfully deleted from the latter's verses – using up the bulk of the time; Harrison ends his own 'Don't Bother Me' with a rather unconvincing shout of 'Oh yeah! Rock 'n' roll now!'

With its feedback intro and complex riff, 'I Feel Fine' was one of the first Beatles songs to be recorded with the instruments first and the vocals later: Lennon had to sing and do the tricky guitar part. Outtakes exist of 'Paperback Writer', 'Day Tripper' and 'Help!' in this pure instrumental state. This may seem curiously pointless (unless you're into riff heaven), but, in the case of 'Rain', hearing the backing track without the vocals both highlights the extraordinary performance from all four Beatles and teases out, as Ian MacDonald has noted, the way that the opening chords are a direct steal from 'Visions of Johanna' – 'Ain't it just like the night to play tricks when you're trying to be so quiet.' Recorded in India before Harrison's vocal was added in the UK, the backing track of 'The Inner Light' stands as a piece of music on its own.

Bar the pre-orchestral take five of 'A Day in the Life', there is little else of any real substance from 1966 or 1967: an elongated 'Flying', called 'Aerial Tour', with a cheesy 'hotcha' music-hall ending, an extended 'It's All Too Much', an early run-through of 'I am the Walrus', which emphasizes its sly tune. There is, however, one masterpiece from the Beatles' most ambitious studio years, and that is the piecing-together of 'Strawberry Fields Forever': takes one to seven, overdub takes 25 to 26. Take one is a simple acoustic guitar/Mellotron run-through, beginning with the verses 'Living is easy with eyes closed'; 'No one I think is in my tree.' The familiar invocation, 'Let me take you down', does not appear until well into the song. There is no exotic instrumentation.

As the song moves through to the final first version, take seven, it adds drums, bass and guitar (the first appearance of Harrison's trademark slide-guitar sound), as well as some extra choruses: the song now begins as it does on the record. This early version is poignant but inconclusive. Leaping to take 26, you get the full freakout: backwards cymbals, Lennon's vocal, pounding drums. This is an overdub for the second version: at the end you can hear why the first fade exists, as the take breaks down. Starr quickly gets a second wind, which sets Lennon off again, with shrieks and cries of 'cranberry sauce'. The take ends on his shout: 'Right, calm down, Ringo.' Take 26 is the same backing track with a different mix: the backwards percussion, Harrison's swarmandal, Lennon's piano. Putting these versions together, you can see what happened in the famous edit: as well as increasing the tempo, Martin added an extra chorus, which adds to the song's relentless downward momentum.

Another find is the first (February 1968) mix of 'Across the Universe' in all its glory, before Spector mushed it up. Instead of rock solemnity, you have three minutes of pure psychedelic pulses which, if they don't rectify the song's shortcomings, precisely locate it in a moment. In this, the backwards guitars work perfectly with Lennon's ego-less vocal and the young fans' voices, which are right up in the mix. A certain tendency to solipsism – 'nothing's gonna change my world' – is transcended by a truly cosmic finale: as Lennon chants 'Jai guru deva', you know that he's speaking to God. Released in early 1968, this would have made perfect sense as the Beatles' last great psych move; by 1970, Lennon had gone through several further incarnations and the song had little relevance.

Apart from an acoustic demo of 'While My Guitar Gently Weeps' – which serves only to accentuate the song's rather doleful, sanctimonious air – there is little available from the *White Album* sessions; many think that some cuts on this daunting, minatory release are little more than outtakes anyway. The Beatles put everything on. As the group began to fray round the edges, so did their quality control: certainly, both *Abbey Road* and *Let It Be* offer different approaches – produced and *faux vérité* – to the problem of how to break up in public. The outtakes from this period only accentuate what George Martin, Phil Spector and the waning magic of the Beatles themselves were attempting to disguise: the dissolution of a group mind.

There is stuff of interest. Heard in full, the May 1969 Glyn Johns assembly of 'Get Back' (with all the *vérité* and adding 'Rocker', 'Teddy Boy' and 'Save the Last Dance For Me') works as a concept album of a lost adolescence. With hindsight, the Beatles should have rushed it out as an 'Unplugged' and moved on. It highlights just how awful 'Dig It' and 'Dig a Pony' are, emphasizes the poor Lennon bass playing on 'The Long and Winding Road', and rescues 'The Two of Us' as a great song – the last attempt to regain the group adolescence that changed the world, complete with McCartney's final 'goodbye'. The

Abbey Road material points to the fact that the group had already dissolved: at the end of 'Something', almost before the take has finished, Lennon is off into a piano jam to work something that's on his mind – the central sequence of 'Remember' from his *Plastic Ono Band* album.

Hearing these outtakes is an ambiguous business. On the one hand, they are often exasperating: unlike jazz giants like Charlie Parker, who can vary from moment to moment, the Beatles were not great improvisers. Most often, they are searching and, later, noodling for something just out of reach. You can know too much and, to some extent, the glitter is tarnished, the idols rendered all too human – an unsettling experience. Yet they are valuable in several ways, if only to show that young groups should not compare themselves to the Beatles lightly: the Beatles worked in the studio with an incredible dynamism and willingness to experiment that, for all the claims made on their behalf, it is not yet possible to hear in Blur or Oasis.

They also help us to read the Beatles better. This is creative nostalgia: these outtakes make it clear that, far from being a fixed, static artefact, each record was an adventure. The fact that each journey could have had a different ending throws up speculation about what would have happened to pop music – if, say, 'Across the Universe' had been released instead of 'Lady Madonna' – such was the Beatles' centrality in 1960s youth culture. In this way, the stripping of the veil is yet another tease for, in spite of all the attempts made upon it, the Beatles' myth remains inviolate: the century's ideal of group adolescence made flesh. This hidden history returns us to the beginning of the story that many of us are compelled to tell and retell.

Mojo, November 1995